The Atmosphere

Richard A. Anthes
Hans A. Panofsky
John J. Cahir

Meteorology Department
The Pennsylvania State University

and

Albert Rango

National Aeronautics and Space Administration/
Goddard Space Flight Center

Charles E. Merrill Publishing Company
A Bell & Howell Company
Columbus, Ohio 43216

Cover photo by Roger Reinking. ©Alistair B. Fraser

Published by
Charles E. Merrill Publishing Company
A Bell & Howell Company
Columbus, Ohio 43216

Type: text—Oracle; display—Eurostile Light Extended, Palatino, Helvetica
Production editor: Linda M. Johnstone
Cover designer: Will Chenoweth

International Standard Book Number: 0-675-08698-1

Library of Congress Catalog Card Number: 75-3539

3 4 5 6 7 8 9 10 — 84 83 82 81 80 79 78 77 76

Printed in the United States of America

preface

Although part of modern meteorology is a sophisticated mathematical science, much of the weather around us can be qualitatively understood and appreciated without the use of mathematics. In writing an introductory meteorology book designed to appeal to the reader who is not mathematically inclined, as well as to the reader who has a strong mathematical and physical science background, we have tried to strike a common responsive chord—curiosity about the atmosphere and its fascinating variety of phenomena.

Rather than emphasize abstract facts or physical concepts, which many people find boring or irrelevant, we have departed somewhat from the traditional introductory textbook format by integrating explanations of basic physical processes into the discussion of atmospheric phenomena or processes that are likely to be of interest to the reader. Thus, instead of studying the Coriolis force in a section by itself, we discuss it when it is necessary to explain the observation that winds tend to blow parallel to, not across, lines of equal pressure.

The Atmosphere is written for the casual, as well as the sophisticated, observer of the weather to explain those atmospheric phenomena that people are likely to encounter, whether driving across the country on an interstate highway, flying 30,000 feet over the ocean, or skimming across whitecaps in a small sailboat on a summer afternoon. Observation is the key to understanding the atmosphere, and we have tried to show the reader what to observe, and how to interpret these observations in a meaningful way.

In addition to explaining how the perceptive observer can infer tomorrow's weather today, we have indicated how weather and climate have shaped human culture throughout history. Included are numerous examples of the direct influence of weather on folklore, literature, music, art, and religion.

The Atmosphere considers several subjects usually not found in elementary meteorology texts. These include the beautiful and complex world of meteorological optical phenomena, including halos, sun dogs, sun pillars, mirages, and the elusive green flash, as well as the rainbow. One chapter is devoted to air pollution, and covers both the effect of the atmosphere on the pollution and the increasingly important effects of pollution on the weather. The controversial, yet vitally important, subject of climate change, both natural and man-made, is considered.

As energy supplies become scarce and expensive, we begin to look more closely at the sun and the wind as sources of energy. Some of the meteorological aspects of these alternative sources are discussed. In keeping with the "weather and man" theme, a chapter on biometeorology (the study of interactions of biological life with meteorology) includes human physiological response to the weather. Finally, the book is summarized with a review of the annual progression of weather over the United States, illustrating how the various physical processes and atmospheric phenomena unite to provide the ever-repeating, yet never-the-same, weather drama.

The authors would like to thank Professor Alistair Fraser, The Pennsylvania State University, for writing the chapter on atmospheric optics and for providing the unique and beautiful photographs of several optical phenomena. We also wish to thank Virginia Ultee, who carefully read the first draft of the manuscript and made many helpful suggestions, and Susan Anthes, who was very helpful in obtaining reference materials and in proofreading the various stages of the manuscript. The production editor, Linda Johnstone, was very cooperative and efficient in transforming the manuscript into the finished product. We thank Peter Black, Theodore Fujita, Joseph Golden, Ronald Holle, Robert McAlister, Frank Schiermeier, Robert Sheets, and William Shenk for providing color photographs. Tanya Sharer and Joyce Sabol typed the manuscript. Finally, we found the prepublication reviews and constructive criticisms by Professors J. M. Wallace, Eugene Chermack, David Marczely, and Stephen Berman very helpful.

Richard A. Anthes
Hans A. Panofsky
John J. Cahir
Albert Rango

contents

1 Rainbow in morning

1.1 "Some Are Weatherwise; Some Are Otherwise" [Benjamin Franklin, *Poor Richard's Almanac,* 1735]

It is not surprising that man has always been fascinated by the capricious nature of the weather. In his early days as hunter and food gatherer, his comfort and supply of food were dependent upon the weather more than any other factor. Even as tools and technology have freed him from scurrying about in search of wild animals and fruits and have provided him with warm clothing, houses, and studded snow tires, man's dependence on the weather has not ceased, but has only become more subtle.

Even if man should try to ignore the weather, he is likely to fail, for sooner or later the weather will directly affect his life. The day-to-day effects of weather shape his psychological outlook slowly, but surely, and changes in the weather may affect his mood, or even his attitude to his business associates, as indicated by the ancient advice: "Do business with men when the wind is in the northwest." In his remarkable book *Mainsprings of Civilization,* Ellsworth Huntington argues that climate and weather play a dominant role in determining a nation's history and shaping its culture. Indeed, climate is a major factor in the advancement or lack of development of civilization.

Besides the subtle molding of character by the climate, misbehavior of the atmosphere can also completely interrupt normal human activities, as anyone whose life has been touched by a flood, hurricane, tornado, or even slippery roads on New Year's Eve can readily attest. Even the most sheltered urban existence is occasionally threatened by varieties of weather, such as traffic-halting blizzards, air pollution events, canceled airplane flights, or power blackouts caused by disruptive weather.

Of course, most of us escape the spectacular weather events that seriously threaten lives. However, we find gray drizzling days depressing, we pay for the corrosion of our cars from road salt used to melt snow, we drag garden hoses around the lawn during droughts to save our grass, and then spend hours behind a mower after the rains resume. On the positive side, we feel exhilarated as we ski down snowy mountain slopes on a bright winter day, or serene as we drowse by the fireside during a cold November rain.

So, the weather is a constant, and occasionally dominating, force in our lives. This inevitable influence, however, is not the only reason for wanting to know how and why the weather acts as it does. Even the casual study of day-to-day weather can be an enjoyable rest from daily pressures. The dramatic evolution of weather events in the middle and high latitudes and the subtle changes in tropical weather patterns can afford the perceptive observer hours of satisfaction that are varied, exciting, and rewarding. Naturally, we may enjoy the bracing north wind of winter or the distant lightning of a faraway summer thunderstorm without understanding the underlying physical processes. As comprehension of the forces that shape these weather events grows, however, so does our enjoyment and appreciation of the weather drama.

Anticipation spices any pleasure, and the weatherwise individual will soon be able to recognize the signs of impending weather changes. For example, great snowstorms in the eastern United States usually give the alert individual warning of a day or two. First, the brilliant cerulean skies of a midwinter cold snap give way to high, delicate cirrus clouds which streak in from the southwest and entangle the sun in a web of icy fibers. The barometer, which has been rising for several days, hesitates, and then slowly begins to fall. Flags which have been rippling from the north now hang limp, and smoke from the fireplace, refusing to rise, drifts slowly westward. The thermometer climbs sluggishly from the teens into the more moderate twenties. Many people might interpret these signs, if they noticed them at all, merely as a welcome respite from the cold, windy weather of the previous two days. However, the perceptive observer of the atmosphere will alert himself to the unmistakable signs of an impending storm.

Twelve hours later, the signs are even more legible. The halo around the moon disappears in a thickening mass of clouds. The barometer, as if making up its mind, falls more rapidly, and bare trees begin to sway in the increasing northeast wind.

In another 12 hours, the first tiny snowflakes flutter across the landscape. The clouds lower, and, soon, thick-falling snow reduces the lights from nearby houses and street lamps to hazy patches of luminescence. The snow falls heavily during the day as the barometer continues to fall. The weather observer monitors the falling pressure and the wind direction, looking for the abrupt upward swing of the glass (barometer) and the shift of the wind vane that will signal the end of the storm. At the same time, he watches the rising thermometer, which now indicates 31°F and hints at the possibility of the snow changing to rain.

Suddenly, the steady wailing of the east wind subsides. Snowflakes hesitate in their westward drive before the wind, and a cold gust from the northwest halts the upward progress of the thermometer. The barometer ceases its downward drive and jumps decisively upward. The heavy snowfall is over. Now, stronger northerly (from the north) winds pile the snow into mountains and canyons, sweeping bare exposed areas and piling high drifts behind every obstacle. The mercury in the thermometer contracts hurridly into the teens, ending the possibility of any melting. Now, the snow flurries diminish gradually, and through the scudding clouds we can catch occasional glimpses of the moon. The storm is over.

To many people, all the preceding signs were completely missed and unappreciated. To the weather observer, however, with $20 worth of instruments and a pair of open eyes and ears, the signs were far more exciting and revealing than the sterile words of the newspaper forecast: "Cloudy and cold today and tonight with snow likely, possibly accumulating four or more inches. Snow ending tomorrow, followed by clearing weather."

Most of the daily weather events that touch our lives are not as spectacular, nor the signs as obvious, as those in the preceding example. Nevertheless, a sensitive observer may catch many small clues to the behavior of the atmosphere that nature provides every day. The purpose of this book is to instill in the amateur this interest in observing and interpreting the weather and to describe the physical processes that are responsible for weather events. In this way, we hope to heighten the amateur's appreciation and comprehension of the atmosphere's seemingly mysterious ways.

1.2 The Ancient Meteorologist

We do not know who first peeked out of the cave, saw a halo around the moon, and called off the mammoth hunt for the next day. Yet, it is certain that the earliest people, whose lives were intimately entwined with the weather, would devote some of their intellectual energies trying to comprehend the meaning of the wind and the sky. Indeed, all of the information (except the barometer) that was available to the modern weather observer watching the evolution of the snowstorm in the previous section was available to early man. (Although objections may be raised about the availability of a thermometer, human skin could have served the purpose. Skin is very sensitive to the temperature, and most people can estimate the temperature within 5°F.) Therefore, intelligent and perceptive people would soon notice the signs that portend major weather events, and then verbally pass their weather wisdom to successive generations. In this way, an increasing inventory of weather folklore could be established.

Prehistoric man's view of meteorology must remain a matter for speculation. Our first real glimpse of the ancient meteorologists begins with the Greeks, who named the study of heavenly phenomena *meteōrología,* and called raindrops, hail, and snowflakes *metéōron,* which means "a thing in the air." In contrast to today's science of meteorology, which includes only atmospheric phenomena, early meteorology was closely associated with astronomy. Thales (640 B.C.) believed in a flat earth, but his views on meteorology were somewhat more sound. He correctly ascribed the four seasons to variations in the position of the sun in the sky. The seasons were separated, as they are today, by the winter and summer *solstices* (sun-standings), and by the autumn and vernal (spring) *equinoxes* (dates of equal night and day).

Because of its unmistakable association with human health and psychology, early meteorology was an important part of medicine. Hippocrates (fifth century B.C.) cautioned that physicians, upon reaching a new city, should first study the meteorology of the region, including the prevailing winds, the relative amounts of sunshine and cloudiness, the amount of rainfall, and the exposure of the town in relation to the winter sun. Today we call Hippocrates' type of meteorology *biometeorology,* and even in a day of technologically controlled human environments, we have essentially confirmed Hippocrates' advice.

1.2.1 *Aristotle's view of meteorology*

Aristotle was the most famous of the ancient meteorologists. His *Meteorologica,* written about 340 B.C., gives a detailed account of Aristotle's view of most aspects of weather and climate. We present Aristotle's ideas here in considerable detail, not only because they are interesting, but also because these views dominated meteorological thinking until the Renaissance.

Aristotle was a philosopher, not strictly a meteorologist, and did not differentiate clearly between the sciences of astronomy, geography, geology, and meteorology. In *Meteorologica,* he talks about shooting stars, the aurora borealis, comets, earthquakes, rivers, springs, and the oceans, as well as rain, clouds, mist, dew, snow, hail, winds, thunder, and hurricanes. Aristotle believed that weather phenomena were caused by mutual interactions of the four elements, *fire, air, water,* and *earth,* and the four prime contraries, *hot, cold, dry,* and *moist.* However, he never clearly explained the nature of these interactions.

Aristotle frequently argued against ideas which were closer to the truth than his own. For example, in considering the cause of hail, he presents first the "wrong" view of Anaxagoras:

> *Some then think that the cause of the origin of hail is as follows: when a cloud is forced up into the upper region where the temperature is lower . . . the water when it gets there is frozen, and so hailstorms occur more often in summer and in warm districts because the heat forces the clouds up farther from the earth.* [Aristotle, *Meteorologica,* with an English translation by H. D. P. Lee (Cambridge, Mass.: Harvard University Press, 1952), p. 81]

As will be discussed in a later chapter, Anaxagoras' theory is amazingly correct. Nevertheless, Aristotle denies this view and offers his own theory:

> *The process (hail) is just the opposite of what Anaxagoras says it is. He says it takes place when clouds rise into the cold air: we say it takes place when clouds descend into the warm air, and is most violent when the clouds descend farthest.* [Ibid., p. 85]

After reading the chapters on precipitation and hail, you should be able to point out Aristotle's folly.

A second example of Aristotle's meteorological deductions is illustrated by his concept of the wind:

> *There are some who say that wind is simply a moving current of what we call air . . . and define the wind as air in motion. The unscientific views of ordinary people are preferable to scientific theories of this sort.* [Ibid., p. 89]

In a later chapter, Aristotle presents his belief that winds are simply dry or moist exhalations from a breathing earth. The dry exhalations are the wind; the moist exhalations, the rain.

Aristotle's winds are classified according to points on the compass:

> *. . . westerly winds are counted as northerly, being colder because they blow from the sunset; easterly winds are counted as southerly, being warmer because they blow from the sunrise. Winds from the sunrise are warmer than winds from the sunset, because those from the sunrise are exposed to the sun for longer; while those from the sunset are reached by the sun later and it soon leaves them.* [Ibid., p. 193]

This discourse is typical of some of Aristotle's meteorological observations, because the statement (that east winds are milder than west winds), over much of the middle-latitude regions, is correct, but the interpretation is not.

Aristotle did contribute many accurate explanations for atmospheric phenomena, however. His reasoning for rain could have been lifted from a modern textbook:

> *The earth is at rest, and the moisture about it is evaporated by the sun's rays and the other heat from above and rises upwards; but when the heat which causes it to rise leaves it, . . . the vapor cools and condenses again as a result of the loss of heat and the height and turns from air into water. The exhalation from water is vapor. The formation of water from air produces clouds.* [Ibid., p. 69]

His views on the formation of dew and frost are also essentially valid, and he correctly notes that dew and frost form on calm nights and in the valleys rather than on the mountain peaks. His observations on rainbows, halos, and mock suns (sun dogs) are quite detailed and accurate. For example, Figure 1.1 shows Aristotle's diagram of the primary and secondary rainbow. It is interesting that he was well aware of the rare, but possible, night rainbow (rainbow caused by moonlight).

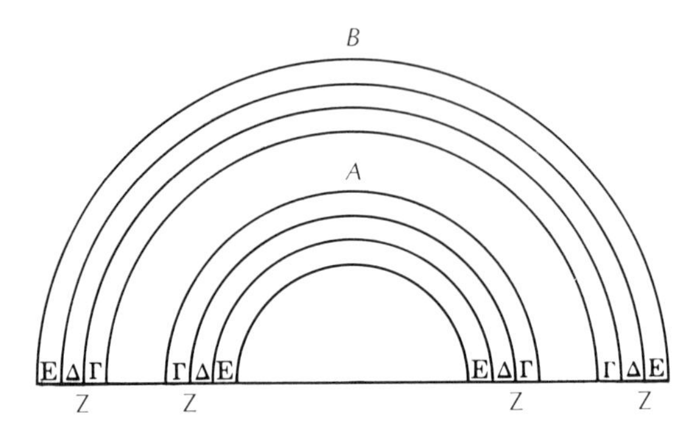

Figure 1.1 *Aristotle's rainbow.* "Let *B* be the outer and *A* the inner, primary rainbows and to symbolize the colors, let us use **Γ** for red, **Δ** for green, and E for purple." (*Aristotle, Meteorologica, p. 266*).

Aristotle noted one of the best-known and most reliable sayings of weather folklore—that halos around the sun or moon are harbingers of rain:

> *This formation is therefore a sign of rain . . . It is reasonable to regard it as a sign of rain, since it shows that a condensation is taking place of the kind, which, if the condensing process continues, will necessarily lead to rain.* [Ibid., p. 247]

We now know that halos are caused by fine ice crystals high in the sky and are frequently associated with the advancing edge of moisture that is racing ahead of an approaching storm.

A concise summary of Aristotle's work in meteorology might be that his observations were mostly correct, but his explanations were frequently erroneous. The chief value of *Meteorologica* is its detailed exposition of the early ideas concerning the causes and effects of meteorological phenomena.

1.2.2 *Archimedes takes a bath*

Although best remembered for his ecstatic "Eureka!" as he stepped into his bath, Archimedes (born 287 B.C.) made an important contribution to the future understanding of cumulus cloud formation. (Cumulus clouds are the familiar isolated clouds which develop vertically in the form of rising domes or towers.) At the time of his bath, however, Archimedes was far more interested in a problem about gold than in the formation of clouds. Hiero, King of Syracuse, suspecting that a lump of gold had been diluted with silver by workmen who were making a crown for him, assigned Archimedes the problem of determining whether the fraud had actually occurred. Archimedes debated unsuccessfully with himself until, while stepping into the bath, he realized that any solid body displaces an equal amount of water. This experience gave Archimedes a method for determining the percentage of gold in the crown and led to the recognition of buoyancy as an important force associated with the motions of fluids. This buoyancy principle was the basis for the design of the hot-air balloon, which is sometimes called the Montgolfier balloon after its French inventors. The Montgolfier brothers launched the first balloon on June 5, 1783. Later, hydrogen and helium were used instead of hot air to provide the necessary buoyancy. Weather observations taken from balloons have contributed much to our knowledge of the vertical structure of the atmosphere. Meteorologists today utilize Archimedes' principle as a basis for theoretical investigations of the buoyant rise of cumulus clouds.

1.3 Renaissance Meteorology

Science in general, and meteorology in particular, made little progress during the Middle Ages. In meteorology, Aristotle's theories were accepted without reservation, as indicated in the following passages, which were written in the tenth century by an unknown Anglo-Saxon:

> *There are four elements in which all earthly bodies dwell, which are, aer, ignis, terra, aqua. Aer is a very thin corporeal element; it goes over the whole world, and extends upward nearly to the moon . . .* [*Popular Treatises on Science*, Thomas Wright, ed. (London: R. and J. E. Taylor, 1841), p. 17]

Concerning the formation of precipitation, the Anglo-Saxons showed little more insight than Aristotle:

> *The atmosphere licks and draws up the moisture of all the earth, and of the sea, and gathers it into showers; and when it can bear no more, then it falls down loosed in rain . . .*
>
> *Snow comes of the thin moisture, which is drawn up with the air, and is frozen before it be run into drops, and so immediately falls.* [Ibid., p. 19]

The Renaissance ended the period of blind acceptance of such meteorological dogma. The spirit of healthy doubt was typified by René Descartes' (1596–1650) condemnation of unquestioning assertion:

> *It is not true to say we know a thing simply because it has been told us . . . to know anything requires much more than this, and unless the reasons for any belief are so clear to our minds that we cannot doubt them, we have no right to say we know it to be true, but only that we have been told so.* [Arabella B. Buckley, *A History of Natural Sciences* (London: Edward Stamford, 1894), p. 103]

The reader of this or any other book should take its assertions with a grain of Descartes' salt, until he understands the reasons behind the facts and thereby discovers the truth for himself.

1.3.1 *The air has weight . . .*

Just as fish are not conscious of the weight of water, we do not normally perceive the weight of the atmosphere. Because air seems so nebulous, it is easy to imagine why the ancients considered it weightless. Nevertheless, the atmosphere presses upon the earth's surface at sea level with a weight of approximately 14.7 pounds on every square inch. We do not feel this weight crushing our shoulders, because the pressure inside our bodies equals that of the surrounding environment, so that there is no net force for us to sense.

Some of the consequences of air pressure were known before it was accepted that air had weight. The siphon and suction tubes operate on the principle that large forces can be achieved if pressure differences are created. For example, if a movable piston is drawn upward in a tube which is set in a tub of water (Figure 1.2), water will rise into the tube. Prior to the recognition of air pressure, the popular explanation for this phenomenon was that nature "abhorred a vacuum," and that the rising of the water was nature's way of preventing this crime.

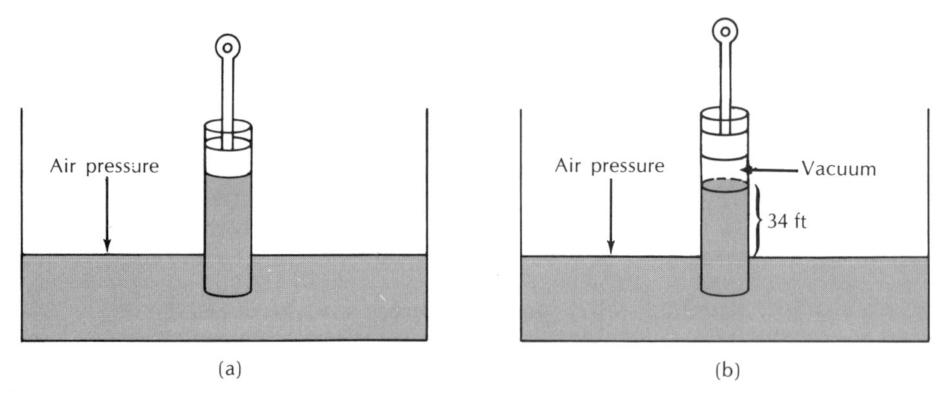

Figure 1.2 *Water barometer.* Air pressure at sea level can support a column of water approximately 34 feet high.

Galileo, however, was puzzled by one aspect of the suction pump. No matter how high the piston was raised, the water could not be made to rise above a height of about 34 feet. It was difficult to explain why nature abhorred a vacuum only to a height of 34 feet, so an alternative explanation was necessary. A friend of Galileo, Torricelli, correctly explained that it was the pressure or weight of the atmosphere pressing upon the surface of the water in the tub that caused water to flow into the tube. When the weight of the column of water equaled that of the atmosphere, an equilibrium was obtained, and no further raising of the water was possible. Torricelli then reasoned that if a denser liquid were used instead of water, the required height for equilibrium would be much less. As predicted, a column of mercury rises only about 30 inches; hence, the mercury barometer was invented.

Soon after Torricelli correctly explained the behavior of Galileo's water barometer and invented the mercury barometer, seventeenth century scientists were probing the lower atmosphere with this new instrument. At the suggestion of Pascal in 1646, Périer carried a barometer to the top of a mountain, the Puy de Dome in France, and showed that air pressure decreases with increasing elevation. Pascal noted the compressibility of air, and that air in the valleys is more compressed than air on the mountain tops. He likened the atmosphere to a huge pile of feathers; the feathers at the bottom of the pile are more squashed than the ones near the top.

Torricelli, in a letter dated June 11, 1644, indicated that air pressure varies not only with elevation at a given time, but also from time to time at a given location. Soon after these variations in time and place were noted, they were associated with changes in the weather. Early references to the barometer as a weather glass were made by Jean Pecquet in 1651 and by Samuel Pepys in his diary. It was much later, however, that systematic analyses of surface pressure were related to weather events associated with storm systems.

1.3.2 . . . and blows hot and cold

It was not some sense of modesty that made early humans in the middle latitudes wrap wooly skins around their bodies on frosty autumn mornings. They did not need a thermometer to tell them that some mornings were cold and others hot. However, quantitative measurement of the property of the air that made people shiver or sweat waited until the thermometer was developed, which occurred at about the same time as the development of the barometer. Galileo had a water thermometer, which must have broken with every freeze, and a Dutchman named Cornelis Drebbel combined science with pleasure by making a wine thermometer. By 1670, mercury was being used as the liquid in the hollow glass. Whether mercury, water, or wine, the principle is the same: the liquid expands with heating and contracts with cooling. The scales on thermometers were arbitrary, and several different ones (for example, the Fahrenheit, Celsius, and Reaumur) were proposed by the eighteenth century. These scales were based on one or two reference points, usually the freezing and boiling temperatures of water. The common Fahrenheit and Celsius scales are compared in Appendix 1.

The invention of the barometer and thermometer enabled measurement of pressure and temperature, two of the basic variables important to the scientific study of meteorology. The discovery of the relationship between these two variables and a third, density, was the beginning of a quantitative rather than qualitative description of weather; the door was opened for a more detailed understanding of the weather.

Much of the foundation for later progress in meteorology was laid by the basic advances in chemistry, physics, and mathematics during the seventeenth and eighteenth centuries. Robert Boyle discovered the relationship between pressure and density (at a constant temperature, pressure = constant x density) in 1661, and, in 1802, Jacques Charles discovered the relationship between temperature and density of gases (at a constant pressure, temperature x density = constant). In 1666, Newton, most famous for his discovery of gravity, developed the so-called method of fluxions, which is similar to differential calculus, and is essential in solving theoretical meteorological problems.

1.3.3 *Heat that melts the ice but does not warm the water*

Latent heat, a physical process which is very important in meteorological processes, was discovered by a chemist, Dr. Joseph Black, in 1760. We know that when ice is melted the temperature of the icewater mixture does not change while any ice remains, no matter how much heat is applied. This lost heat, which is required to melt the ice but does not warm the water, is called *latent,* or *hidden, heat.* This hidden heat is given back to the surrounding air if the water is refrozen. A similar hiding of heat occurs when water is evaporated, with an equal amount of heat being liberated when the water vapor subsequently condenses. Because evaporation, condensation, freezing, and melting occur frequently in the atmosphere, these latent sources of heat greatly influence many meteorological processes. The principle of latent heat was explained by the experiments of Count Rumford in 1798 and Sir Humphry Davy in 1799, which proved that heat consists of molecules in motion. Since molecules locked in the rigid structure of ice crystals have less motion than molecules in the liquid phase at the same temperature, extra heat is required to turn ice to water.

1.3.4 *Invisible heat rays*

While chemists were melting ice cubes, the nature of radiation was being investigated by Sir William Herschel. In 1800, Herschel passed a thermometer through the colors that had been separated from light passing through a prism. As expected, the temperature rose as the thermometer passed from the violet into the yellow. Surprisingly though, the temperature continued to rise as the thermometer moved beyond the red into the darkness. Thus, an important form of energy transfer, the invisible *infrared* ("beyond the red") *rays,* was discovered.

1.3.5 *Kite flying in the eighteenth century*

Benjamin Franklin was a versatile genius of many passions, one of them meteorology. His most dangerous meteorological pursuit was kite flying in thunderstorms, in which he was engaged in 1752 in order to demonstrate the electrical nature of lightning. By sending his kite into the vicinity of thunderstorms, he was able to induce sparks to jump from a key to his finger. Miraculously, he was not electrocuted; unfortunately, others seeking to duplicate his experiment were not so lucky. A year later, a Russian scientist, Georg Wilhelm Richmann, was struck on the forehead and killed by a lightning stroke while kite flying in a thunderstorm. In spite of this, and other accidents, kite flying was a common procedure for exploring the lower atmosphere until the late 1800s.

1.3.6 Storms that move

Prior to the eighteenth century, it was commonly believed that storms were born and died at the same location. With improving communications, however, it gradually became evident that storms move about in a more or less orderly fashion, retaining their identity for several days. Much later, when simultaneous observations could be transmitted rapidly by telegraph, the extrapolation of the storm's motion became an important basis for forecasting. Even so, the realization that storms were only a part of the huge complex weather machine came slowly.

Daniel Defoe, referring to a British storm of November 27, 1703, remarked that the same storm had occurred earlier on the shores of America.[1] Benjamin Franklin, in his pre-kite-flying days, also deduced that storms move. Franklin was ready to observe a 1743 eclipse of the moon in Philadelphia when dense clouds moved in just as the eclipse was about to begin. He was probably even more frustrated when he found out that similar clouds in Boston did not obscure the moon until after the eclipse, so that his friends there obtained a splendid view. In a letter to one of these friends, Franklin noted that the storm must have moved from Virginia to Connecticut, and so reached Philadelphia before Boston.

1.4 Nineteenth-century Meteorology

If progress in the seventeenth century could be described by advances in basic science, and in the eighteenth century by a vague awareness of the nature of storms, then the nineteenth century represented the beginnings of our modern-day understanding of large-scale atmospheric circulations and the typical weather patterns associated with them. During the nineteenth century, controversies on the relationship of the wind, precipitation, and pressure in moving cyclones (large-scale wind circulations about a center of low pressure) arose and were resolved. At the beginning of the century, it was barely recognized that storms move; toward the end of the century, the typical wind, temperature, and rainfall patterns associated with the moving cyclones were documented, and the U.S. Weather Bureau was making forecasts based on simultaneous weather reports. A complete history of meteorological progress during this time is outside the scope of this brief survey. Instead, we mention just a few of the problems that occupied meteorologists during this time.

1.4.1 The rotary wind controversy

It is now such a well-established fact that hurricanes and the larger, but less intense, cyclones associated with our typical rainy days are all whirlwinds that it is difficult to imagine that barely 100 years ago the

[1] Donald R. Whitnah, *A History of the United States Weather Bureau* (Urbana, Illinois: University of Illinois Press, 1961), p. 3.

question of the wind distribution around low-pressure centers was a lively topic of controversy. Much of the motivation for determining the correct distribution stemmed from marine interests which desired interpretation of the behavior of the barometer in terms of the probable effect on the wind.

In the early 1800s, there were two contradictory theories which described the wind flow around storms. One theory, with adherents William Redfield and Colonel William Reid, maintained that the northeast storms and hurricanes were giant whirlwinds, with the wind blowing around the center of low pressure (Figure 1.3a). A contrasting view was held by James Espy, who believed that the winds associated with these storms blew directly toward the center of low pressure (Figure 1.3b). Espy even cautioned mariners against following Redfield's rules on avoiding storms by following the circular wind law:

> *I earnestly recommend to gentlemen who embrace the whirlwind theory of storms, to abstain from laying down rules to the practical navigator, founded on this doctrine, until it is better established than it is at present.* [*The Philosophy of Storms* (Boston: Little and Brown, 1841), p. 253]

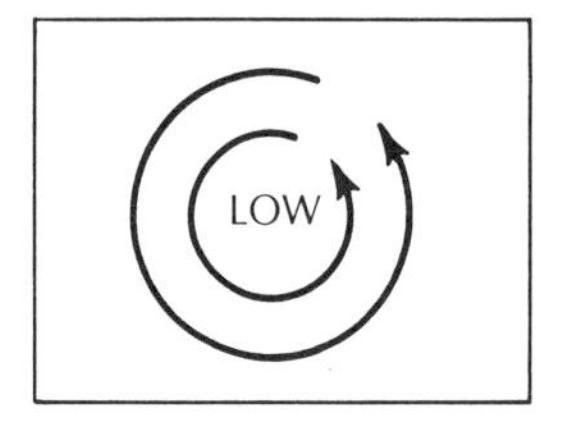

(a) Redfield's storms

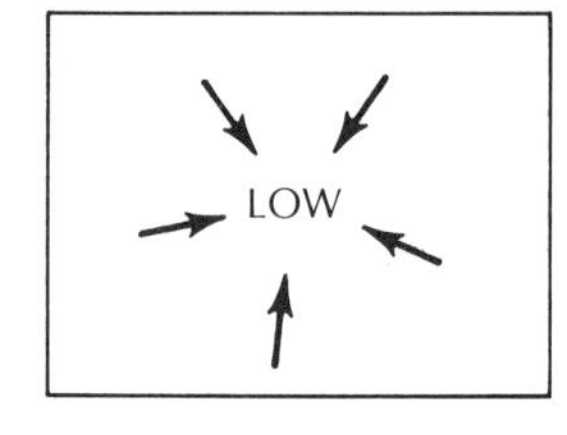

(b) Espy's storms

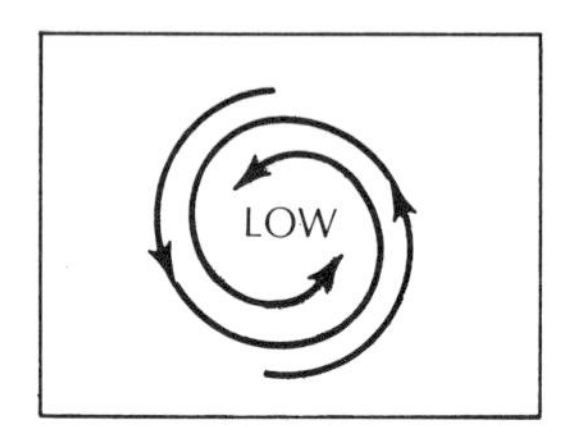

(c) Loomis' storms

Figure 1.3 *Three views of storm circulations in the nineteenth century.* Arrows show the direction of air movement.

Redfield's and Reid's theory of the rotary nature of winds around lows was based on hundreds of observations from ships and land stations. Strangely enough, Espy interpreted many of the same observations as supporting his inflow theory. The reasons for these conflicting interpretations of the same data are understandable when it is realized that both theories are partially correct. At the surface of the earth, the wind spirals inward toward the low center, and therefore has both a circular and an inflow component (Figure 1.3c), as correctly concluded by Professor Elias Loomis in 1859 after a detailed analysis of an 1836 storm:

> *For several hundred miles on each side of the center of a violent storm, the wind inclines toward the area of least pressure, and at the same time circulates around the center contrary to the motion of the hands of a watch.* ["On Certain Storms in Europe and America, December, 1836," *Smithsonian Contributions to Knowledge,* Vol. II (Washington, D.C.: Smithsonian Institution, 1860), p. 25]

This fact, plus the paucity of observations, which made imagination an important ingredient of the analyses, explain the opposite conclusions drawn by the proponents of the two theories.

1.4.2 The philosophy of storms

Although his pure inflow theory was carried a bit too far, Espy made a very important contribution by explaining how the release of latent heat associated with condensing water vapor maintains the necessary warmth of ascending air as it rises to great heights. As we shall see in Section 2.2.4, without the extra heat of condensation, ascending bubbles of air would soon become colder and denser than their environment, and would therefore sink. Because clouds and their offspring, rain, snow, sleet, and other forms of precipitation, depend on deep currents of upward-moving air, it is difficult to overemphasize the importance of latent heat.

Consideration of latent heat led Espy to many correct explanations of phenomena associated with the cumulus-type clouds. These theories, with observations to back them up in many cases, were explained in his *The Philosophy of Storms*. Espy showed why cumulus clouds have relatively flat roots (bases), all at one level. He described the atmospheric conditions favorable for tornado formation and gave the correct explanation for the formation of the visible funnel of tornadoes and waterspouts. His theory that rising air is necessary for clouds and precipitation formation solved an old riddle which argued that clouds should be more frequent at night. According to that theory, nights are colder than days, and cold air cannot hold as much water vapor; therefore, condensation should be heaviest at night. Espy showed that while the nocturnal cooling is frequently conducive to fogs, a more efficient way to produce condensation is the lifting of air in cumulus clouds.

1.4.3 Early weather maps and fractured forecasts

Prior to the nineteenth century, weather records were kept sporadically, and the techniques, time, and frequency of observation were entirely at the whim of the individual recorder. Early observations were taken for Wilmington, Delaware (1644–1645) and Boston, Massachusetts (1738–1750). Thomas Jefferson (1743–1826) carefully recorded the daily weather at Monticello, in Charlottesville, Virginia, in the late 1700s.[2] However, taking observations on a regular basis did not start in the United States until 1812, when an order from the Surgeon General of the Army was given to hospital surgeons. These army doctors were required to take weather observations and keep climatological records. By 1853, 97 army posts were recording the daily weather.

The mushrooming of weather stations made it possible to study the spatial variation of wind, pressure, and precipitation associated with important storms by plotting the information on a map. Because of slow communications, the first maps were produced long after the storms occurred. Espy post-

[2] Ibid., p. 9.

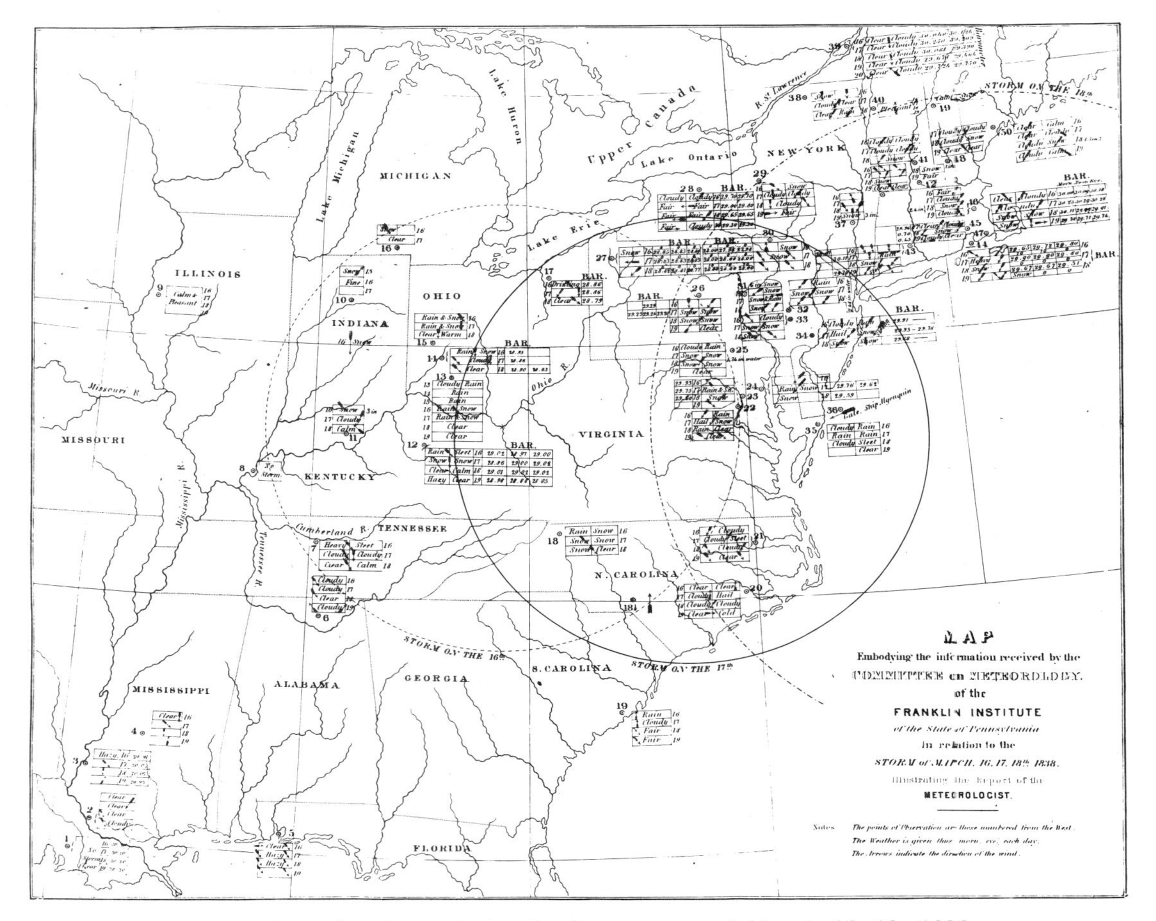

Figure 1.4 *Espy's analysis of winter storm of March 16–18, 1838.*

analyzed many storms during the 1830s and 1840s: one example from *The Philosophy of Storms* is shown in Figure 1.4, which summarizes a late-winter storm which brought heavy snow to the northeastern United States.

The map shown in Figure 1.4 appears slightly naked to today's meteorologists because of the absence of the familiar web of isobars (lines connecting points of equal pressure) and the fronts (boundaries between cold and warm air masses). In fact, Espy's map is little more than a list of the sequence of weather events at each station during the three-day period of March 16–18, 1838. No attempt was made to analyze the conditions over the entire map at a given time in the storm's history. We may note, however, the typical sequence of events associated with a major snowstorm at many locations. For example, the weather sequence during the three days at station 45 (near Boston) was clear, cloudy, snow, cloudy, and finally clear again.

Espy's map contained too much information to be analyzed in a way that would show significant details of the storm's structure, mainly because the plotted data spanned three days. A much simpler, and more revealing, procedure was to plot all the data for a given time on a separate map. An early set of maps of this type, still barren of fronts, was analyzed in color by Professor Loomis in 1860.[3] These maps covered the evolution of a major storm in December, 1836. An example is shown in Plate 1. Professor Loomis' analysis does not show the pressure and temperature structure associated with the storm in the manner usually seen on today's weather map. Rather than draw isobars and isotherms (lines connecting points of equal temperature), Loomis analyzed departures from the mean of pressure and temperature at each station. Thus, every point on the +10°F temperature departure line is 10°F above the mean temperature of that station. Loomis also indicated the wind direction by an arrow, and depicted the state of the sky (clear or cloudy) and rain and snow areas by colors.

Even though Loomis' analyses show no fronts (fronts were not discovered until the twentieth century), distinct evidence of a strong cold front is present. The modern meteorologist, schooled in the frontal theory of storms, would not hesitate in drawing the characteristic heavy blue line (to designate a cold front) in the trough of minimum pressure departures that extends along a north-south line through Michigan, Ohio, Kentucky, and Georgia. Ahead of this front, in the warm air, temperatures are 10°F above the mean, and winds are from the southeast. Behind and to the west of the front, the winds are from the northwest, and the temperature departures fall to –30°F. The cloud and precipitation pattern also supports the cold frontal system, with rain ahead of the front changing to snow behind the front. Far to the southwest, the clear skies are indicative of a high-pressure center moving in from the west.

The maps of Professor Loomis, along with other analyses, gradually led to a qualitative model of the typical middle-latitude cyclone. One such early model, devised by the Reverend W. Clement Ley, is reproduced in Plate 2. To interpret this model, it is helpful to imagine the sequence of weather

[3]Elias Loomis, "On Certain Storms in Europe and America, December, 1836," Smithsonian Contributions to Knowledge, Vol. 11 (Washington, D.C.: Smithsonian Institution, 1860), pp. 27–39.

events that would be experienced by people at different points along the storm's track. To a person on the southeast side of the storm track, the *cirrostratus* clouds (layer of high, thin clouds) gradually thicken to the *nimbus* (rain) clouds as the storm approaches. As the center passes, there is a sudden clearing as the winds shift around to the northwest. An observer on the northwest side of the track would see a more gradual clearing as the storm passed.

It is interesting that Ley's model also recognizes the existence of the not-yet discovered cold front. We quote Robert H. Scott's description, written in 1887, of the Ley cyclone model:

> *One of the most striking characteristics of a cyclonic storm is the sudden shift of wind which takes place between S.W. and N.W., accompanied frequently by a heavy squall and a shower, together with an almost instantaneous fall of temperature.* [Robert H. Scott, *Weather Charts and Storm Warnings* (London: Longmans, Green and Co., 1887), p. 65]

This description of the weather changes concurrent with the passage of a cold front would stand up well in any modern meteorological text.

The early weather maps were attempts to summarize in an orderly way the many observations about major storms and served as research tools to determine the relationship of wind and weather to the low-pressure centers. In 1849, however, the first weather observations were transmitted by telegraph, making possible the rapid collection and analysis of data. Now, weather maps could be produced during the storm's lifetime and utilized for forecasting purposes. By 1860, 500 stations were reporting the weather, and forecasting based on more than local observations became possible. The increased data spurred optimism that truly accurate weather forecasting was imminent, and the number of aspiring weather forecasters suddenly multiplied.

Weather forecasting is not an easy job, however, and, unlike in other endeavors, it is almost impossible to conceal mistakes. From the beginning, and continuing to the present, weather forecasters have been remembered for their failures, or their fractured forecasts. Any forecaster today can sympathize with the poor wretch lambasted by President Lincoln:

> *It seems to me Mr. Copen knows nothing about the weather, in advance. He told me three days ago that it would not rain. . . . It is raining now, and has been for 10 hours. I cannot spare any more time to Mr. Copen.* [Donald R. Whitnah, *A History of the United States Weather Bureau,* p. 15]

But weather forecasters, like economic forecasters, do not give up because of such fiascos, for, if they did, there would be no forecasting at all. In 1870, a national meteorological service was created by Congress under the direction of the Army Signal Corps. By 1872, national forecasts were being issued on a daily basis. These early forecasters claimed an accuracy of 75 percent, a figure not much below what is claimed today.

Weather forecasting in the late 1800s was based almost entirely on the surface charts, which included pressure, wind direction and velocity, speed of movement of high- and low-pressure systems, and amounts of fallen precipitation. Forecasting consisted of locating the weather-producing systems, determining their direction and speed of movement, and extrapolating forward to some future time. And even today, although meteorologists are aided by a bountiful supply of upper-air charts, computer forecasts, and satellite information, extrapolation of features on the surface chart is one of most important methods for short-range forecasting.

1.5 Emergence of Modern Meteorology

Progress toward understanding the atmosphere has continued the acceleration begun in the nineteenth century into the twentieth. Here, we will touch on a few of the major developments during the 1900s. However, first we should update the weather maps discussed in Section 1.4.3 by introducing the concepts of cold and warm fronts. A typical weather map today (see Chapter 5 for examples) will show several fronts, while prior to 1930, maps contained no fronts whatsoever. The atmosphere, of course, did not suddenly develop cold fronts in the 1930s, as indicated by the indirect references to fronts in Loomis' maps and Ley's model. The frontal concept was born of the Norwegians' ideas of the existence of distinct air masses, which were advanced shortly after World War I. Jacob Bjerknes stressed the conflict between masses of hot and cold air, with the boundary separating the two antagonists appropriately called the *front*. If the cold air was marching southward, routing the warm air, the front was termed a *cold front*. When the forces of the south regrouped and overran the weakened and now retreating cold air, the front was renamed a *warm front*. Stationary fronts marked indecisive boundaries at which neither side was making much progress.

Besides the introduction of the frontal concept in surface analyses, there have been three major technological advances during the 1900s that have made significant impacts on forecasting. The first was the beginning of routine upper-air observations in the late 1930s by balloon-carried instruments called *radiosondes*. These radiosondes radioed the temperature, pressure, and humidity of the upper atmosphere back to the ground stations. With this information, the three-dimensional structure of the atmosphere could be studied, and forecasting became more complicated than simple extrapolation of the movement of surface weather systems.

The second revolutionary advance in weather forecasting came in the 1950s when high-speed computers were developed. These computers are able to solve the equations that describe the wind, temperature, and moisture behavior over the entire atmosphere. The equations are predictive in the sense that they can be solved for a future state of the atmosphere, given the present conditions. This concept of forecasting by numbers will be described in more detail in a later chapter.

Finally, the third revolution in weather forecasting began on April 1, 1960, when the first weather satellite, *Tiros I*, was launched. This satellite provided television coverage of storm systems from an altitude of 450 statute miles, and gave meteorologists a fresh look at old weather patterns. Subsequent satellites were more sophisticated and could be used to deduce the winds, temperature, and humidities of the atmosphere below. One of the major difficulties yet to be solved satisfactorily is the meshing of these abundant data with the forecasting phase of the computer, for no forecast can be better than its initial input. The satellite revolution is still in progress. Pictures from some of the more recent satellites will be used in the discussion of various weather phenomena in later chapters.

2 Explaining what we observe

We are aware of many aspects of the physical laws that govern the behavior of the atmosphere even if we do not formally understand the details of these laws. Thus, we "know" that sunny days feel warmer than cloudy days, even if the temperatures are the same; that clear, clean skies are blue, but dusty, hazy skies are white; that cloudy nights are warmer than clear nights; that hot-air balloons rise; that friction between moving objects creates heat; and that it is harder to set a heavy object in motion than a lighter object, but once in motion, the heavy object is harder to stop. We also know that mountains are generally cooler than valleys, that boiled eggs take longer to cook in Denver, Colorado than in Tallahassee, Florida, and that basements are cooler than attics in the summer. The amazing thing is that these, and many more apparently unrelated, phenomena can be explained by a relatively small number of general physical laws or principles. This is the beauty of once learning the general principles—these laws may be applied over and over again in different contexts to explain phenomena that are new to the observer. Thus, when we understand that expanding gases become colder and compressed gases become warmer, we have the correct explanation for why aerosol sprays, when released from the compressed can, feel cold, why suddenly pressurized air in a hand tire pump feels hot, and why air at a mountain resort is generally cooler than the air in the desert below.

For understanding most meteorological processes, it is sufficient to have a qualitative understanding of the following topics:

(1) *Radiation.* All bodies emit and absorb heat energy through radiation; the hotter the body, the more heat energy emitted.
(2). *Newton's second law.* The acceleration of an object (or a parcel of air) is proportional to the net force applied to the object; force = mass X acceleration. For a given mass, the stronger the force, the greater the acceleration. For a constant applied force, the greater the mass, the smaller the acceleration.
(3) *The first law of thermodynamics.* Temperature changes of a gas are produced by expansion or compression, as well as by application or extraction of heat.
(4) The ability of air to hold water vapor increases rapidly with increasing temperature. An approximate rule is that an increase of temperature of 10°C allows the air to hold twice as much water vapor.
(5) *Latent heat.* Condensation of water vapor to liquid releases heat; evaporation of liquid water absorbs heat.

Items (4) and (5), dealing with the properties of moisture in the atmosphere, are discussed in detail in Chapter 8. Although the other items are discussed throughout the book when they directly relate to the meteorological phenomena, a somewhat more general explanation is presented in this chapter.

2.1 Radiation in the Atmosphere

To most people, the word *radiation* is associated with atomic processes such as those in nuclear power plants or atomic bombs. However, radiation is not confined to intense nuclear reactions of this kind, but instead represents the transfer of energy associated with a wide variety of familiar experiences, including radio, television, microwave ovens, and, most importantly, sunlight. Radiation can be generally defined as the transfer of energy by the rapid oscillations of electromagnetic fields in space. These oscillations may be considered as traveling waves, with a characteristic wavelength (distance between successive crests or troughs), frequency (number of wave crests passing a point per time, or the number of oscillations per time), and speed (rate of travel of an identifiable part of the wave, such as the crest). The wavelength, frequency, and speed of a wave are illustrated in Figure 2.1.

It may seem surprising that radio waves, television waves, microwaves in ovens, sunlight, and X rays are basically the same phenomenon—electromagnetic waves which are capable of transmitting energy through a vacuum. The basic difference between these forms of radiation is the wavelength, as indicated in Figure 2.2. Also, because the speed of all these waves is the same [3×10^8 meters per second (m/s), equal to the speed of light], the

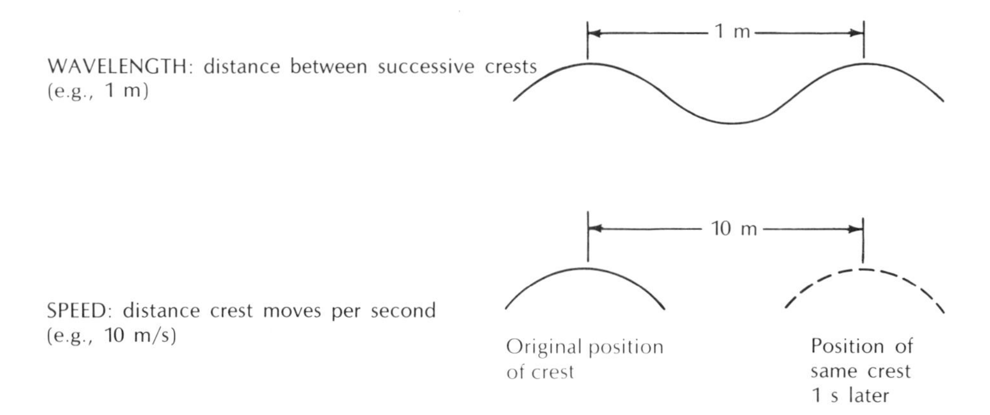

Figure 2.1 *Illustration of wavelength, frequency, and speed of a wave.*

frequency of each type of wave varies. For example, the frequency of AM radio waves of wavelength 500 m is 6×10^5 oscillations per second. The frequency of visible radiation, whose wavelength is 0.5×10^{-6} m, is 6×10^{14} oscillations per second.

2.1.1 *Propagation of energy through a vacuum*

Even though we are all familiar with the propagation of waves on lakes and oceans, it is admittedly difficult to visualize waves propagating through a vacuum. Indeed, it is tempting to envisage, as did some early scientists, a sort of mysterious plasma or ether, filling the voids of space as water fills an ocean, in order to explain the propagation of electromagnetic radiation. However, if we accept the fact that gravitational forces act through a vacuum, or that a magnet can move a compass needle which is isolated in a vacuum, we can more readily visualize the concept of oscillating electromagnetic forces which are generated by vibrations of electrical charges at a point and act through empty space. The propagation of electromagnetic radiation through a vacuum and its subsequent effect on terrestrial matter are illustrated in Figure 2.3. The oscillatory force field, which is the propagation of electromagnetic waves, is capable of causing charged electrons or other particles to move, thereby transferring energy to these particles. The rapid vibrations of the extremely hot [about 6000 kelvins (K)] molecules in the solar atmosphere create an oscillating electromagnetic force field, which propagates through empty space at the speed of light. When the electrical charges in the molecules of a gas, the ground, or our skin are subjected to this oscillating force, they begin to vibrate faster, increasing their *kinetic energy* (energy of motion). (This effect is somewhat analogous to waving a magnet back and forth near a compass needle that is isolated in a vacuum within a glass jar; the oscillating magnetic field causes the compass needle to swing, thereby transferring energy through the vacuum.) We feel this increase in kinetic energy as a rise in temperature. Thus, radiation, acting through a vacuum, is able to transmit heat energy from one group of molecules (the sun) to another group (the atmosphere, ground, or our bodies).

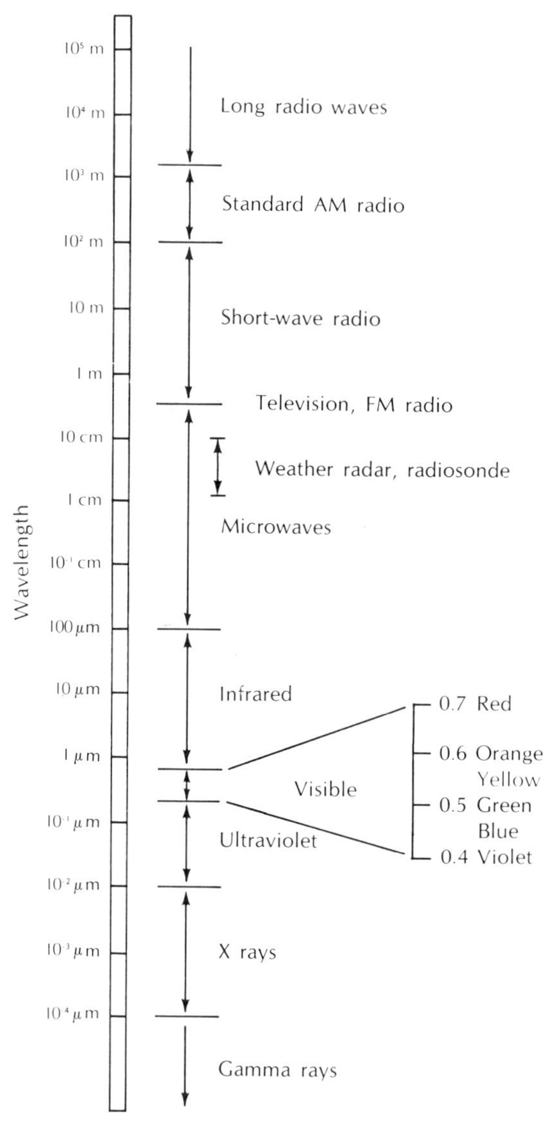

Figure 2.2 *Electromagnetic spectrum.*

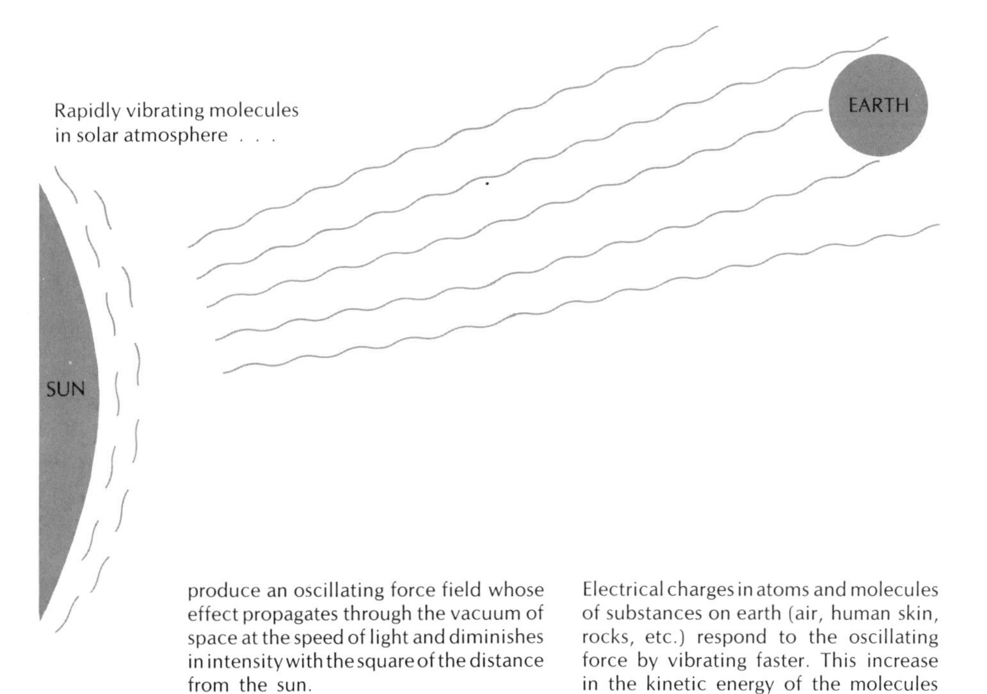

Figure 2.3 *Propagation of radiation through a vacuum.*

2.1.2 *Intensity of radiation*

All objects are sources of radiation, but the amount and kind of radiation they emit depend upon factors such as their temperature and *emissivity,* which is a measure of how efficiently the object emits radiation. The intensity of the radiation emitted by an object is strongly dependent upon the temperature of the object. The amount of radiation emitted increases as the fourth power of the temperature, which means that if the absolute temperature doubles, the emitted radiation increases sixteenfold. [The absolute temperature, in kelvins (K), is the number of degrees above absolute zero, which is the hypothetical temperature at which all molecular motion ceases.] This relation between the fourth power of temperature and radiation is commonly known as the *Stefan-Boltzmann law.* Thus, a unit area on the sun, at an average temperature of 6000 K, emits 160,000 times as much radiation as a unit area on the earth's surface, which has a temperature of about 300 K.[1]

Fortunately, less than one part in a billion of the energy of the sun reaches the earth's surface. Therefore, the earth is able to achieve a radiative

[1] The number 160,000 is obtained by raising (6000 K/300 K) to the fourth power.

equilibrium (amount of radiation received by the sun equal to the radiation lost to space, averaged over a year) at a temperature which is suitable for the development of life as we know it.

The sun is a large spherical body which emits radiation in all directions into space. As the distance from the sun increases, the spherical area over which a given amount of energy is distributed becomes larger and larger, so that the energy received per unit area and per unit time decreases as the square of the radius from the sun (see Figure 2.4). Earth, at a radius of 150 million kilometers (km), or 93 million miles, from the sun, receives about two calories of heat per square centimeter per minute (the *solar constant*) at the top of its atmosphere on a surface that is perpendicular to the line between the earth and the sun. This constant means that a square measuring one centimeter (cm) on a side would receive two calories (cal) of heat every minute (min), or enough to raise the temperature of water one centimeter deep over this square one degree Celsius every minute. The planet Mercury, located at a radius of only 58 million kilometers from the sun, receives about 6.7 X $(150 \text{ million}/58 \text{ million})^2$ as much energy per unit area and time as the earth, while Jupiter, at a radius of 778 million kilometers, receives only about 0.037 X $(150 \text{ million}/778 \text{ million})^2$ as much.

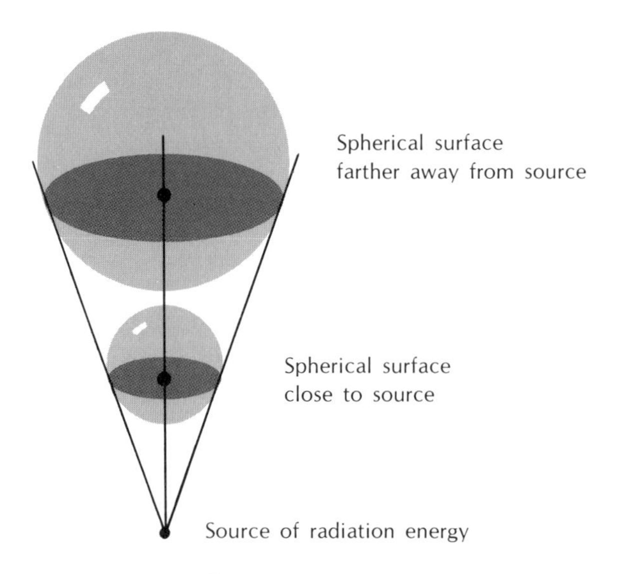

Figure 2.4 *Illustration of inverse square law.*

Besides the absolute temperature, the emissivity of a body is important in determining the intensity of the radiation emitted by the body. The emissivity is defined as the ratio of the actual radiation emitted by a body to the maximum possible amount of radiation that could be emitted at the same temperature. Another way of describing the same concept is by using the *absorptivity,* which is a measure of the efficiency of a body in absorbing radiation. According to *Kirchhoff's law,* the absorptivity and emissivity of a body are equal; a good emitter is a good absorber. In the visible wavelengths, black coal dust is a good absorber (reflects very little visible sunlight). White snow, on the other hand, is a poor absorber (and emitter) of radiation in the visible range. A perfect absorber/emitter is called a *black body,* and has an emissivity of 100 percent. A perfect reflector would have an emissivity of 0.0 percent. In many applications, it is assumed that radiation is emitted by a black body; however, the emissivities of most objects and gases vary from about 10 to 95 percent.

An object or substance that is a good (poor) absorber of certain wavelengths is not necessarily a good (poor) absorber of all wavelengths. Thus, fresh snow is very nearly a white body (emissivity 5 percent) in visible wavelengths, while almost a black body (emissivity 98 percent) in infrared wavelengths.

2.1.3 *Variation of wavelength of maximum radiation with temperature*

Besides Kirchhoff's law and the Stefan-Boltzmann law, a third aspect of radiation is given by the *Wien law,* which describes the relationship between the wavelength at which the maximum amount of radiation is emitted and the temperature of the body (Figure 2.5). The wavelength of maximum radiation is inversely proportional to the absolute temperature, so that the higher the temperature, the shorter the wavelength of maximum radiation. The outer regions or layers of the sun which emit sunlight have a temperature of about 6000 K. The wavelength of the maximum solar radiation corresponds to visible light [0.4–0.7 micrometer (μm). A micrometer is 0.0001 centimeter]. Because temperatures on the earth are much lower than the solar temperature, the maximum terrestrial radiation occurs at longer wavelengths. Thus, the earth, the atmosphere, and we ourselves emit the maximum amount of radiation in the wavelengths of the *infrared* (8–15 micrometers) (Figure 2.5). It is important to note that although the maximum earth radiation occurs at around 10 micrometers, while the maximum solar radiation occurs at about 0.5 micrometer, the sun emits more radiation at all wavelengths than the earth. (Note the difference of six orders of magnitude—one million—in the scale of the radiation intensity for the solar and the terrestrial radiation curves.)

It is interesting that our eyes have evolved to have maximum sensitivity at the wavelengths of maximum solar radiation. Had early man been more dependent upon hunting warm-blooded animals rather than on the diversity (and visual acuity) demanded by cave dwelling, evolutionary changes might have developed for him "eyes" capable of seeing in the infrared, as some snakes have. Although we have not developed a natural visual sensitivity to the infrared wavelengths, we have artificially developed a variety of infrared detectors which enable us to observe the long-wave radiation of the atmosphere. These infrared sensors, carried aloft by earth-orbiting weather satellites, are now used to estimate atmospheric temperatures over the entire globe.

An example of the use of infrared sensors on satellites is shown in the infrared photograph of the eastern United States on May 20, 1973 (Figure 2.6), which may be contrasted with the normal photograph taken in the visible

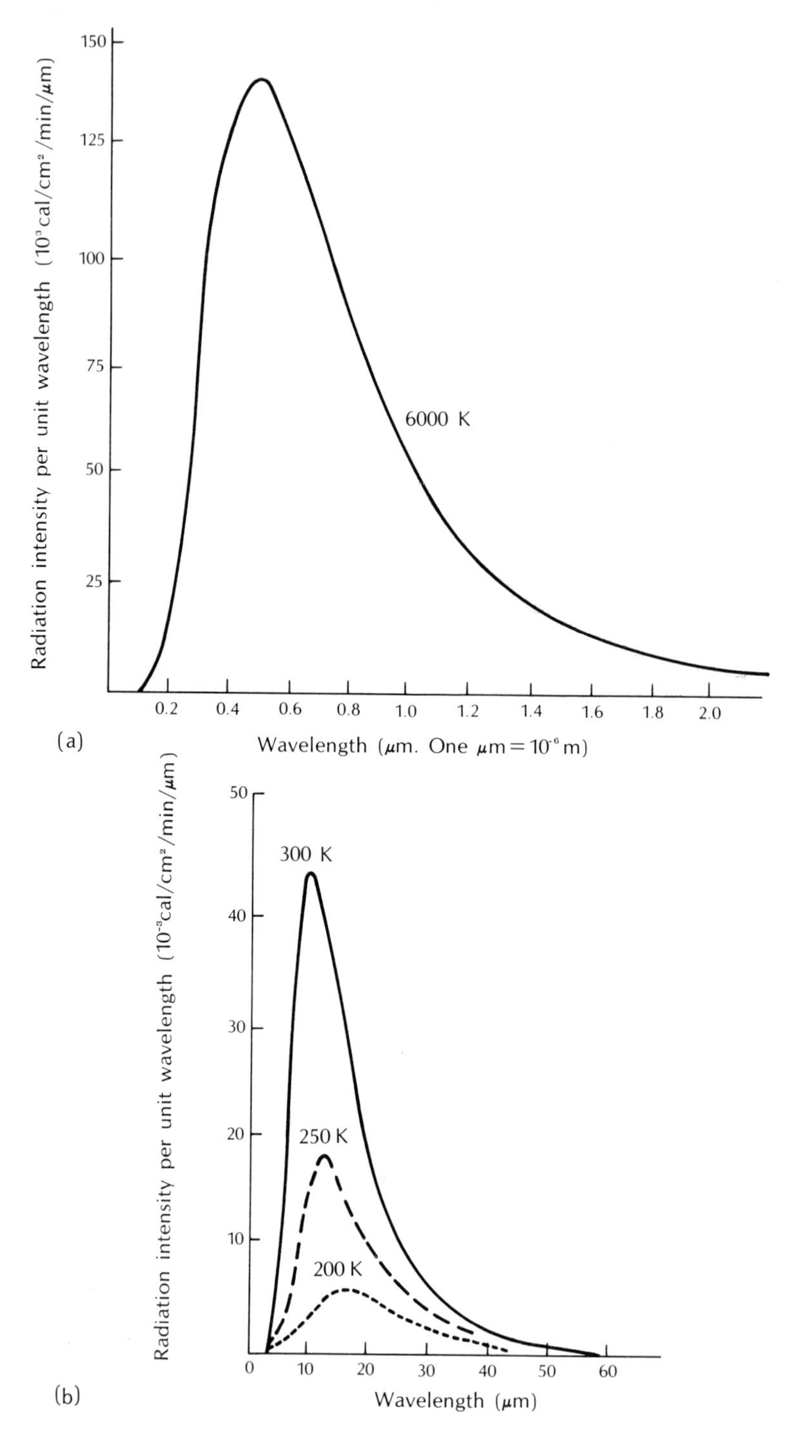

Figure 2.5 *Variation of black-body radiation intensity as a function of wavelength for (a) 6000 K, which corresponds to the solar temperature, and (b) 300, 250, and 200 K, which correspond to temperatures found in the earth's atmosphere.*

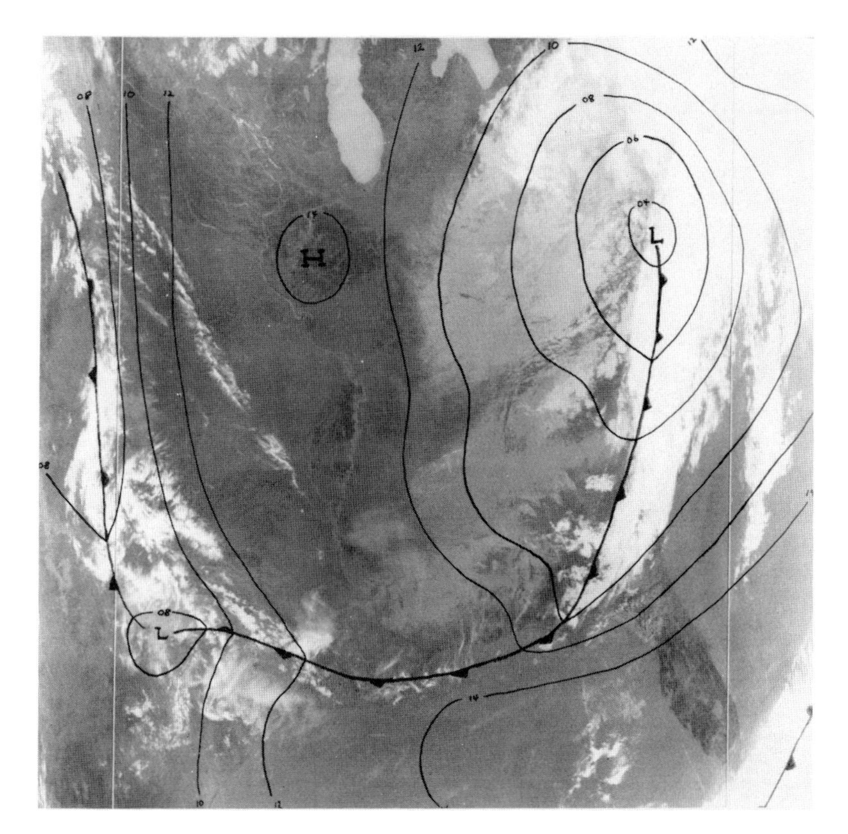

Figure 2.6 *Infrared satellite view of cold front along the East Coast for May 20, 1973.*

wavelengths (Figure 2.7). In the infrared photograph, cold objects (ground, water, or clouds) appear white and warm objects appear dark. In May, the Great Lakes are colder than the surrounding land and therefore appear white. High cloud tops are colder than low cloud tops, and so a comparison of the infrared photograph with the visible picture enables us to differentiate between high, thick clouds, which may be producing precipitation, and shallow, low clouds, which can produce, at most, drizzle. Thus, the white clouds along the cold front, which dips southward along the East Coast, are high, while the clouds to the southwest of the low-pressure center are much lower.

The seemingly esoteric concepts of emissivity and variation of the wavelength of maximum radiation with temperature according to Wien's law have some very important consequences on our weather. For example, snow absorbs very little incoming solar radiation, which is strongest in the visible wavelengths. However, at the temperature of snow, the maximum radiation occurs in the longer wavelengths of the infrared, a region of the spectrum in which snow absorbs (and emits) very well. Thus, snow gains little heat energy during the day by absorption of solar radiation, but loses energy rapidly at night by effectively emitting infrared radiation to space. This difference in emissivity at long and short wavelengths is the major reason why snow has such a cooling effect on the earth and atmosphere.

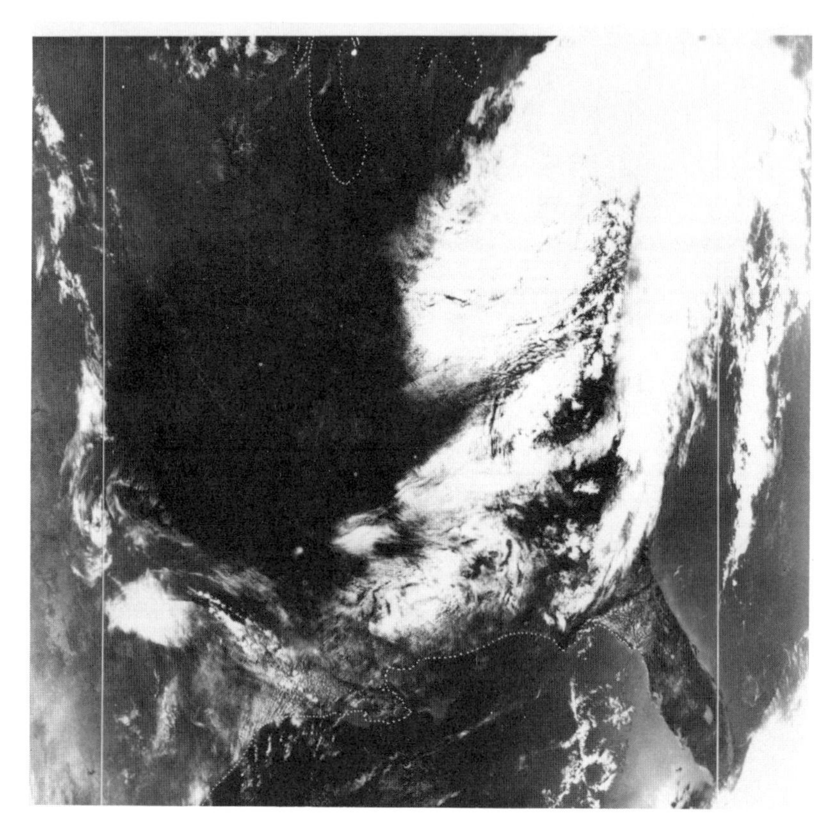

Figure 2.7 *Visible satellite view of cold front along the East Coast for May 20, 1973.*

2.1.4 Scattering, absorption, and reflection of solar radiation by the atmosphere

Radiation does not pass unimpeded through the atmosphere; part is reflected back to space, part is scattered or absorbed by the atmosphere, and part reaches the ground. Both the scattering and absorption of radiation in the atmosphere depend directly upon the electronic properties of the gas molecules, and upon the size of the molecules and any natural or man-made dust particles present in the gas.

Oxygen (O_2) and ozone (O_3) are gases which most effectively absorb, and thus deplete, the sun's radiation as it passes through the atmosphere (Figure 2.8). In the ultraviolet wavelengths, the absorption efficiency is so high that virtually no radiation is transmitted. The process by which ozone absorbs dangerous ultraviolet radiation in the stratosphere is discussed in detail in Section 3.3.1.

The scattering of solar radiation by gas molecules is the reason for the brightness of the daytime sky; planets without atmospheres have black skies, even during the day. Scattering molecules distribute energy in all direc-

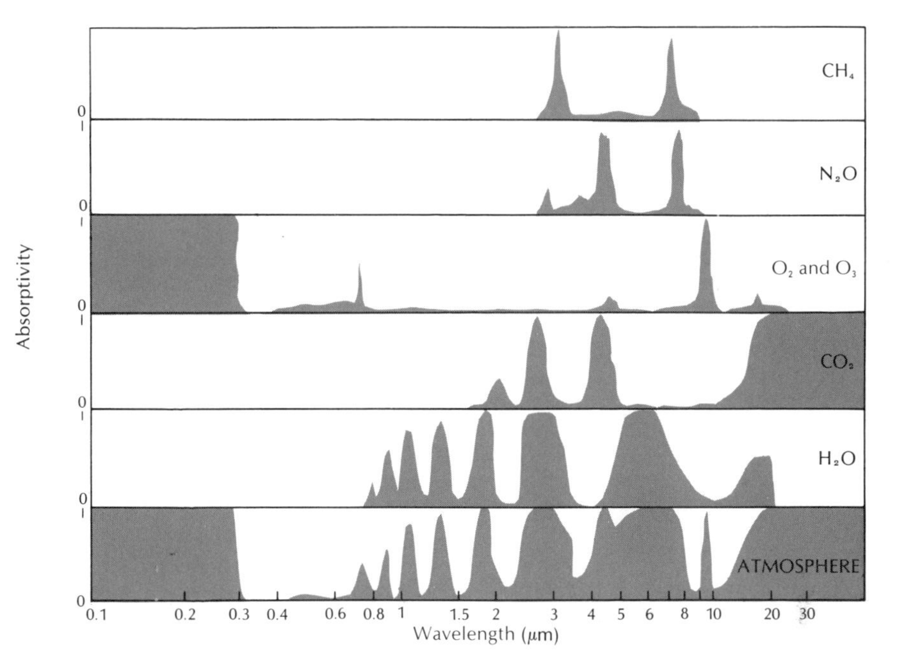

Figure 2.8 *Absorptivity as a function of wavelength for methane [CH_4), nitrogen oxide (N_2O), oxygen (O_2), ozone (O_3), carbon dioxide (CO_2), water vapor (H_2O), and the entire atmosphere [after Robert Fleagle and Joost Businger, An Introduction to Atmospheric Physics (New York: Academic Press, 1963)].*

tions, forward and backward. The amount of scattering is very strongly dependent upon the size of the scattering particle or molecule. Particles that are much larger than the wavelength of light (0.5 micrometer) do not scatter; instead, they partially absorb and reflect the radiation. Particles or molecules much smaller than the wavelength of light produce a scattering of the sunlight that is inversely proportional to the fourth power of the wavelength, which means that shorter wavelengths are scattered much more than longer wavelengths.

Atmospheric molecules are much smaller than the wavelengths of visible light. Therefore, on clear days, the shorter wavelengths of the blue and violet colors are scattered more than the red and yellow colors from the incoming beam of "white" sunshine (Figure 2.9). Thus, the clear sky appears blue. Haze, fog, and cloud droplets, however, have diameters as large as or larger than the wavelength of sunlight, and they do not show a preference for scattering any particular wavelength; all are scattered equally. Thus, when the atmosphere is hazy or cloudy, the sky appears white.

The efficiency of scattering is increased when large amounts of natural or artificial dust particles are in the air. The explosion of the volcano

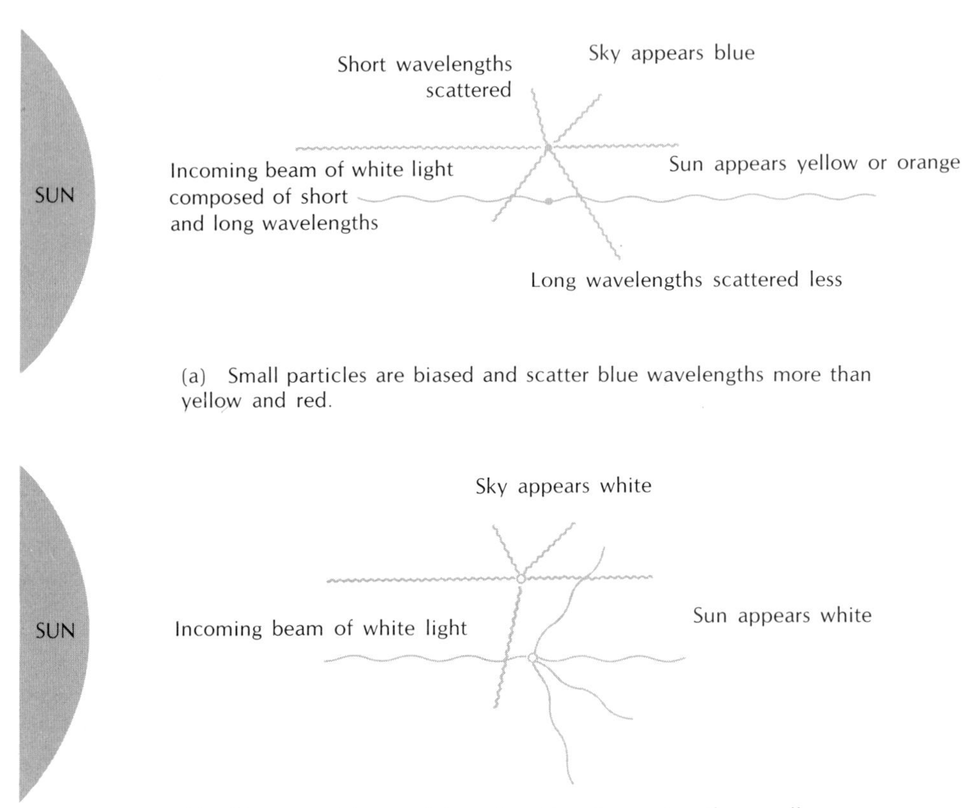

Figure 2.9 *Scattering of sunlight by small and large particles.* (a) Scattering by molecules in the atmosphere produces blue sky. (b) Scattering by larger dust, haze, or cloud droplets produces white sky.

Krakatoa in the East Indies in 1883 produced beautiful sunsets for several years over the entire globe. Frequently now, in highly polluted air, so much scattering and absorption occur that the astronomical sunset is not red or dirty yellow, but, in fact, ceases to exist. For the earthbound observer, the sun may disappear in the murky sky as much as 5 or 10 degrees above the true horizon.

2.1.5 Terrestrial radiation

On the average, the earth emits very nearly as much radiation to space as it receives from the sun; otherwise, the temperature of the earth and the atmosphere would not remain so constant. The earth receives most of its radiative energy in the visible wavelengths, but emits most of the energy in the longer infrared wavelengths (4–70 micrometers).

Figure 2.8 shows the absorptivity as a function of wavelength for some of the important gases in the atmosphere. For example, we have already seen that ordinary oxygen and ozone absorb strongly in the short wavelengths (between 0.1 and 0.3 micrometer) but are relatively transparent (have small absorptivities) in the longer wavelengths. The sum of the absorptivities by the various gases in the atmosphere determines the absorptivity as a function of wavelength for the entire atmosphere, which is shown in the bottom of Figure 2.8.

While the atmosphere is relatively transparent to radiation in the visible wavelengths, the absorptivity in various bands in the infrared range can be very large. The gases which absorb significant amounts of terrestrial radiation include methane (CH_4), nitrous oxide (N_2O), carbon dioxide (CO_2), and water vapor (H_2O). Most of the important absorption in the infrared range is accomplished by carbon dioxide, which is a relatively constant component of the atmosphere, and water vapor, which is highly variable. The net effect of these gases is to absorb the radiation in most of the long wavelengths. On the other hand, certain wavelength bands, called *windows*, exist in which essentially all the infrared radiation is transmitted through the atmosphere. These windows occur around 8 and 11 micrometers (see Figure 2.8).

The high absorptivity of radiation in the infrared by water vapor and the huge variability of water vapor in the atmosphere have a dramatic effect on the amount of radiation lost to space, and consequently, on the nocturnal cooling. Even on perfectly clear nights, a humid atmosphere greatly restricts the radiation of heat to space, and the lowest minimum temperatures occur with very dry air. Thus, the temperature at Miami (which has a very high humidity) during a clear June night may fall to only 20°C (68°F) from a high of 30°C (86°F). At El Paso, Texas, however, where the atmosphere is usually very dry, the temperature may fall from 30°C to 10°C (50°F) by morning.

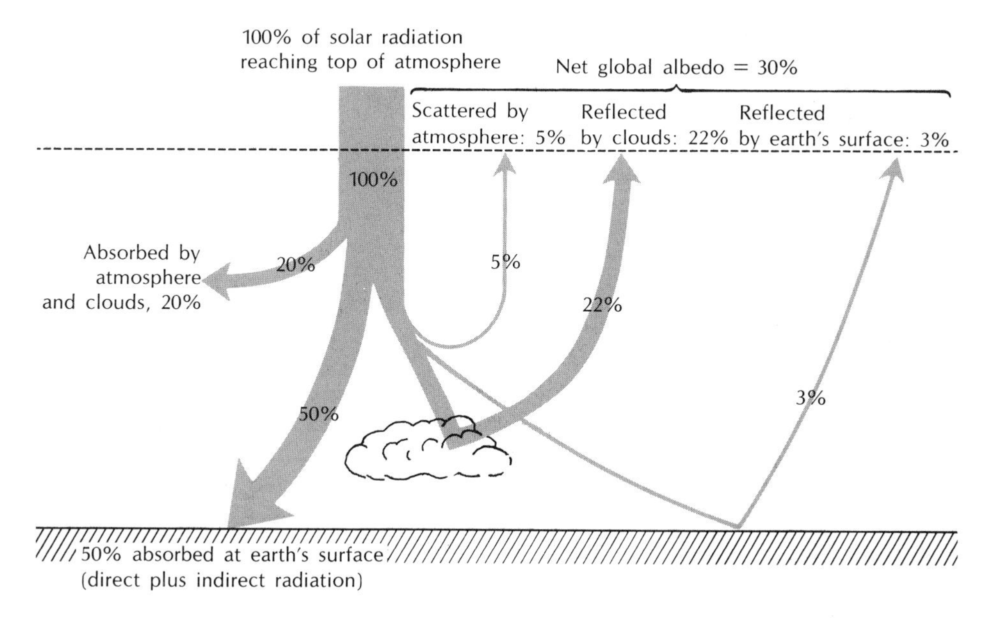

Figure 2.10 *Solar radiation budget of earth and atmosphere.*

2.1.6 The radiation budget

The radiation budget of the earth and its atmosphere shows quantitatively the relative amounts of radiation reflected, scattered, or absorbed by the clear skies, clouds, and earth. Figure 2.10 shows approximately what happens to the solar radiation that reaches the top of the earth's atmosphere.[2] About 20 percent of the radiation is absorbed by the atmosphere and used to warm the air. A large amount, 22 percent, is reflected back into space by clouds and lost, which indicates the very important role clouds play in determining the climate and weather. An additional 5 percent is lost by scattering in clear air, and only about 3 percent of the incoming radiation is reflected from the earth's surface, which is relatively dark.

The reflectivity (percentage of the incident radiation reflected) of a substance is called the *albedo*. Albedos of various terrestrial surfaces are listed in Table 2.1. The reason for the small overall reflection of radiation of sunlight from the earth is that water, which covers about 75 percent of the earth's surface, has a very low albedo during most of the day when the sun is high above the horizon. Only near the sunrise and sunset does water reflect an appreciable amount of radiation.

Table 2.1 *Typical albedo of terrestrial surfaces*

Surface	*Percent of visible radiation reflected*
Fresh snow	90
Sand	25
Earth	15
Forest	7
Water (sun high in sky)	4
Water (sun near horizon)	50
Thick clouds	75
Thin clouds	25

The remaining 50 percent of the solar radiation reaches the ground, either by direct transmission through the atmosphere to the surface, or indirectly as diffuse radiation from clouds and clear sky. It is interesting that, because of the significant amounts of diffuse radiation, it is possible for a surface to receive more radiation than the solar constant (2 cal/cm^2/min), which, at first, is hard to accept, because the solar constant is the amount of radiation received at the top of the atmosphere, and would appear to be the

[2] The percentages of the incident solar radiation and the emitted terrestrial long-wave radiation used in the various processes in the energy budgets shown in Figures 2.10 and 2.12 are only estimates. The actual percentages may vary by ± 2 percent from the values given. The important points are the relative orders of magnitude and that the budgets balance.

maximum possible amount that any surface could receive. However, if the sun is directly overhead and the atmosphere is clean and clear directly above the surface, nearly the entire amount of radiation given by the solar constant reaches the ground (Figure 2.11). And, if there are scattered cumulus clouds, the solar radiation reflected and diffused from the clouds may contribute another 0.4 cal/cm²/min. Thus, the surface at the ground can receive more solar radiation than the same surface at the top of the atmosphere, a situation which would be obviously ideal for rapid suntanning (or, more likely, burning).

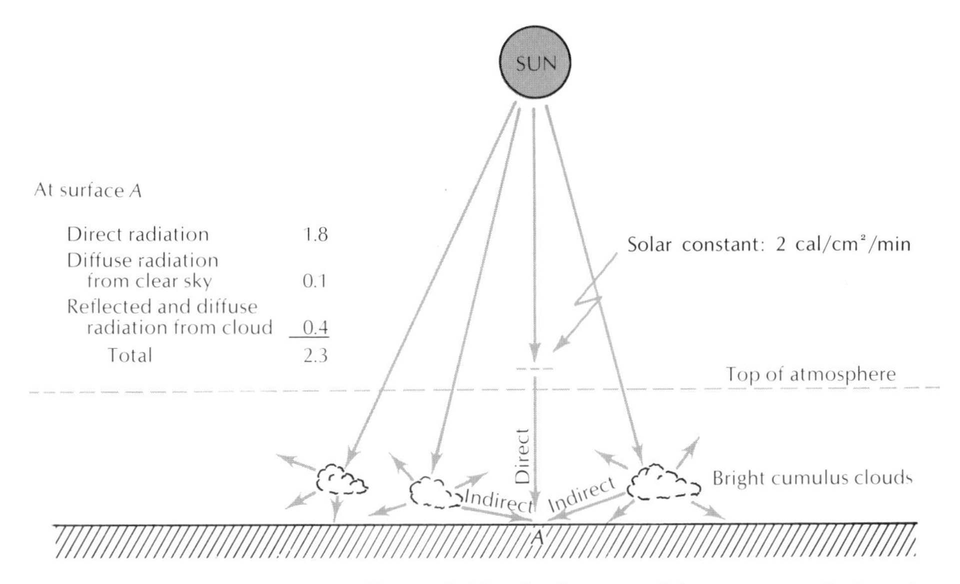

Figure 2.11 *Surface receiving more radiation than the solar constant.* Because of indirect radiation from the clouds and scattered radiation from the sky, a surface at the ground can receive more radiation per unit time than the solar constant.

The approximate long-wave energy budget is shown in Figure 2.12. About 96 units of long-wave radiation from the atmosphere are added to the 50 units of solar short-wave radiation, yielding a total of 146 units absorbed by the ground. In order for the radiation budget at the ground to balance, the same 146 units must be lost. The greatest loss, about 114 units, occurs through the emission of long-wave radiation. About 20 units are used to evaporate water at the earth's surface, adding latent heat to the atmosphere which may later be released in condensation. The remaining 12 units are utilized to warm directly the air near the ground by *conduction* (direct transfer of heat energy in a medium by the collision of rapidly moving molecules) and *convection* (vertical transfer of heat by eddies that are much bigger than the molecular scale).

Most (104 units) of the 114 units of infrared radiation emitted by the ground is absorbed by carbon dioxide and water vapor in the atmosphere.

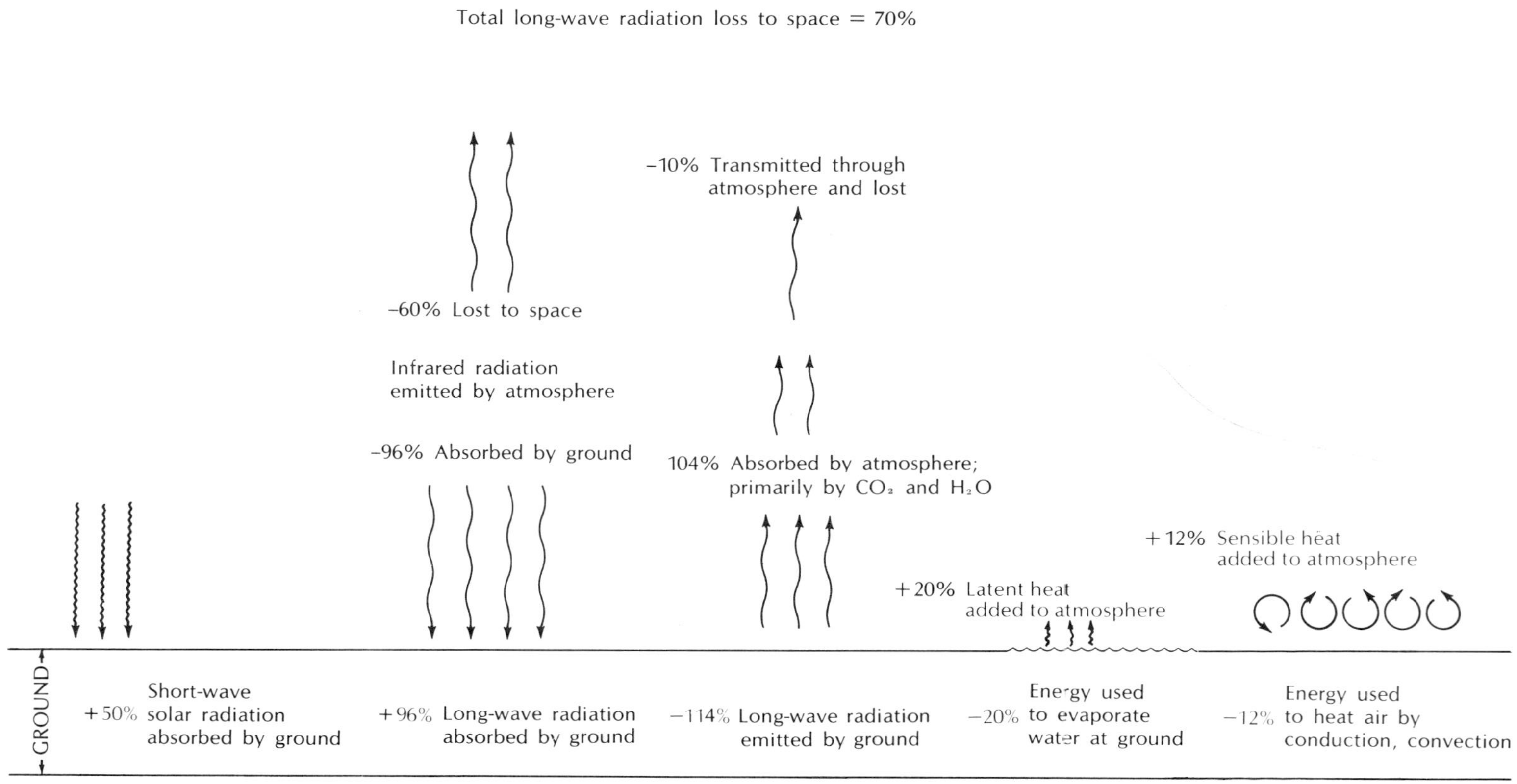

Total long-wave radiation loss to space = 70%

−60% Lost to space

−10% Transmitted through atmosphere and lost

Infrared radiation emitted by atmosphere

−96% Absorbed by ground

104% Absorbed by atmosphere; primarily by CO_2 and H_2O

+12% Sensible heat added to atmosphere

+20% Latent heat added to atmosphere

GROUND

+50% Short-wave solar radiation absorbed by ground

+96% Long-wave radiation absorbed by ground

−114% Long-wave radiation emitted by ground

−20% Energy used to evaporate water at ground

−12% Energy used to heat air by conduction, convection

Figure 2.12 *Long-wave radiation budget of earth and atmosphere.*

Part of the infrared radiation absorbed is reradiated back to the earth. This trapping and recycling of terrestrial radiation, which makes the earth warmer than it would be otherwise, is called the *greenhouse effect,* because it was once thought that greenhouses remain warm by the same process. Glass allows the short-wave energy (sunlight) to pass into the greenhouse because glass is transparent to short-wave radiation. At the same time, glass is opaque to infrared radiation and traps the long-wave radiation in the greenhouse. However, it has been shown that this radiation effect is not the major reason for the warmth in a greenhouse; instead, the glass simply prevents mixing of the warm air with the colder air outside. Thus, ironically, the earth remains warmer by the greenhouse effect (transmitting short-wave sunlight, absorbing and reradiating long-wave terrestrial radiation), while greenhouses do not.

In summary, the disposition of the incoming solar radiation is determined by scattering, reflection, and absorption by the clear atmosphere, clouds, solid particles, and surface of the earth. The infrared radiation which is emitted by the earth is largely absorbed by carbon dioxide and water vapor in the atmosphere. The result of all this scattering, reflection, absorption, and reradiation is a complex equilibrium energy budget which determines the mean temperature of the earth. Because of the importance of the highly variable amounts of clouds and water vapor in this equilibrium, small changes in either or both could have large effects on the climate at the earth's surface.

2.2 Behavior of Gases—Changes in Temperature, Pressure, and Density

In the range of temperatures and pressures in the atmosphere, the gases which make up the atmosphere (mainly nitrogen and oxygen; see Table 2.2) behave as *ideal gases*. Ideal gases are by definition those which obey the relatively simple relationship between pressure, density, and temperature known as the gas law, which is discussed in Section 2.2.2. Although necessary in quantitative calculations, this law has only limited use in qualitatively explaining how temperature changes are caused by expansion or compression because of the presence of three variables, all of which may change simultaneously. To explain many important atmospheric phenomena, we need a relationship between pressure and temperature alone. We find this more revealing and useful dependency by making use of the *first law of thermodynamics,* which tells us what will happen to the temperature of a gas if it is heated, cooled, or the pressure is changed by processes such as the lifting of air over a mountain.

Before discussing the behavior of gases that make up the atmosphere, let us introduce the useful concept *parcel of air* and the environment of this parcel.

Table 2.2 *Composition of the atmosphere*

Gas	*Percentage*
Nitrogen (N_2)	78.08
Oxygen (O_2)	20.95
Argon (A)	0.93
Carbon dioxide (CO_2)	0.03
Water vapor (H_2O)	0.00–3.0 (variable)

2.2.1 *Air parcels and their environment*

Meteorologists are always talking about rising parcels of air, squeezing the water vapor out of parcels of air, or the acceleration of a parcel of air. What is this nebulous parcel of air? A parcel of air usually refers to a very small volume of the atmosphere with horizontal dimensions of a few meters in which the temperature, pressure, humidity, density, dustiness, etc. are the same throughout. This homogeneous parcel is viewed as a sort of invisible balloon, which can be followed as it moves up and down or sideways over a period of time. The parcel may expand, stretch, shrink, or even twist, but the air molecules which made up the original parcel always stay together; that is, the parcel does not break in two, with half of it heading south and the other half north. We might imagine coloring all the air molecules in a volume the size of a classroom green and then observing what happens when this green volume of air is subjected to lifting, heating, cooling, condensation, or any other process we might wish to consider. We may visualize parcels of air in other ways. By riding along in a weightless balloon, we follow the sample of atmosphere in the immediate vicinity of the balloon as it drifts around the earth. A smoke puff rising from a chimney identifies a parcel of air, as does a cloud of tiny gnats, rising and falling with each turbulent eddy near the ground.

We find it useful to talk about parcels rather than the entire atmosphere when we want to isolate certain processes or forces. A parcel is simply a convenient increment of the atmosphere which behaves as a unique element during the process under discussion.

The air parcel concept is very useful when considering the formation of cumulus clouds (to be discussed in detail in Chapter 6), for the condensation of water vapor into cloud droplets makes this particular parcel of air visible (Figure 2.13). Imagine a spherical parcel of warm, moist, buoyant air near the ground. Because of its high temperature and moisture content, it is slightly less dense than the surrounding air, which is called the *environment*. It is important to realize that the properties of the parcel and the environment are not usually the same, although they could be. As the parcel of air rises through its environment (much like a bubble rises through water), its temperature decreases, but so does the temperature of the surrounding air (which is not moving). Therefore, the parcel may remain warmer than its environment and continue to rise. Figure 2.13 shows a plot of temperature versus height above the ground in this example and emphasizes that the rising parcel and its environ-

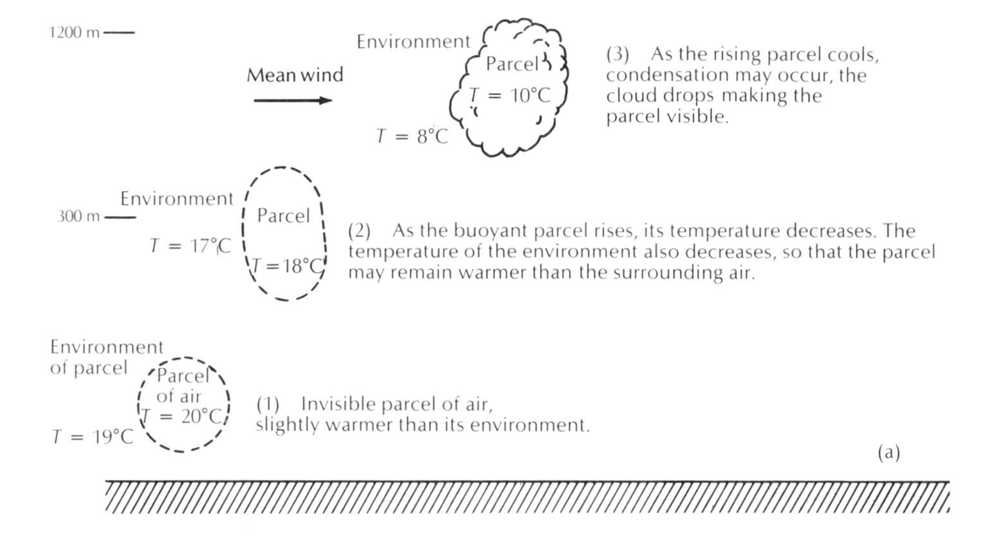

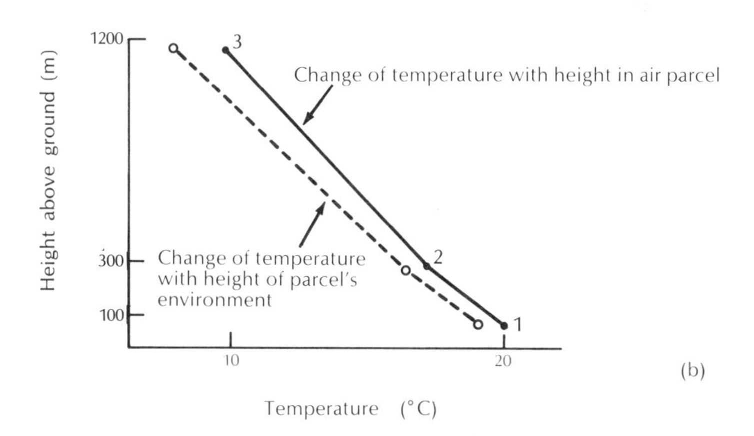

Figure 2.13 (*a*) *Rising parcel of air and its environment.* (*b*) *Temperature change with height in a parcel of air compared to the temperature change with height of the parcel's environment.*

ment are quite different. Finally, as the parcel continues to rise, condensation may occur, making the parcel visible.

2.2.2 *The gas law: equation of state*

The *gas law,* or the *equation of state,* says that the pressure of a gas is proportional to the product of the absolute temperature and the density (mass of the gas per unit volume); that is,

$$\text{Pressure} = \text{constant} \times \text{density} \times \text{temperature} \tag{2.1}$$

Thus, in a closed container holding a fixed amount of gas (so that the density is constant), increasing the temperature increases the pressure, and decreasing the temperature decreases the pressure, as we can demonstrate by heating the air in an open metal can on a stove and then removing and quickly capping the can. As the gas in the sealed container cools and the inside pressure falls, the can will be crushed by the higher atmospheric pressure outside the can.

The gas law, by itself, has only limited applicability in explaining the behavior of the atmosphere. We know, for example, that at the earth's surface, pressure and temperature are not directly proportional. In fact, during the winter over the continents, the lowest temperatures frequently occur with the highest surface pressures, and the temperature usually rises as the pressure falls. However, these observations do not invalidate the gas law; the difficulty lies in the fact that the density of the atmosphere is not constant because the atmosphere is compressible. Thus, when the temperature falls as the pressure rises on a cold winter day, the density rises also, and the gas law is still satisfied.

Let us try to apply the gas law in one more way. We hear repeatedly how the temperature of rising air decreases, while the temperature of sinking air increases. Can the gas law be used to explain this important observation? If air rises to a greater elevation, its pressure must fall, and if the density were constant, the temperature would have to fall also. Thus, the gas law seems to explain the phenomenon of rising air becoming cooler. However, the density of rising air does not stay constant; in fact, it decreases, as does the pressure. Looking back at the equation of state, we see that if both the pressure and the density fall, we cannot say whether the temperature will rise, fall, or remain the same; it could do any of these things, depending on which decreases more, the pressure or the density. Again, the problem is that we are working with three variables. To explain what happens as air is lifted and the pressure falls, we would much prefer to have a relationship between temperature and pressure alone, so that we do not have to worry about what the third variable, density, is doing.

2.2.3 Temperature changes of a gas under heating, cooling, or changes in pressure

One of the most useful laws in explaining the behavior of a gas (or a mixture of gases such as the atmosphere) under various conditions of pressure, temperature, and heating or cooling is the empirical (experimentally determined) first law of thermodynamics, which says that the temperature of a gas may be changed by addition (or subtraction) of heat, a change in pressure, or a combination of both:

$$\text{Change in temperature} = \text{constant X heat added (subtracted)} + \text{another constant X pressure change} \quad \textbf{(2.2)}$$

The effect of adding or subtracting heat is simple to understand; when the atmosphere is heated by solar radiation, by long-wave radiation from the earth, or by contact with the warm ground, its temperature rises. When the atmosphere loses radiation to space, or is cooled by evaporation of rain falling through a dry layer, the temperature decreases.

In spite of the ultimate importance of heating and cooling processes (called *diabatic* processes), there are many atmospheric processes, usually involving time scales of a day or less, in which the amounts of heat added or subtracted are small. In these cases, we can neglect the heating or cooling term in the first law of thermodynamics, and are left with a very simple relationship between small changes in temperature and pressure:

Change in temperature = constant X change in pressure **(2.3)**

Because this latter relationship is valid if no heat is added or subtracted, it is called the *adiabatic* form of the first law of thermodynamics. Note that the temperature may change without heating or cooling; thus, heating a gas is not synonymous with a temperature increase, nor is cooling (removing heat) synonymous with a temperature decrease. An adiabatic process is simply one in which no heat energy (calories) is added or subtracted from the gas during the process. Thus, the process of a parcel of dry air flowing over a mountain may be adiabatic, but the temperature will decrease as the parcel flows up the mountain slope and the pressure falls, and will rise again as the parcel descends the opposite side and the pressure increases.

It is difficult to overemphasize the importance of the adiabatic form of the first law of thermodynamics—that temperature changes are directly proportional to pressure changes when no heat is added or subtracted from the gas. Because the pressure changes so rapidly over small vertical distances, large temperature changes result as the atmosphere sinks or rises. For example, the adiabatic form of the first law of thermodynamics predicts a 10°C temperature fall for a decrease in pressure corresponding to an increase in height of 1 kilometer (or 5½°F/1000 ft). This special rate of decrease of temperature with height (10°C/km) is called the *dry adiabatic lapse rate*.

Air flowing over tall mountains, rising in vigorous thunderstorms, or being lifted in the complex cyclones of middle latitudes may change elevation by tens of thousands of feet. Thus, if a parcel of dry air at the surface, with a comfortable temperature of 25°C (77°F) were lifted to 6 kilometers (about 20,000 feet), the temperature would be a frigid −35°C (−31°F). Conversely, if air at a typical temperature of −20°C (−4°F) at 6 kilometers sank to the surface, its temperature would be a roasting 40°C (104°F).[3] All of these changes in temperature could be produced without any heating or cooling whatsoever, so powerful is the effect of compression or expansion on temperature changes.

2.2.4 Atmospheric stability

The details of the vertical temperature structure of the atmosphere are important in many meteorological processes. For example, the change of temperature in the vertical may determine whether morning cumulus clouds will grow into afternoon thundershowers, or whether

---

[3] The warming effect of compression has a surprising consequence. Air conditioners, not heaters, must be used in aircraft flying at low pressures (high altitudes), even though temperatures of the air outside may be −30°F. A compression of this air at a height of 30,000 feet to sea level pressures in the aircraft would produce a temperature of 130°F if the air conditioners were not used to extract heat from the air.

pollution from a smokestack will rise thousands of feet and be dispersed over a wide area instead of turning back downward toward the ground. The vertical temperature (and to a lesser extent, moisture) distribution characterizes the static stability, or more simply, the *stability* of the atmosphere. The concept of stability is illustrated in Figure 2.14.

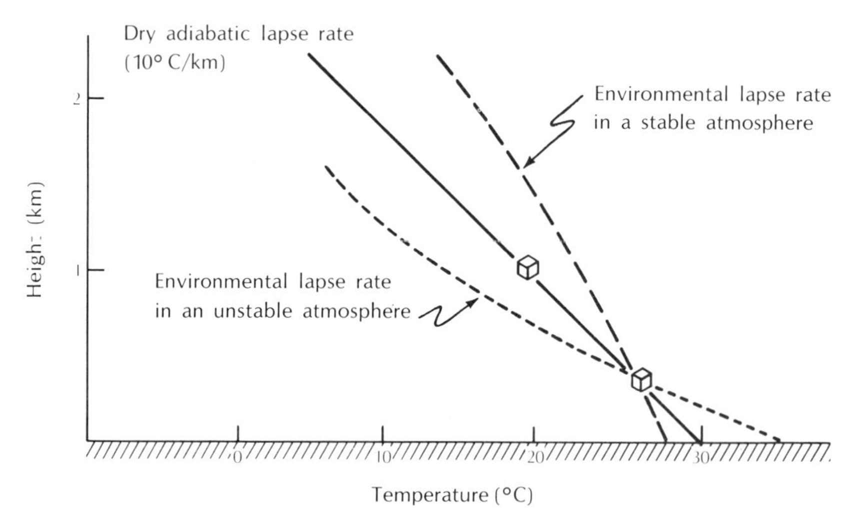

Figure 2.14 *Illustration of stability in the atmosphere.* When no sources or sinks of heat are present, the temperature of a rising parcel of air decreases at a rate of 10° C/km. If the rate of temperature decrease with height in the environment is less than 10° C/km, the parcel will become colder than its environment and will sink back toward its original level, a condition termed *stable.* If the environmental rate of temperature decrease exceeds 10° C/km, however, the parcel will remain warmer than its environment and accelerate away from its original level, a condition termed *unstable.*

Small-scale irregularities in terrain or uneven heating or cooling of the ground are always producing small perturbations in the large-scale flow of air. In a stable atmosphere, parcels of air which are displaced small vertical distances by these irregularities in the flow will tend to return to their original level. If the parcel moves upward, for example, and no sources or sinks of heat are present, its temperature decreases at the dry adiabatic rate of 10°C/km. If the environmental rate of temperature decrease is less than this rate, the parcel soon becomes cooler (and more dense) than its environment and sinks back toward its original level. This atmospheric condition is termed *stable*. In contrast, the atmosphere is called *unstable* if the environmental rate of temperature decrease is greater than 10°C/km; then, the parcel becomes warmer (and less dense) than its environment and accelerates away from its original level.

The vertical distribution of moisture also affects the stability of the atmosphere. Because water vapor is less dense than dry air, moist air underlying dry air favors instability. If condensation occurs in the rising parcel of air, the latent heat of condensation is added to the parcel and reduces its rate of cooling to approximately 6°C/km. If the rate of decrease of the environmental temperature is less than this *moist adiabatic lapse rate,* the saturated (100 percent relative humidity) parcel of air may continue to rise. In this case, the atmosphere is unstable with respect to vertical motions of saturated air parcels. This process, which leads to the development of cumulus clouds and thunderstorms, is discussed in Chapter 6.

Except near the ground on hot sunny days, the environment is normally stable for dry vertical displacements of air; that is, the rate of temperature decrease with height is less than 10°C/km. The environmental lapse rate frequently exceeds the moist adiabatic rate of about 6°C/km, however; otherwise, we would not have so many thunderstorms.

2.3 Forces, Accelerations, and the Production of Motion

The previous sections have considered the radiative and thermodynamic properties of the gaseous atmosphere. Now, we turn briefly to the dynamics of gases—the laws that explain the motion of the ever-restless atmosphere.

Starting the wind in motion, changing its direction, or causing it to cease requires acceleration (change in velocity), and producing acceleration requires a net force. The basic law relating force to acceleration is Newton's second law:

$$\text{Net force} = \text{mass} \times \text{acceleration} \tag{2.4}$$

Thus, changes in velocity are produced by unbalanced forces acting over time; a balance of forces (zero net force) means no acceleration and, therefore, constant velocities.

Although the important forces which accelerate (and decelerate) the atmosphere will be discussed in Chapter 4, let us briefly preview them here in Table 2.3 and make the point that each force is normally counteracted by another force of opposite sign, but nearly the same magnitude. Therefore, the net forces (and, therefore, accelerations) acting on parcels of air are small (Figure 2.15). Note that even though there is little acceleration, there can still be rapid motion, just as a bicycle will coast without being pedaled.

In general, large-scale atmospheric motions are nearly balanced, which means that net accelerations are small. Exceptions occur in very violent weather such as tornadoes, or on small scales such as in the gusts and eddies which form in the vicinity of tall buildings on a windy day. In the balanced, large-scale flows, the vertical pressure gradient force balances gravity, and the horizontal pressure gradient force is balanced by the Coriolis force.

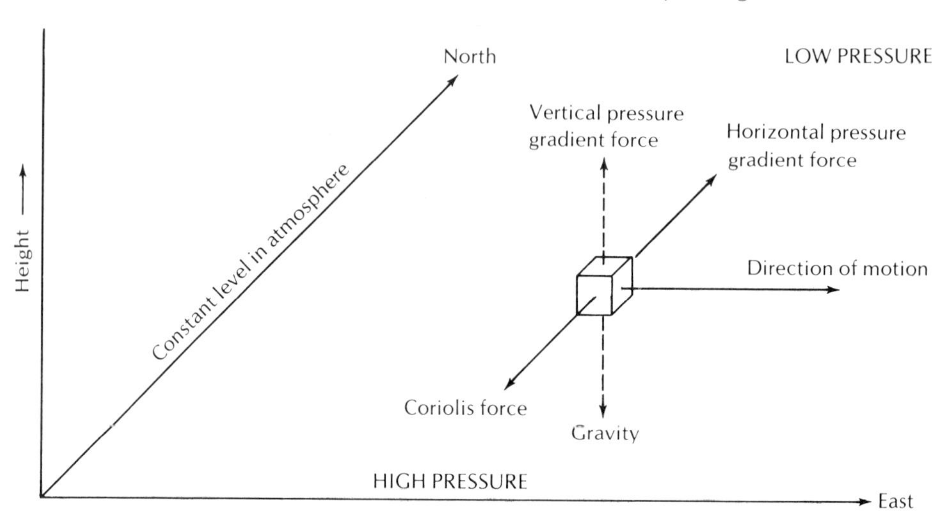

Figure 2.15 *Balance of forces on an air parcel traveling eastward.* The parcel is experiencing no accelerations because all the forces are balanced. The vertical pressure gradient force (directed upward) is balanced by gravity (directed downward). The horizontal pressure gradient force (directed northward toward low pressure) is balanced by the Coriolis force (directed southward, at right angles to the direction of motion).

2.4 Scales of Meteorological Phenomena

The equations and laws described thus far are very general. In a sense, they are too general in that they can be used to describe phenomena as large as cyclones or as small as the vortex of water draining out of a bathtub. Obviously, to avoid confusion, we do not want to study all sizes of motion at the same time. In practice, we select our principal time and space scales of motion. The *time scale* is determined by the lifetime, or period, of the phenomenon; the *space scale* is determined by the typical size or wavelength.

Figure 2.16 gives the time and space scales for several atmospheric features. The concept of scale is somewhat imprecise, and, therefore, the actual sizes or time periods of phenomena which belong to the same scale may vary considerably, perhaps by as much as an order of magnitude (factor of ten). Thus, in a sample of 100 thunderstorms, individual storms may vary in diameter between one and five miles, and may last from 30 minutes to several hours. And yet, all of these thunderstorms have a time scale of about an hour and a horizontal space scale of a few miles. The important point is that thunderstorms never go through a life cycle in a second, nor does a single thunderstorm

Table 2.3 *Important forces in the atmosphere*

Force	*Remarks*
(1) Gravity	Acts downward; is nearly balanced by the vertical pressure gradient force (see 2a).
(2) Pressure gradient force	Acts toward low pressure.
(a) Vertical pressure gradient force	Component of the pressure gradient force in the vertical; acts upward; is nearly balanced by gravity.
(b) Horizontal pressure gradient force	Component of the pressure gradient force in the horizontal; acts from high to low pressure; is much smaller in magnitude than the vertical pressure gradient force, but is very important in producing horizontal air motions (winds). On large scales, tends to be balanced by the Coriolis force (see 3).
(3) Coriolis force	Force caused by the rotation of the earth; acts at right angles to the direction of motion; is proportional to the speed of motion; is strongest at the North Pole; vanishes at the equator. For large-scale motions, balances the horizontal pressure gradient force.
(4) Friction	Generally acts in direction opposite to the direction of motion, and slows the speed of the moving air. Friction is usually important only near the ground, in the lower 5000 feet of the atmosphere.

ever cover an area the size of the United States. Thunderstorms have their own niche in the spectrum of time and space, even if individual thunderstorms show variations within this niche.

As another example of a class of weather phenomena which is identifiable with particular time and space scales, consider the extratropical cyclones (low pressure-systems) and anticyclones (high-pressure systems) that affect many states over a period of a few days. To forecast the evolution of these weather-producing systems, we deal with horizontal scales that can be resolved on the weather map. These *synoptic,*[4] or large, scales determine the density of the observing stations, and range from several hundred miles to several thousand miles in the horizontal. The vertical scale of these weather systems is quite a bit smaller, about 10 miles.

On a very small scale, ordinary ocean waves can be understood through the same physical laws described in this chapter. For these waves, the space scale is about 10 meters, and the time scale is a few seconds.

[4] In meteorology, the term *synoptic* means "coincident in time," so that a synoptic weather map shows meteorological conditions over an area at a given moment. It is also frequently used to denote large-scale, as opposed to small-scale, atmospheric patterns.

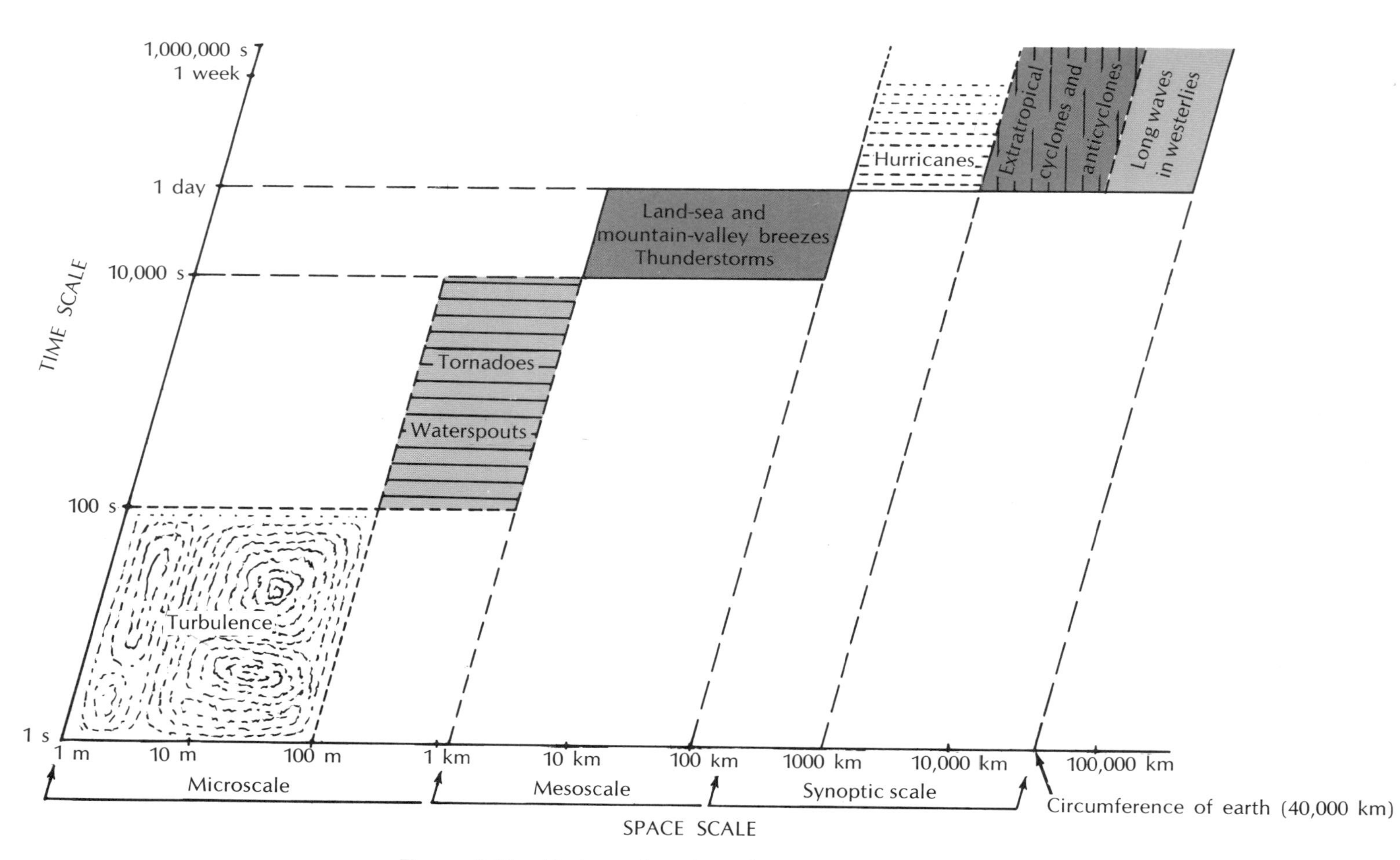

Figure 2.16 *Horizontal scales of atmospheric motions.*

Rather than attempting to comprehend the entire atmosphere at once, we simplify the problem by isolating an individual weather phenomenon for study. Depending on its scale, we can then make some important simplifying approximations valid for that scale only. For example, when we work with large-scale motions and try to understand the meandering currents of air that circle the globe, we can assume that gravity exactly balances the vertical pressure gradient force, which means that vertical accelerations are zero. This condition of balance is called *hydrostatic equilibrium*. Furthermore, we will see that, on this scale, the Coriolis force and horizontal pressure gradient force nearly balance (*geostrophic balance*), at least above the lowest kilometer or so, and away from the equator.

Neither the hydrostatic nor the geostrophic approximation is valid for an individual cloud climbing in the sky, a gust of wind on a spring afternoon, or a wave in the air flowing over a small mountain. On these small scales, however, we can simplify the physical laws in another way by neglecting the effect of the earth's rotation, and, therefore, the Coriolis force. Thus, each scale of atmospheric motions has its own set of simplifying assumptions, as well as its own complexities.

2.4.1 *The synoptic scale*

Probably the most important scale for determining tomorrow's weather is the *synoptic,* or weather-map, scale, which includes atmospheric phenomena with typical horizontal scales of 800–8000 kilometers (500–5000 miles) and lifetimes from one day to a week. The basis for prediction on this scale is the weather-map analysis, which consists of reducing the vast amounts of meteorological data (pressures, temperatures, winds, humidities, clouds, precipitation) from hundreds of weather stations into meaningful patterns that can be interpreted by the meteorologist.

A familiar example of a weather analysis is the sea level pressure chart shown in Figure 2.17 for December 25, 1973. By plotting the sea level pressure of each station and drawing lines (isobars) through points which have equal pressures (estimating values between stations, if necessary), a large number of data that would be virtually meaningless in tabular form is organized into patterns which reveal much about the weather. One purpose of drawing isobars is to summarize the enormous amount of information available on the map and produce a visualization of large-scale pressure patterns; the second purpose is to isolate the large-scale features and separate them from features of smaller scales, which is accomplished by drawing smooth lines. In Figure 2.17, the isobars were required to fit exactly each individual observation; the lines are irregular, with many wiggles, especially in the Ohio Valley and the Rocky Mountains. One reason for the wiggles are inaccurate observations, but small-scale motions also contribute to the irregularity. Smoothing eliminates both errors and unwanted small scales. In Figure 2.18, small-scale variations in the pressure pattern, which are unimportant for synoptic-scale analyses, are removed.

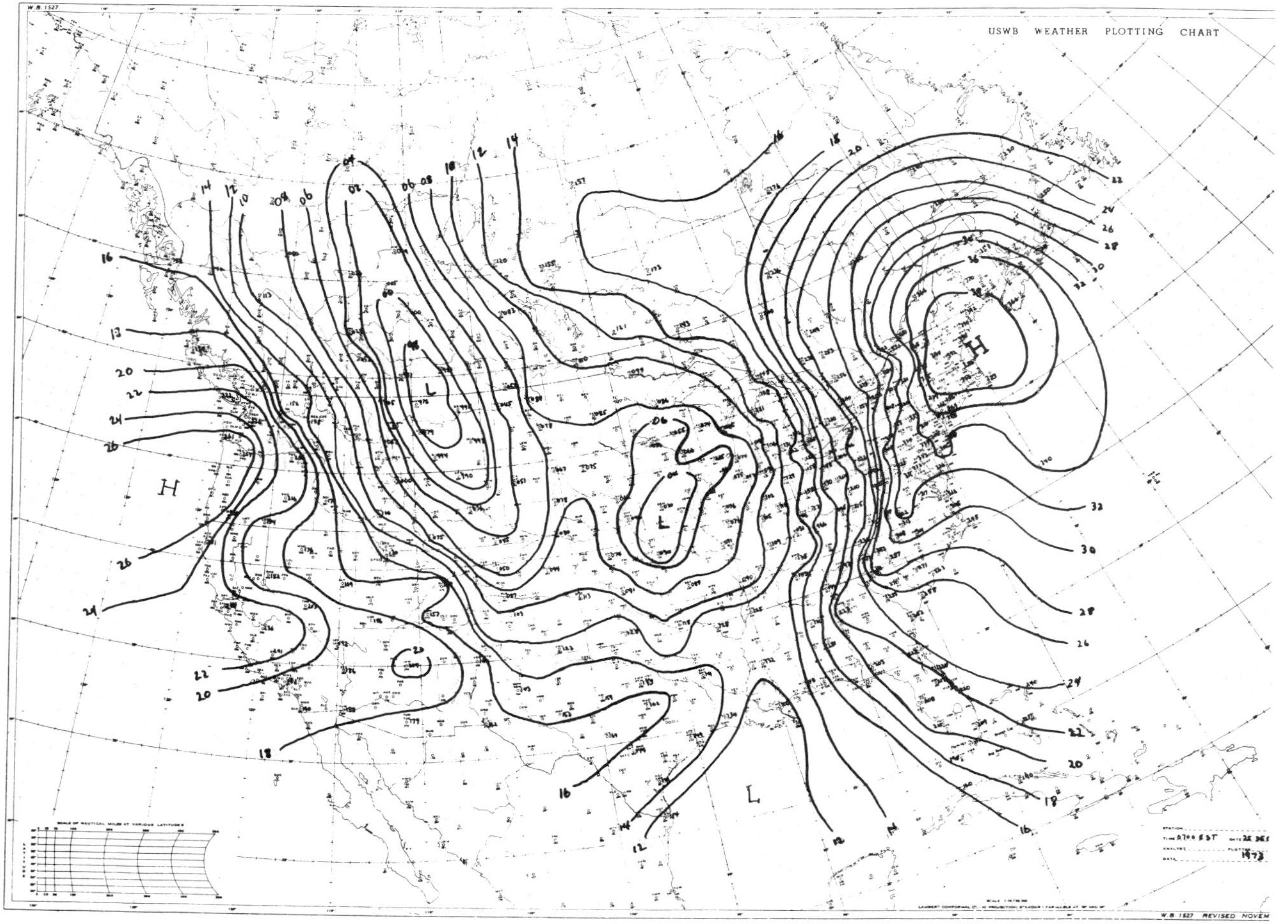

Figure 2.17 *Unsmoothed sea level pressure map for December 25, 1973.*

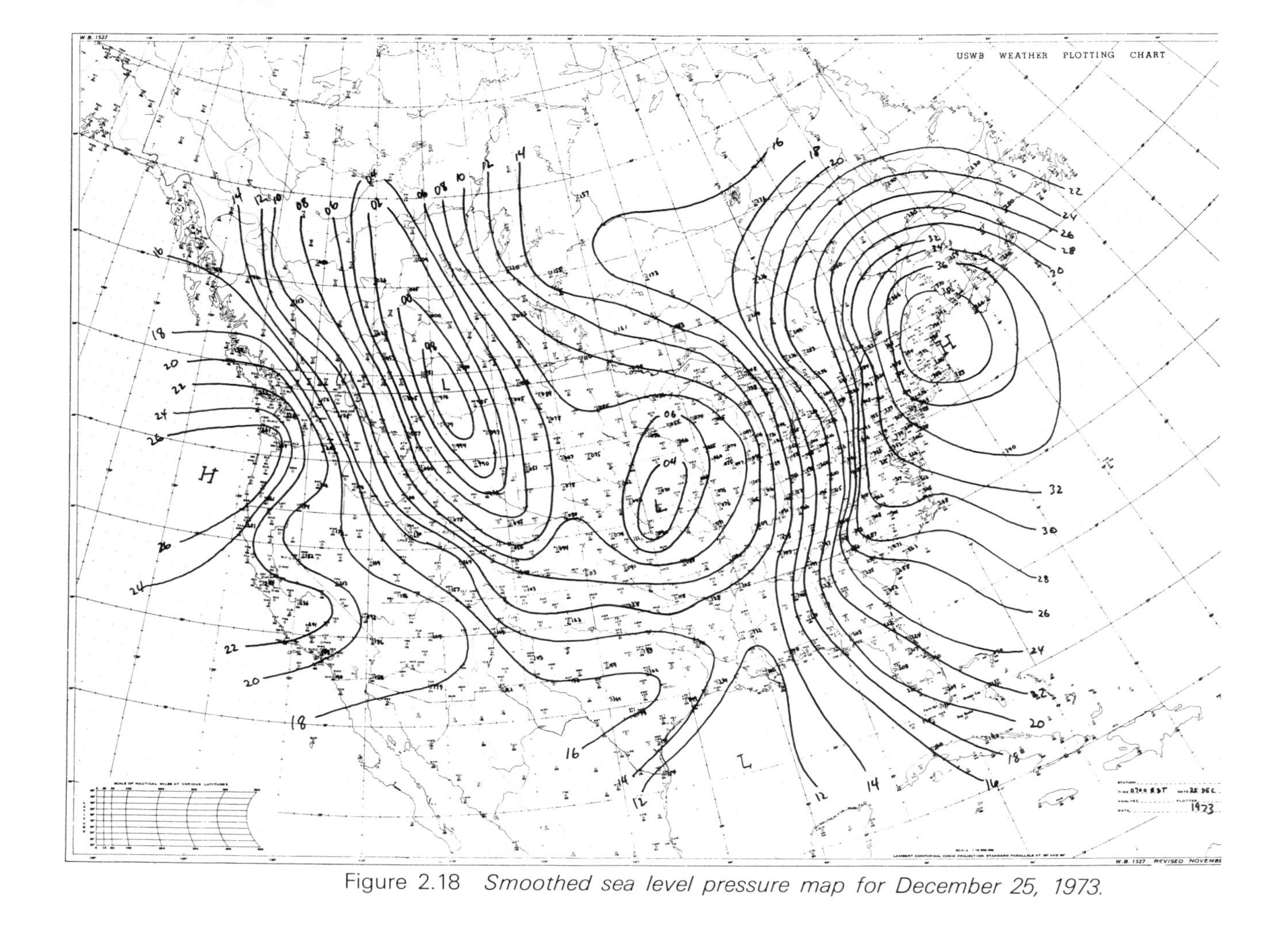

Figure 2.18 *Smoothed sea level pressure map for December 25, 1973.*

In the smoothed sea level pressure chart, we see several examples of synoptic-scale features. A region of high pressure covers New England. The diameter of this "high" is roughly 30 degrees of latitude (3330 kilometers). A minimum of pressure occurs over Iowa, and the scale of this "low" is about 10 degrees of latitude (1110 kilometers). Another elongated low, of about the same scale, is centered near the Montana-Canada border.

If we were to look at the surface pressure map for an hour later than the map shown in Figure 2.18, we would see very small changes in the large-scale patterns, indicating that pressure systems with horizontal scales of 1000 to 3000 kilometers have time scales longer than an hour. On the other hand, if we looked at the surface map a month later, the highs and lows of December 25, 1973, would have disappeared, indicating that the time scale is less than a month. Finally, if we looked at hourly pressure maps and followed the movement of each particular high and low, we would find that they persist as identifiable features for two to five days. Thus, the time scale of highs and lows with horizontal scales of 1000 to 3000 kilometers is of the order of a few days.

After seeing the isobar pattern at sea level, we might wonder what the isobars look like at other elevations in the atmosphere. Do highs and lows exist aloft, and if they do, have they the same time and space scales as the surface pressure systems? We could answer these questions by plotting the pressures at another fixed elevation in the atmosphere, for example, 18,000 feet. In practice, however, most upper-air weather data are analyzed at a constant pressure level rather than at a constant height level. A constant pressure level is a two-dimensional surface (something like a sheet or a rug) on which the pressure is identical everywhere (see Figure 2.19). Constant pressure surfaces are very nearly parallel to constant height surfaces, but they can bend and buckle a little. The typical variation in height of the 500-millibar surface is from 4.8 to 5.8 kilometers (16,000 to 19,000 feet).

There are several reasons for making analyses on constant pressure charts rather than on constant height charts, which would seem more straightforward. First, aircraft tend to fly on constant pressure surfaces rather than on constant height surfaces since the aircraft altimeter is really a barometer. Second, and more important, the equations used in meteorology are simpler if the variables are all defined on constant pressure surfaces. Neither of these reasons directly concerns us in this book; for our purposes, it would be just as easy to work with analyses on constant height surfaces. However, all the upper-air weather maps we are likely to encounter will be on constant pressure surfaces. Therefore, we need to understand how these analyses relate to their counterparts on constant height surfaces.

Actually, the interpretation of constant pressure surface analyses is not really difficult. The main thing to remember is that pressure variations on a constant height surface correspond to height variations on a constant pressure surface. In Figure 2.19, for example, where the height of the constant pressure surface is high (above 5.5 kilometers), the pressure of the constant height surface is also high (greater than 500 millibars). Likewise, where the height of the 500-millibar surface is low (below 5.5 kilometers), the pressure of the

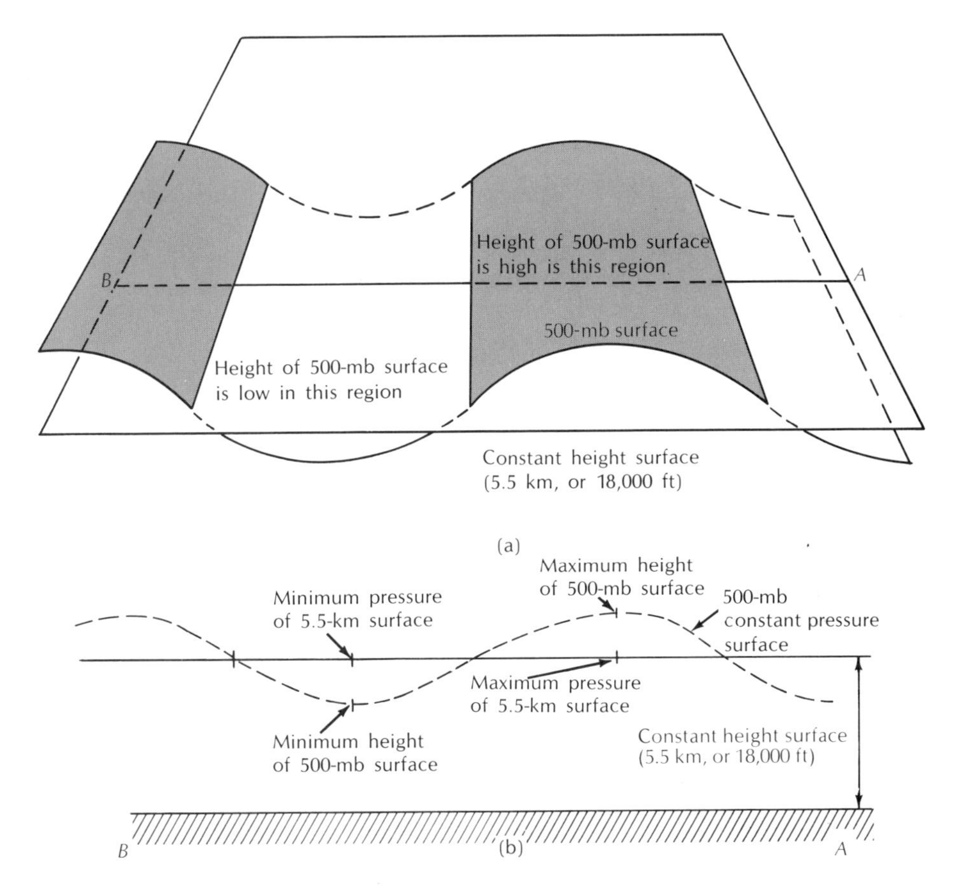

Figure 2.19 *Relationship of constant pressure and constant height surfaces.*

constant height surface is also low (less than 500 millibars). Thus, instead of drawing *isobars* on a constant pressure surface, we draw *contours* of equal elevation. Thus, highs and lows on a 500-millibar map are really regions of high and low elevations of that surface, but they behave and look exactly like high- and low-pressure centers on a constant height map.

Figure 2.20 shows the 500-millibar height contour analysis for the same time (December 25, 1973) as the surface pressure analysis of Figure 2.18. The lines of equal elevation are labeled in feet. A comparison of the upper-air analysis with the surface analysis shows two important points. First, the height contour patterns have some similarity to the surface pressure patterns (note the relatively high heights—a *ridge*—over New England and the low heights—a *trough*—over the Midwest), but are generally smoother and more wavelike, with fewer closed centers of high or low heights. Second, the wave patterns in the height contours are about the same scale as the surface pressure patterns; the wavelength of the trough-ridge system over the United States is about 30

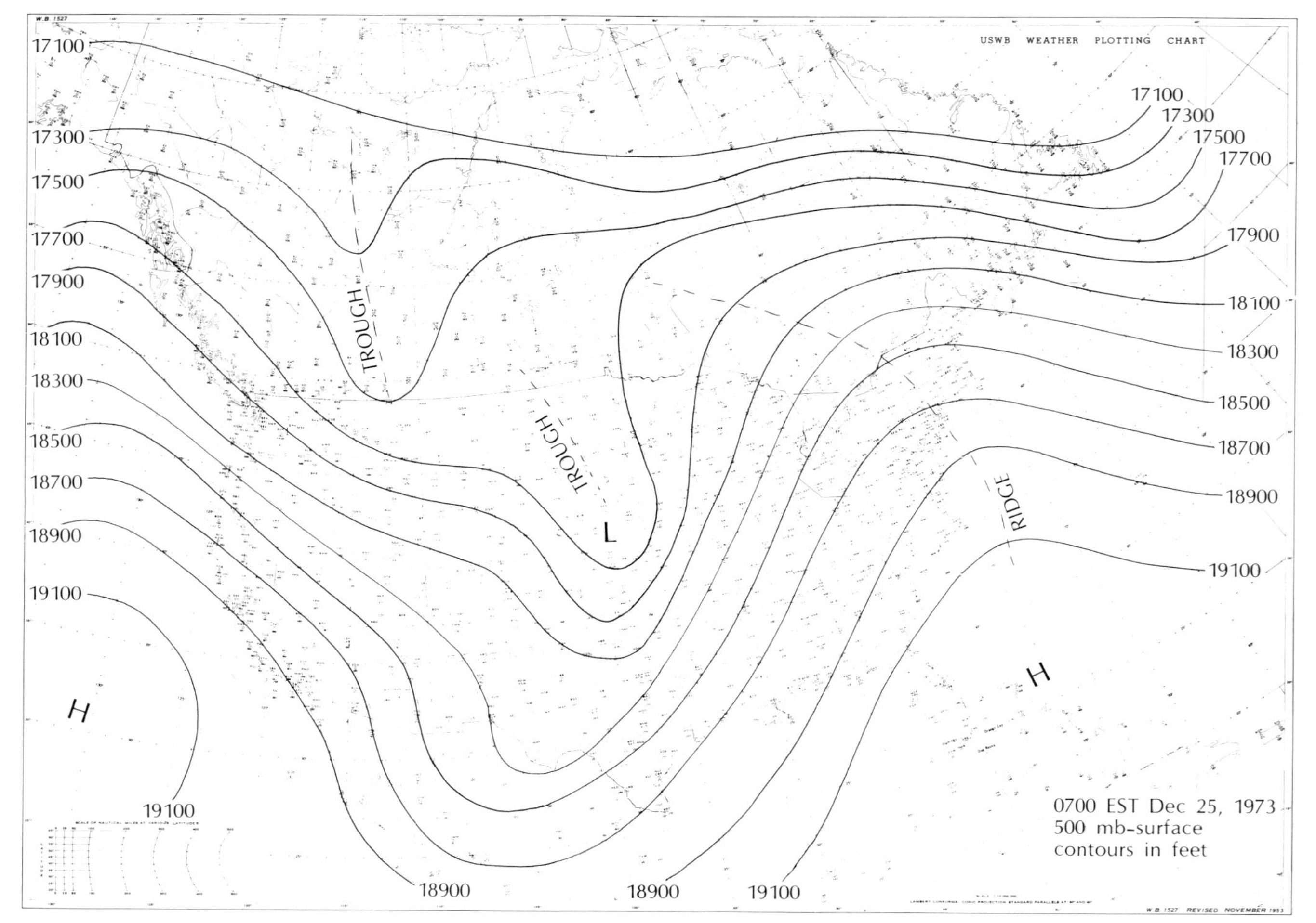

Figure 2.20 *Height contours of 500-millibar pressure surface for 0700 EST December 25, 1973.*

degrees of latitude, or 3330 kilometers. Again, if we were to watch these patterns from hour to hour, we would be able to follow individual troughs and ridges for several days before they lost their identity. So, the time and space scales of the upper-air features are roughly the same as the smoothed highs and lows on the surface map. The evolution of weather systems of this scale is thus easily studied by having weather stations separated by about 150 miles and taking observations every 12 hours; in fact, these are the space and time densities of upper-air observations over the United States.

2.4.2 *The microscale: turbulence*

Even if they have been artificially removed from the large-scale weather map by smoothing, small-scale motions do exist. These small-scale motions (eddies) are frequently chaotic, with dimensions from a centimeter to several hundred meters, and with time scales varying from a second to several minutes. This particular class of wildly fluctuating motions is called *turbulence*. One type of turbulence, *thermal* turbulence, is caused by heating from below, which leads to rising columns of air called *thermals,* or *convection currents* (*cells*). When these convection cells have enough moisture in them, condensation may occur, and we see cumulus clouds. Another form of turbulence is *mechanical* turbulence caused by air flowing over rough terrain. This turbulence is associated with the rapid change of wind with altitude which occurs just above the ground. The gusty nature of surface winds is caused by either mechanical or thermal turbulence.

The importance of small-scale turbulence lies in its ability to gradually change the large-scale conditions, especially close to the ground. It warms the air by bringing up heat from the surface, it adds moisture, and it slows down the average wind by mixing it with the slower-moving air from lower levels.

Even though the physical relationships discussed earlier can theoretically be used to describe the motions of each turbulent eddy, we are clearly incapable of observing, visualizing, and analyzing the millions of eddies forming and dying each minute over the earth. The representation of the effect of the rapidly varying small-scale motions on the slowly varying large scale is a problem which has not yet been solved satisfactorily. Because it is impossible to describe each individual eddy, we attempt to relate the total small-scale transport of heat, moisture, or velocity to some easily measured property of the large-scale weather system. For example, consider a bathtub with hot water on one side and cold on the other. If the water is stirred, heat is transferred by small-scale eddies from the warm to the cold side. It makes sense to assume that the amount of heat transported by these eddies depends on the difference of temperature of the hot and cold water; the larger the temperature contrast, the larger the heat transport. This simplifying assumption, which reduces a very complex physical problem to a single parameter (the difference in temperature between the hot and cold water), is an example of *parameterization*.

Meteorologists frequently use similar approximations, or *parameterizations*. For example, it is assumed that the evaporation rate at the surface of the ocean is proportional to the difference between the amount of water vapor just above the ocean and the amount of water vapor in the air higher up.

Unfortunately, not all small-scale motions can be treated as chaotic turbulence. For example, sea-breeze circulations, tornadoes, or local circulations caused by cities do not act as turbulence does, but instead are quite organized. These systems belong to a third general class of scales, the *mesoscale,* or middle scale.

2.4.3 The mesoscale

Between the large (weather-map, or synoptic) scale and the smaller turbulent scale (microscale) lies the mesoscale. A typical example of motion on this scale is the sea breeze, which has horizontal dimensions of about 20 kilometers and a time scale of a day. Air near the ground flows toward the land; 2 kilometers or so above the ground, the air flows from the land back to the sea. Another mesoscale circulation is the flow created by the differences in temperature between urban and rural areas. Still another example is the *squall line,* which is a system of large thunderstorm clouds oriented in a line perhaps 200 kilometers long.

Mesoscale motions are especially difficult to study. For large-scale flow, we can ignore vertical accelerations, while in mesoscale flow, we cannot. For small-scale flow, we need not consider the earth's rotation; for mesoscale flow, we must. These problems in studying the mesoscale quantitatively, along with the difficulties in obtaining observations on this scale with conventional radiosonde balloons, have hampered research in this area, so that we know more about both the synoptic scale and the microscale than we do about the mesoscale. However, important weather phenomena, such as air pollution episodes, thunderstorms, and tornadoes are mesoscale circulations. With advanced remote sensing equipment to provide the detailed data necessary to resolve these mesoscale features, and with bigger and faster computers to process the vast amounts of data, there is hope for understanding this important scale.

2.4.4 Interaction of scales of motion

At a given time, meteorological observations may be influenced by many scales of motion. For example, the wind at a station may be controlled by turbulence, mountain waves, sea breezes, or convection cells, as well as by the much larger map structures—cyclones, anticyclones, and fronts. Thus, although we frequently consider only one scale at a time for simplicity, in reality, all of the scales of motion are interacting, the small-scales affecting the large scale and vice versa. As an example of scale interaction, consider a large body of cold, dry air moving over much warmer water, as happens frequently off Cape Hatteras, North Carolina in winter. When cold air flows over a warm surface, the air becomes unstable. Convection cells which are quite small, perhaps 500 meters in horizontal diameter, develop. These cells bring dry, cold air down

A lightning stroke darts from the side of an evening thunderstorm over Miami, Florida (photo by Peter Black).

Ice crystals assume the shape of a bird in an unusual cirrus cloud formation. The "beak" is a small funnel cloud (*photo by Richard Anthes*).

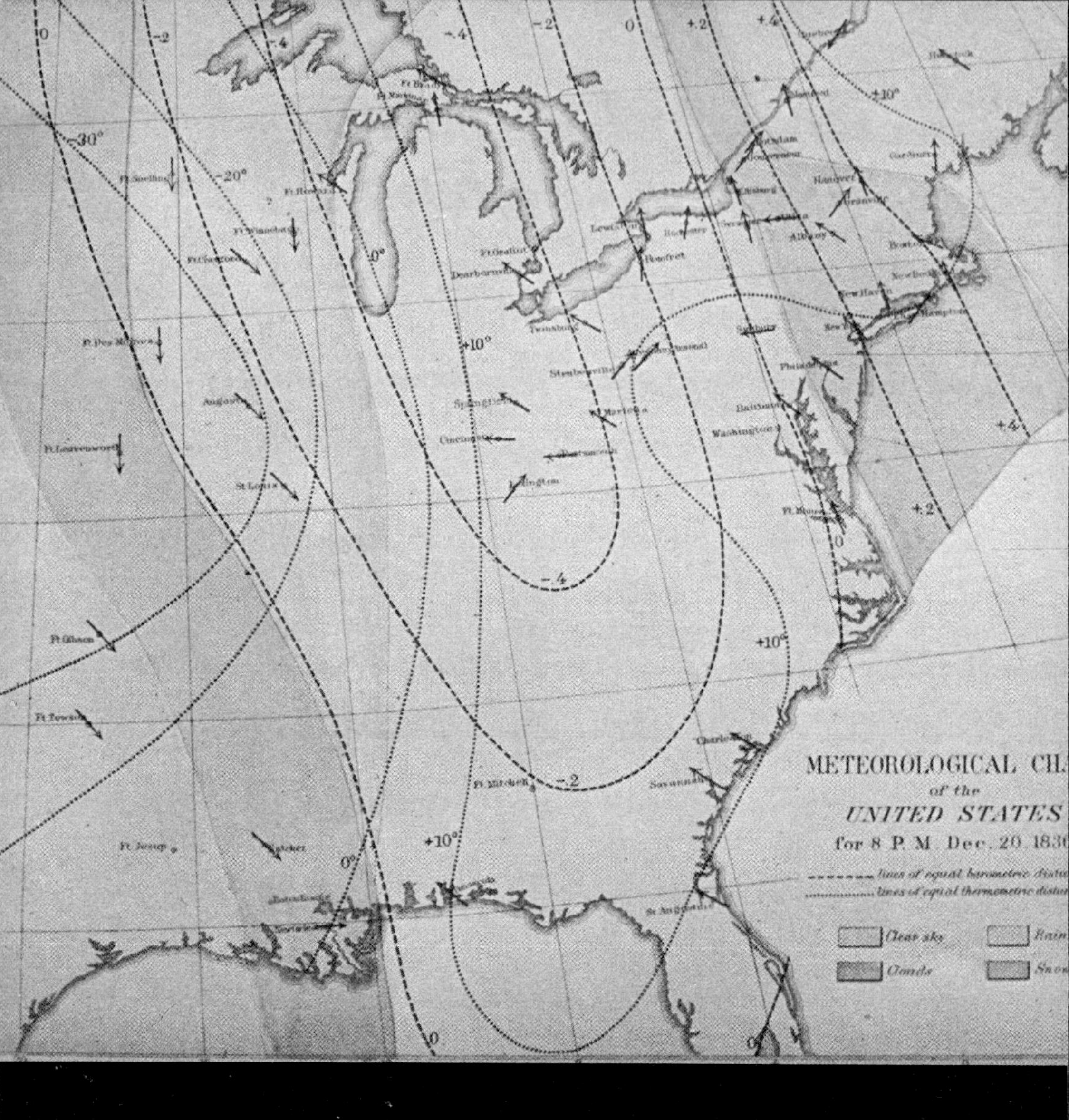

Plate 1 (above) *Loomis' analysis of winter storm of December, 1836.*

Plate 2 (below) *Ley's model of the extra-tropical cyclone [from Robert H. Scott, Weather Charts and Storm Warnings (London: Longmans, Green and Co., 1887)].*

and moist, warm air up, with the result that the initially dry and cold air is warmed and filled with water vapor. The small-scale motion has affected the large-scale distribution of temperature and humidity.

The interaction of different scales of motion is one of the most difficult problems of quantitative meteorology, for it is impossible to treat numerically all relevant scales, which range from a centimeter to thousands of kilometers in space, and from a second to months in time. Because only one particular scale at a time can be considered in detail, the combined effects of all other scales must be approximated in some manner. Sometimes these approximations work, and the forecast rain arrives on time. At other times, they do not work very well, and the forecast rain becomes 12 inches of snow.

3 Climate

3.1 Climate Is More Than Average Weather

The climate of a region describes the overall character of the daily weather that prevails from season to season and from year to year. It does not describe the weather on a particular day; rather, it provides a synopsis of the typical weather over a long period of time. Thus, we say that the tropical climate is generally hot and humid, and that the Siberian winter is dry and cold.

But climate not only describes average or typical values of meteorological quantities; it also includes such statistics as the annual range of temperature and the extreme values likely to be reached in any given time period. For example, the largest amount of rainfall in a single 24-hour period in 100 years, or the strongest wind ever recorded at a station, is part of climate. Climate is the total of all statistical weather information that helps describe the variation of weather at a given place or region.

We need to understand the climate of the various regions of the earth, because climate affects all biological processes and human activity, as will be discussed in Chapter 10. For example, if we wish to build a house in a certain location, we need to consider several climatic factors. First, the coldest and hottest temperatures likely to be reached during the year should be used to determine the insulation required, the power of the heating system, and the desirability of air conditioning. Wind speed would influence these decisions, too, because wind increases the cooling power of the atmosphere. The likelihood of flooding should be considered in selecting sites near bodies of water or in low-lying areas. The possibility of occasional very strong winds, such as the winds over 100 mi/h that sometimes blow down the slopes of the Rocky Mountains, has to be considered in determining the strength of the walls. Wind directions may also be important, but are often ignored. For example, in much of the northeastern United States, the prevailing wind is from the southwest in summer and from the northwest in winter. From this consideration alone, most windows should face the southwest to take advantage of the summer breezes. There should be few, if any, windows to the northwest to minimize the effect of the chilling winter winds.

Other variables that comprise the climate may also be considered. The amount of snowfall determines the kind of roof that can be used in a certain location without danger of collapse. Thus, the houses in alpine villages have steeply sloped roofs to allow the snow to slide off harmlessly, whereas completely flat roofs are common in Miami. The amount of sunlight, which is strongly influenced by meteorological variables, influences the choice of color for the home. White and other strongly reflecting paints are preferable in climates with strong, hot sunshine, whereas deeper hues may be used effectively for cloudy, cool climates.

Climatological information can be obtained in a number of ways. First, meteorological data are collected at many weather stations, and statistics, such as averages, ranges, and extremes, that are useful in planning our activities are computed. Such statistics are available for many places in the world. In the United States, the National Oceanographic and Atmospheric Administration (NOAA) publishes climate summaries by states and by individual cities. Table 3.1 is an example of such a summary for State College, Pennsylvania for the years 1926–1962. This summary gives the monthly average maximum and minimum temperatures, and the extremes. The mean heating *degree-days* are also given. (The number of heating degree-days in a day is obtained by subtracting the average daily temperature from 65°F. Therefore, a day with a high of 60°F and a low of 40°F has 15 heating degree-days. As might be expected, fuel consumption is very closely related to the number of heating degree-days.) The climatological summary in this table also gives the average and extreme amounts of precipitation. In describing precipitation at a point, it is especially important to know the extreme amounts that are possible. Thus, the mean snowfall in March is about 11 inches, but nearly 50 inches fell in March, 1942. Appendix 2 gives a monthly summary of the climate of 16 cities in the United States. These summaries include values for mean monthly temperatures, percentages of cloudiness, precipitation, and snowfall.

One trouble with climatological summaries is that they apply only to the weather station where the data were observed. However, since we do not build houses or grow gardens right at the weather station, we must understand the factors that cause climate to vary over small distances in order to estimate the climate between. The difference in climate between a location on a southern slope near the foot of a mountain and another location only a mile away,

Table 3.1 *Climatological Summary: Means and Extremes for 1926–1962*

Latitude: 40°48′
Longitude: 77°52′
Elevation (ground): 1175 feet

Station: State College, Pennsylvania

	Temperature (°F)								Precipitation totals (in.)								Mean number of days				
	Means			Extremes								Snow, sleet						Temperatures			
																		Max		Min	
Month	Daily maximum	Daily minimum	Monthly	Record highest	Year	Record lowest	Year	Mean degree-days*	Mean	Greatest daily	Year	Mean	Maximum monthly	Year	Greatest daily	Year	Precip. 0.10 in. or more	90° and above	32° and below	32° and below	0° and below
(a)	37	37	37	37		37		37	37	37		37	37		37		37	37	37	37	37
Jan	35.5	20.9	28.2	71	1950	−14	1936	1141.	2.66	1.60	1952	8.9	28.7	1936	14.0	1936	7	0	12	27	1
Feb	37.5	21.2	29.3	73	1954	−17	1934	1007.	2.39	1.47	1961	9.7	23.6	1961	16.5	1961	6	0	8	25	1
Mar	45.7	27.8	36.7	82	1938	1	1943	876.	3.55	2.10	1936	10.8	47.5	1942	17.5	1942	8	0	3	22	0
Apr	58.6	38.0	48.3	89	1942	+17	1950	509.	3.67	2.08	1937	2.3	18.1	1928	17.3	1928	8	0	††	9	0
May	70.9	48.3	59.6	92	1962	+28	1931	207.	4.06	2.66	1953	T	0.2	1947	0.2	1947	9	††	0	1	0
Jun	78.8	56.7	67.8	96	1952	+35	1929	49.	3.69	2.28	1957	T	0.3	1954	0.3	1954	8	2	0	0	0
Jul	83.1	60.7	71.9	102	1936	40	1929	8.	3.71	2.11	1933	0.0	0.0	----	0.0	----	7	4	0	0	0
Aug	80.9	59.2	70.0	101	1930	39	1934	20.	3.40	3.95	1955	0.0	0.0	----	0.0	----	6	3	0	0	0
Sept	73.9	52.3	63.1	98	1953	28	1947	132.	2.69	2.15	1939	0.0	0.0	----	0.0	----	6	1	0	††	0
Oct	62.8	42.2	52.5	90	1941	18	1936	396.	3.03	3.42	1954	0.1	1.0	1962	1.0	1962	6	††	††	4	0
Nov	48.8	33.0	40.9	81	1950	1	1929	723.	3.04	3.21	1950	3.2	14.0	1953	10.0	1953	6	0	2	15	0
Dec	37.3	23.2	30.2	65	1933	−6	1942	1077.	2.60	1.52	1941	8.0	24.4	1950	10.0	1960	6	0	10	26	1
Year	59.5	40.3	49.9	102	July 1936	−17	Feb 1934	6145.	38.49	3.95	Aug 1955	43.0	47.5	Mar 1942	17.5	Mar 1942	83	10	35	129	3

(a) Average length of record, years.
* Base 65°F.
T = trace, an amount too small to measure.
†† Less than one-half.

but near the top of the mountain facing north, may be as great as the difference in climate between two weather stations separated by hundreds of miles. Therefore, we need to study local as well as global variations in the climate.

Climate is determined by many factors, such as the intensity of the sunlight, the proximity of an ocean, the origin of air masses arriving at the locality, and topography.[1] Lakes, mountains of all sizes, and even cities influence the climate. Because there are so many factors influencing climate, we will first separate them into large-scale and local causes. Large-scale controls include the seasonal distribution of sunlight and the global distribution of continents, oceans, and the major mountain chains such as the Rockies or Alps. These global features produce climatic characteristics that dominate large regions of the earth; therefore, their effects may be interpolated easily between weather stations. As was already pointed out, local geographic features, for example, lakes, cities, and hills, vary greatly over small distances and cause considerable differences in the local climate from one place to another. For example, on a cold night, there may be a difference in temperature of 30°F between a valley and the adjacent ridge only a few miles away. On a cold night in State College, Pennsylvania, the authors measured −18°F at the bottom of a hill and 0°F at the top, merely two blocks away. Tabulated data at weather stations are not much help in estimating local variations in climate if such differences occur frequently. To interpret properly these local variations in climate, we must understand the small-scale meteorological processes that produce them.

3.2 Variation in Solar Radiation—The Basic Cause of Weather and Climate

We know by the drift of smoke across a room, or the rustling of curtains, or the feel of a faint draft on our faces that hot-water radiators produce weak circulations of air in a closed room. The air in contact with the hot surfaces of the radiator is heated, expands, and rises, and cooler air moves in to replace it. The cause of this very small-scale wind circulation is *differential* heating, which means that a portion of the air in the room is heated more than the rest, thereby producing temperature and pressure differences that lead to acceleration of the air and the production of winds. On a much larger scale, the complicated circulations of the atmosphere also owe their existence to differential heating, where the sun is the radiator and the entire atmosphere is the room.

3.2.1 *Variation of temperature from north to south*

The earth is almost a sphere; the diameter across the equator is only 0.3 percent larger than the diameter from the North Pole to the South Pole. Usually, we can ignore this difference. Because the earth is nearly a

[1] An *air mass* is a body of air that covers a wide area and is characterized at any constant level by relatively uniform temperatures and humidities.

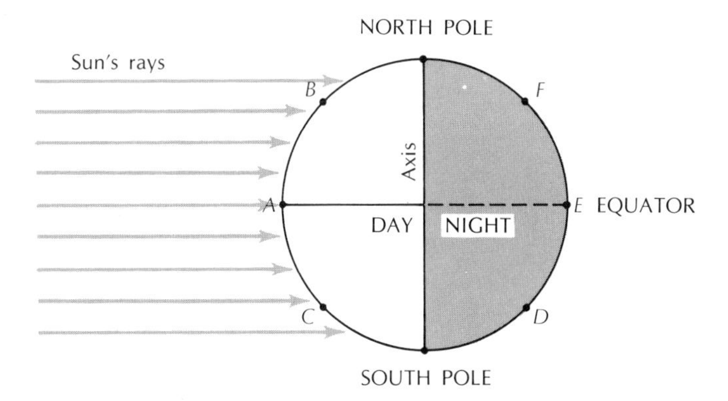

Figure 3.1 *Variations in sunlight received on the surface of the earth as a function of latitude.*

sphere, the average angle between sunlight and ground changes from latitude to latitude. In Figure 3.1, it is noon at places on the earth's surface designated by *A, B,* and *C;* these points are receiving the maximum intensity of sunlight they will receive on this day. The stations on the right, *D, E,* and *F,* are receiving no sunlight at all; they are experiencing local midnight. But do *A, B,* and *C* get the same amount of light and solar energy? Indeed, they do not. At point *A*, the sun is overhead and the light intensity is great; at *C*, the rays of the sun just graze the surface, and the amount of light received is very small. This figure is drawn for March 21 or September 23, at which times the equator gets most of the light at local noon and the poles receive the least. This distribution of the sun's energy is true only for March 21 and September 23. At other times, as we shall see, other latitudes may get more radiation than the equator. Over the year, however, the equator gets the most radiation, and the poles the least. This simple difference is the basic cause of all atmospheric motions, and, therefore, of weather.

3.2.2 *Diurnal variations in temperature, clouds, and wind*

Many meteorological variables undergo daily, or *diurnal,* cycles which are caused by the earth's rotation. The earth rotates on an imaginary axis that connects the North and South poles, and the time from local noon to the next local noon is 24 hours, on the average. The rotation is counterclockwise, as seen from above the North Pole (see Figure 3.2). The result of this rotation is obvious. As points *A–C* on Figure 3.1 move out of the figure after noon, the angle between the surface of the earth and the rays of the sun becomes smaller, and the incoming radiation decreases. As everybody knows, sunlight increases in intensity during the morning, reaches its peak at noon, and decreases in the afternoon. However, the temperature reaches its peak after the radiation does, because it takes time to heat the ground, which in turn heats the overlying air. The temperature is usually highest two to three hours after noon; it begins falling in the late afternoon, continues through the night, and reaches its lowest point about sunrise. This daily cycle is called the *diurnal variation of temperature*.

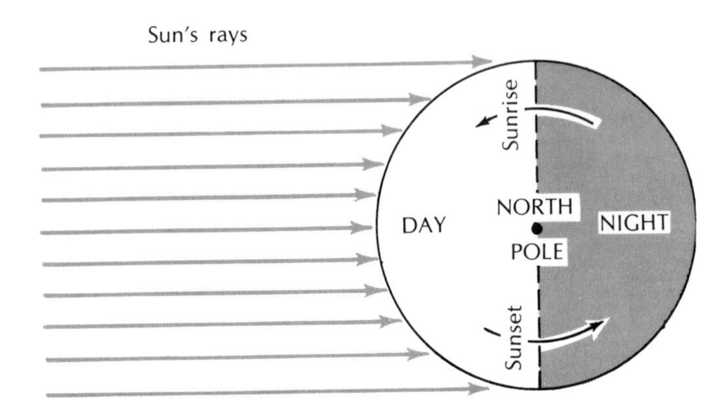

Figure 3.2 *Rotation of earth as seen from above the North Pole.*

The diurnal variation of temperature is influenced by many factors: latitude, season, cloudiness, wind speed, and proximity of water bodies. Cloudiness, of course, reduces the temperature variation, since clouds block incoming radiation during the day, making the daytime temperatures cool. In contrast, at night, the clouds stop the radiation loss by the ground and reradiate heat earthward, helping to keep nighttime temperatures mild. We might consider clouds as insulators which shelter us from extreme temperatures.

Cloudiness can also change the time of the temperature maximum. Normally, as we have seen, it is warmest between about 2 P.M. and 3 P.M. In Denver, Colorado, however, on typical summer days, it is clear in the morning only. By 11 A.M., it becomes quite hot, perhaps 90°F. The hot ground starts convection currents over the mountains to the west, which form cumulus clouds. After noon, these clouds drift eastward and cover most of the sky, preventing the sunshine from reaching the ground. The air begins to cool, so that the temperature reaches its peak about noon.

The variation of cloudiness just described is quite typical; clouds have their characteristic diurnal variations, too. For instance, cumulus clouds are always an indication of strong convection currents, both updrafts and downdrafts, which are usually strongest about midday. Flying through a region of such clouds is likely to be quite bumpy. These strong vertical motions (typically 3–10 mi/h) have another important effect in that they mix air near the ground with air aloft. Under this vertical mixing, polluted air near the ground is exchanged with the cleaner air aloft, so that the pollution near the ground is diluted. Generally, air quality is best in the afternoon at the time of maximum convection.

The vertical convection currents also produce a diurnal variation of wind speed. As we shall see, the horizontal air motions (winds) are normally fast aloft and slow on the ground. If there is strong vertical mixing in the middle of the day, the fast air is carried to the ground, and the wind speed increases. Therefore, we observe fast winds in the daytime and slow winds at night.

All of these diurnal variations are basically caused by the diurnal variation of temperature. Because the magnitude of the diurnal temperature change varies over the earth, some places have stronger diurnal cycles than others. For example, as seen from Figure 3.1, points near the poles keep the same low angle of the sun all day, and so have no noticeable diurnal variation. In general, the diurnal variation decreases from the equator toward both poles.

The oceans have almost no diurnal variation. It takes a great amount of heat energy to warm the ocean, and the oceans warm less than one degree Celsius in a day. There are several reasons for the slow warming of the seas. First, the water in the ocean can overturn vertically, so that warm surface waters mix with colder water below. Second, the ocean is partly transparent, so that a thick layer of water has to be heated. Third, some of the incoming heat is used to evaporate rather than warm the water. Finally, water has a greater heat capacity than land, which means that it takes more heat energy to warm water by one degree than an equal mass of land. For these reasons, diurnal variations of temperature, clouds, and wind are quite small over the open ocean. More generally, in maritime climates, where the air blows mostly from the sea, diurnal variations are small. (See also Chapter 8.)

Since the diurnal variation of temperature is caused by absorption of radiation near the ground, it decreases with increasing height. Above about 1 kilometer, there is almost no diurnal cycle. Paradoxically, in the stratosphere, the daily temperature variation becomes large again, with perhaps a 10°C difference between maximum and minimum temperatures. The variation at this level is caused by a peculiar form of oxygen, called ozone, which exists above an altitude of 25 kilometers and absorbs ultraviolet radiation from the sun during the daytime.

3.3 Variation of Climate with Height

As discussed in Section 2.1, the cloudless atmosphere is quite transparent to visible sunlight, so that the ground is warmed more directly than the atmosphere. Only a small portion of the sun's radiation is trapped by the atmosphere. Instead, the air is warmed indirectly by the upward transport of heat from below. As the ground is heated, convection currents are initiated which transport heat up into the atmosphere.

Another heat source for the atmosphere is the release of latent heat when water vapor condenses. Some of the energy of the sun is used for evaporation of water at the ground. The invisible water vapor is then carried into the atmosphere, where it condenses into water droplets which we see as clouds. This process releases heat to the air.

In general, the indirect heating processes of the atmosphere decrease with height up to 10 kilometers or so. The reduction of the heating processes with height in the troposphere, and the cooling of the air by long-wave radiation (see Section 2.1), which reaches a maximum in the upper troposphere, are responsible for the decrease of temperature with height. The rate of decrease of temperature with height is called the *lapse rate*. Its average value is 6.5°C/km, or about 3.5°F/1000 ft.

However, frequently there are layers in the atmosphere of limited depth, in which the temperature increases rather than decreases with height. These layers are called *inversions*. Inversions form under rather special conditions. For example, close to the ground, on clear nights with little wind, inversions are common, because the ground loses heat more effectively than the air above, resulting in rapid cooling of the ground at night. On a winter night, the ground temperature may be −15°C, while the temperature on top of a skyscraper is 0°C. As we shall see, inversions play special roles in the dispersion of air pollutants, since they inhibit vertical mixing, with the result that pollution remains highly concentrated near the ground.

The general decrease of temperature with height has been known for a long time, and early scientists believed the temperature decreased uniformly to absolute zero at great heights. It was a great surprise when, around 1900, balloonists found that this decrease stops abruptly just above 9 kilometers (30,000 feet) or so in middle latitudes. Above this height, the temperature remains constant. At first, meteorologists thought that this observation was in error, for the constant temperatures would require a heat source at this altitude, and no such heating mechanism was known. However, as we noted earlier, the presence of ozone above this height causes heating by the direct absorption of sunlight.

The region of the atmosphere nearest to the earth, in which the temperature generally falls with height, is called the *troposphere* ("weather sphere"). The next higher region, where the temperature no longer decreases, is the *stratosphere,* and the surface separating the two "spheres" is called the *tropopause* ("end of troposphere"). As the name implies, almost all the "weather"—rain, clouds and snow—occurs in the troposphere. Only occasionally will a violent thunderstorm break through the tropopause into the stratosphere. In contrast, the stratosphere is relatively quiet and dry.

Figure 3.3 shows vertical cross sections of the mean temperatures averaged around latitude circles for the months of January and July. The heavy lines denoting the tropopause separate the troposphere and the stratosphere. This figure shows that the tropopause and the stratosphere have quite complicated properties. For example, the tropopause is higher and colder at the equator than at middle and high latitudes. Further, it usually has breaks in it. The cold equatorial tropopause has a surprising consequence. At 15 kilometers (50,000 feet) and just above, the equator is actually much colder than the regions farther north or south, just opposite the situation close to the ground.

3.3.1 *The stratosphere and ozone*

The stratosphere is relatively stable, with warm air overlying cool air, and for this reason, the vertical motions tend to be weak. In addition to the lack of vertical motion, an extremely low moisture content, typically three parts per million, characterizes the stratosphere. Therefore, clouds occur only in very restricted regions, e.g., in the antarctic during winter. Further-

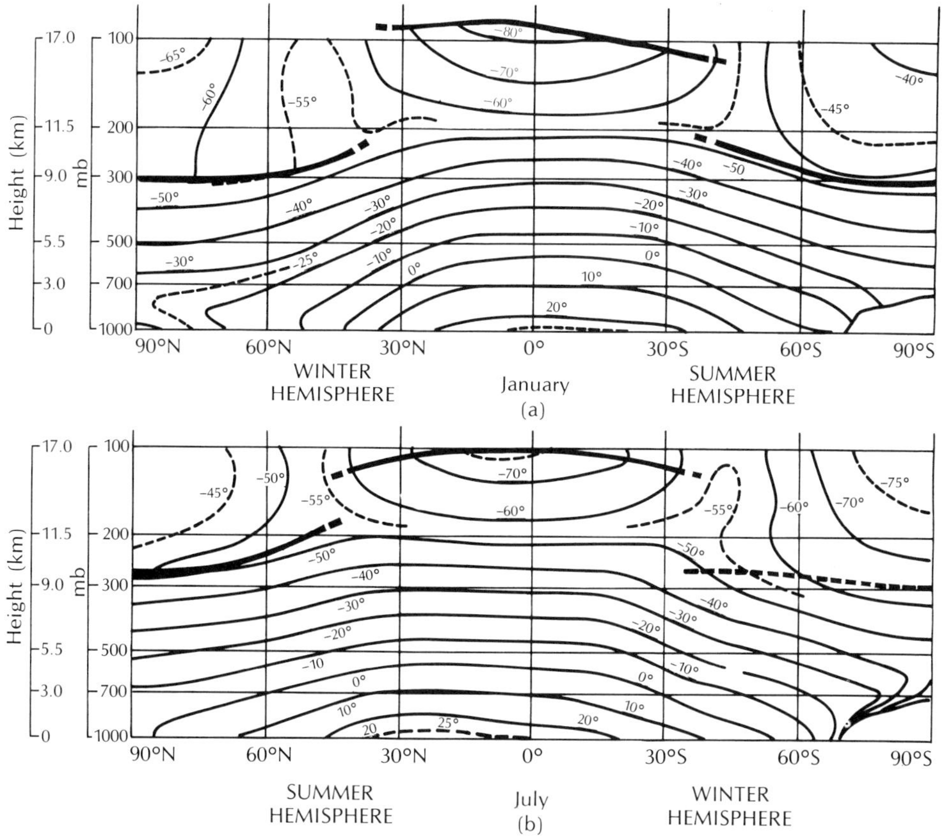

Figure 3.3 *Average vertical temperature distribution in (a) January and (b) July [after Eric Palmén and Chester W. Newton, Atmospheric Circulation Systems: Their Structure and Physical Interpretation (New York: Academic Press, 1969)].*

more, there is never any precipitation. However, the stratosphere is not motionless. There are strong horizontal winds in the stratosphere, particularly near the bottom, and at higher levels in the polar regions during the winter. Winds of 300 km/h (186 mi/h) are not unusual.

The temperature no longer decreases upward in the stratosphere because of the presence of ozone, which is most dense between 20 and 32 km elevation. The total amount of ozone is actually very small. If all the ozone were taken down to the ground from the stratosphere, where it would be compressed by the weight of the overlying atmosphere, it would form a layer less than a centimeter thick. Ozone is so active, however, that a concentration of as little as one part in a million may have important effects. Unlike other gases in the atmosphere, ozone has the ability to absorb ultraviolet radiation. This absorption of energy heats the stratosphere, so that it is much warmer than it would be without ozone. At the same time, the ozone shields animals, plants, and people from dangerous radiation which otherwise would make the world uninhabitable for the familiar forms of life. In fact, the small amounts of ultraviolet radiation that do penetrate the ozone blanket cause sunburning and appear to be an important cause of skin cancer in man.

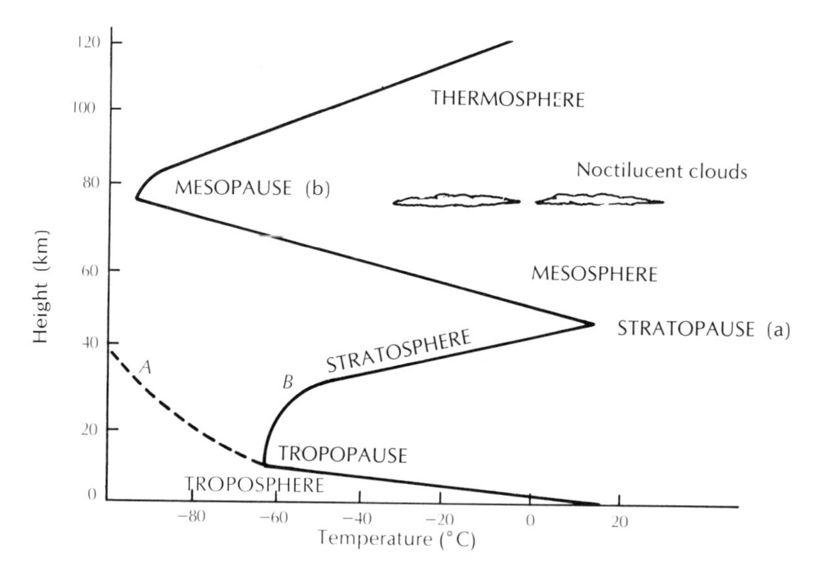

Figure 3.4 *Vertical distribution of temperature to a height of 120 kilometers.* Heavy line: observed; dashed line: without high-level heating.

If no ozone were present, the average vertical temperature distribution would resemble curve *A* in Figure 3.4. The actual average distribution looks like curve *B*, with the temperature peak at point (a) because of ozone heating. This point is called the *stratopause,* the pause (or end) of the stratosphere. Here the temperature is almost as high as at the ground, attesting to the efficiency of ozone in absorbing solar energy.

The warm stratopause has the peculiar property that it can reflect sound waves, to such an extent that powerful explosions can often be heard hundreds of miles away. In fact, this warm region was discovered because the guns at Queen Victoria's funeral in 1901 were heard in Germany, but not between England and Germany.

Since ozone is so important to our well-being, we will try to explain why it is only formed at some distance above the ground. Ozone is a form of oxygen in which three *atoms* of oxygen combine to form one *molecule* of ozone. An atom of oxygen is denoted by the symbol O, and a molecule of ozone is represented by O_3. In the lower atmosphere, practically all the oxygen occurs in molecules consisting of two atoms, and is denoted by O_2. Very high up, say above 110 kilometers, O_2 is divorced by intense ultraviolet radiation and occurs mostly as single oxygen atoms, O.

To produce ozone, we have to combine atomic oxygen, O, and ordinary oxygen, O_2. If O and O_2 merely collide, however, they just bounce off each other and do not form O_3. To make ozone, O_3, and satisfy all physical requirements, we need a third molecule, M, involved in the interaction, represented as

$$O + O_2 + M = O_3 + M$$

This molecule M acts like a minister at a wedding; O and O_2 get married, and M walks away after the wedding is completed. Chemists call the role of M that of a *catalyst*. In other terms, the formation of ozone requires a triple collision, which is rare at very high levels (above 60 kilometers), where there are so few molecules that it is extremely unlikely that three of them will arrive at the same place at the same time. Thus, there is almost no ozone above 60 kilometers. On the other hand, there is almost no atomic oxygen, O, below approximately 16 kilometers, so that the formation of ozone is unlikely there, too. These are the reasons why most of the ozone is found in the stratosphere.

3.3.2 *The atmosphere above the stratosphere*

Figure 3.4 shows that the temperature drops again above the stratopause (above 50 kilometers), marked (a). This region of decreasing temperature is called the *mesosphere*. If there were no special heating mechanism above the stratopause, the temperature would keep on dropping for great distances. Actually, however, the temperature reaches a minimum at an elevation of about 80 kilometers, a region called the *mesopause* (marked (b) in Figure 3.4), which is the coldest region in the atmosphere. The temperature at the mesopause may fall to −100°C (−150°F) at the North Pole in summer. The mesopause, in fact, can be so cold that the tiny amount of water vapor in this region forms ice clouds, called *noctilucent* clouds, which can be seen when the sun hits them after sunset.

The region above the mesopause is called the *thermosphere* ("hot sphere"), where the temperature may actually go up to several thousand degrees. One reason for these hot temperatures is that ultraviolet sunlight can be absorbed by oxygen in the thermosphere. Also, there are so few molecules to be heated (the density is quite low) that a little energy absorbed can produce a large temperature increase. Because there are quite a few loose electrons and positive ions in the thermosphere, this highest part of the atmosphere is also called the *ionosphere*.[2]

By this time, you may wonder why the distant thermosphere is discussed at all in a survey book on meteorology, and how, if at all, it affects you. As far as we know, the thermosphere has very little effect on our daily weather. The only contact most of us are likely to have with the thermosphere is through its effect on ordinary AM radio reception. Loose electrons have the ability to reflect radio waves, which explains why stations far away can be heard at night. During the daytime, long-distance radio transmission is not as good

[2]An *ion* is an electrically charged atom or particle.

because there is too much ionization (and, therefore, electrons) in the lower thermosphere. These electrons absorb the radio waves on the way up to the reflecting layer.

Because radio waves can be reflected and absorbed in the thermosphere, scientists can devise experiments with radio reception to determine the nature of the thermosphere. From such experiments, we know that the thermosphere has huge diurnal variations, both in terms of temperature and number of electrons. Also, the thermosphere becomes most active every 11 years, when the sun is disturbed by large solar storms, called *sunspots*. These storms, which consist of relatively cool gases in the solar atmosphere, appear as dark spots on the face of the sun. Around the sunspot areas, bright flares of short duration appear and disappear. Associated with the flares, the outer portions of the solar atmosphere become extraordinarily hot. These flares also disturb radio propagation on earth. A day or so after the appearance of solar flares, particles arrive from the sun which collide with particles of the upper atmosphere. These collisions excite the atoms, which then emit light. Thus, the *northern* and *southern lights* (also called the *aurora borealis* and *aurora australis,* respectively) usually appear a day or so after a major solar disturbance.

Auroras occur at high latitudes because the earth's magnetic field deflects the incoming beams of particles toward the poles. During the bombardment of the earth's upper atmosphere by the solar particles, the magnetic field of the earth is greatly disturbed. Even at the ground, a compass needle undergoes irregular, and apparently mysterious, movements, which are associated with upper-atmosphere electric disturbances. These disturbances are called *magnetic storms*. The main practical implication of solar disturbances and associated upper-atmosphere phenomena is that communications over long distances become difficult. Not only is radio propagation impeded, but also, irregular currents are induced into cables under the oceans, making difficult communication by wire as well.

3.3.3 *Vertical distribution of density and pressure*

So far, we have mainly discussed the rather complex vertical distribution of temperature. In a general way, the vertical distributions of density and pressure are much simpler: both decrease rapidly with height. For example, at the height of flight of conventional jet aircraft (10 kilometers), the weight of the overlying atmosphere is only about one-fourth of what it is at the ground; the pressure there is only about one-fourth of sea level pressure.

There is a convenient rule about variation of pressure with height, accurate to an altitude of about 60 miles: For every 10 miles of elevation, the pressure decreases by about a factor of ten. Because the pressure at sea level is about 1000 millibars, at 10 miles up, it is about 100 millibars; 20 miles up, 10 millibars; 30 miles, up 1 millibar; and so forth. The same approximate rule holds fairly well for air density. Thus, in the lower thermosphere (see Figure 3.4), both density and pressure are only about 1/1,000,000 of their surface values.

The very small pressure and densities in the thermosphere have an important implication. There have always been scientists who have thought that solar disturbances should affect weather because they profoundly affect the thermosphere. But because the thermosphere contains only about 1/100,000 of the total mass in the atmosphere, it is easy to see that it can be violently disturbed and still have little effect on the weather near the ground.

3.4 Global Distribution of Winds—The General Circulation

To understand the average behavior of wind, we must consider the horizontal variation in pressure. In comparison to the large variation of density and pressure with height, the horizontal variation is quite small. Near the ground, a typical variation of pressure over 1 mile in the vertical is 150 millibars, but the variation is more likely to be only 1 millibar in 50 miles in the horizontal. Even though the vertical change of pressure is much larger than the horizontal, the small horizontal pressure variations are extremely important to atmospheric motions, for it is these variations that cause air to move horizontally and produce winds. Small horizontal pressure differences can produce strong winds, while much larger vertical differences in pressure exist with weaker vertical motions because the upward-directed vertical pressure gradient force is nearly balanced by gravity, which acts downward.

3.4.1 *Why the west winds?*

To understand the reason for the westerly winds that prevail throughout the middle and upper troposphere in middle latitudes, we need to understand the relationship between horizontal pressure differences and horizontal temperature differences. The key to this relationship is that the pressure decreases with height more rapidly in cold air than warm air, as illustrated in Figure 3.5. Because cold air is denser than warm air, it is more compressed in the vertical. Thus, if the surface pressures of a warm column and a cold column ot air are equal, at any level above the surface the pressure will be higher in the warm column.

In the troposphere, the warmest air is located near the equator; the pressure several miles up is high over the equator and decreases toward both poles.

It is important to note that this horizontal pressure variation aloft results from the horizontal temperature variation at lower levels. In general, the larger the horizontal temperature difference below, the larger the pressure difference above. Again, this horizontal pressure variation is not large, but is sufficient to produce strong winds.

To see how the large-scale winds develop, let us consider what might have happened after the creation of the atmosphere. The air near the equator would first become warmer than the air near the poles, and horizontal pressure differences would develop in response to this temperature difference. If the atmosphere were initially at rest, the pressure gradient aloft would cause air in the northern tropics to accelerate toward the low pressure at the North

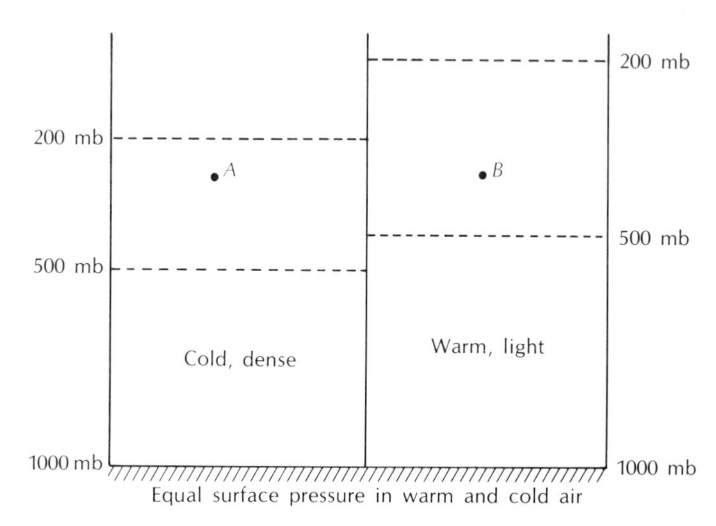

Figure 3.5 *Schematic diagram showing that pressure decreases more rapidly with height in cold than in warm air.* Note that the pressure is higher at point *B* in the warm air than it is at the same level, at point *A*, in the cold air.

Pole, and the air in the southern tropics toward the South Pole (as shown in Figure 3.6). However, the rotation of the earth produces a deflection of the air to the *right* in the Northern Hemisphere and to the *left* in the Southern Hemisphere. The force which causes this deflection is called the *Coriolis force* (see Section 4.5.1). Therefore, in both hemispheres, the air flowing toward the poles would be deflected toward the east, and west winds would prevail all over the earth, except right around the equator. The strength of these winds is proportional to the magnitude of the pressure gradient, which, as we have seen, depends on the temperature gradient below. Therefore, the larger the temperature gradient, the stronger the west winds aloft. Also, the more the temperature varies horizontally from one place to another, the more the pressure gradient and the wind change with height. This is a general meteorological principle which has many applications.

There are many consequences of the general westerly flow aloft in middle latitudes. Clouds generally move from west to east, as do cyclones and anticyclones. Airplane trips are shorter from west to east than from east to west. With a good tail wind, a flight from Los Angeles to New York takes four hours, but the return trip takes over five hours.

We have seen the fundamental result of west winds existing aloft because the temperature decreases from the tropics toward the poles. This increase in westerly wind speed continues up to the tropopause. However, the horizontal temperature gradient generally reverses in the stratosphere, where cold air is located over the equator (consider Figure 3.3 at 150 millibars). Thus, west winds in the stratosphere generally decrease upwards, which means that somewhere near the tropopause the west winds reach a maximum, averaging

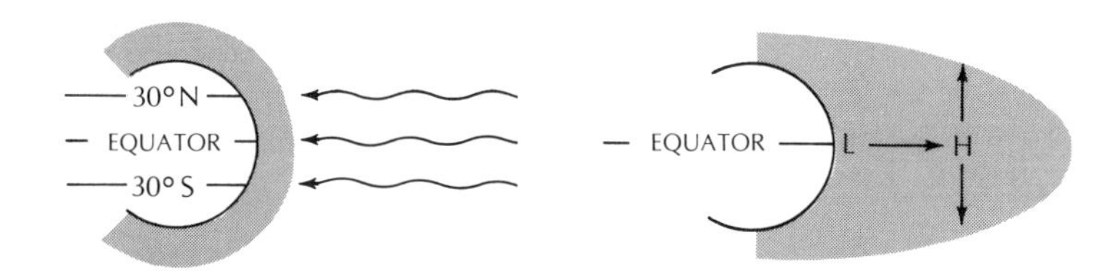

(1) Solar radiation reaching the spherical earth produces differential heating, warming the equatorial regions more than the poles.

(2) Warm air over the equator expands upward creating higher pressure aloft and lower pressure at the surface.

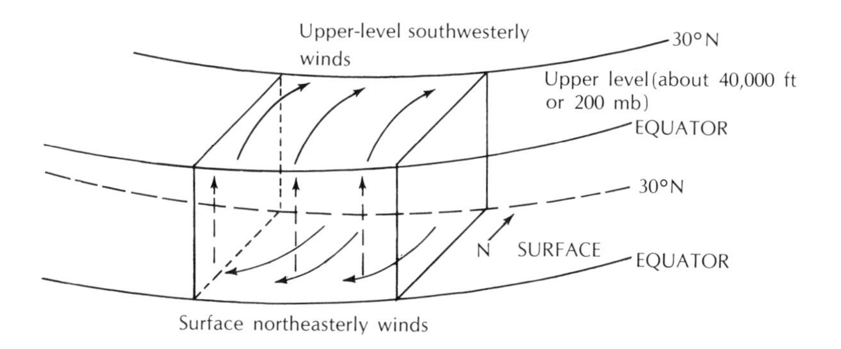

(3) Air rises near the equator and returns poleward aloft. Rotation of the earth deflects poleward-moving air to the right in the Northern Hemisphere (Coriolis effect), producing southwest winds aloft. Air flows toward the equator in low levels. The Coriolis effect deflects air toward the right of motion, producing northeast winds near the surface.

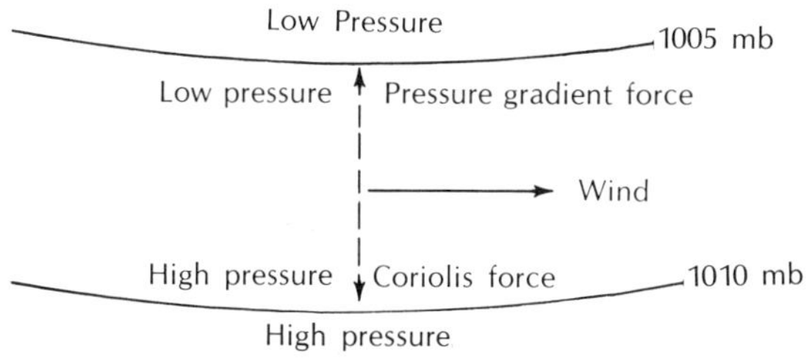

(4) A balanced state of motion is achieved when the Coriolis force exerts a force equal in magnitude but opposite in direction to the pressure gradient force.

Figure 3.6 *Schematic diagram showing differential heating, expansion of atmosphere, horizontal accelerations, and resulting geostrophic balance.*

about 80 mi/h in winter and 40 mi/h in summer. These strong west winds are by no means uniformly distributed from tropics to poles. They occur in relatively narrow ribbons of especially strong winds, called *jet streams*, which may be only a hundred miles wide. In such bands, the west winds may blow over 200 mi/h. These jet stream winds were discovered by accident in World War II, at which time American aircraft had maximum air speeds of about 200 mi/h. Planes on a bombing mission to Japan flew from Pacific island bases toward the west, into the wind. With a 200-mi/h head wind, however, the planes found themselves making absolutely no progress over the ocean, and so had to turn around.

Why do the winds blow in such concentrated jets? We have seen that the wind velocities are determined by horizontal pressure differences, which in turn are governed by the temperature distribution. If strong winds are concentrated in narrow bands, temperature gradients must also be concentrated in narrow bands. In other words, in some places, the temperature varies rapidly over relatively small horizontal distances. Such regions of strong horizontal temperature contrasts are called frontal regions. As introduced in Chapter 1, fronts are boundaries between air masses of differing properties; one is usually cold and the other is relatively warm. They have been studied in great detail and have important applications for forecasting, not only because the temperature changes rapidly across the fronts, but also because much of the world's bad weather occurs near fronts. Later chapters will deal more thoroughly with weather fronts. We mention them here because fronts and jet streams go together and are important characteristics of the world's climate.

We might wonder why the temperature does not fall gradually from equator to poles, but instead changes slowly over most of the distance, and rapidly over quite short distances between. It is difficult to explain this behavior in simple terms, but any good theory of average global winds and temperatures (the *general circulation*) must account for this behavior.

3.4.2 The general circulation

General circulation theories start with air at a uniform temperature. We imagine that the sun is suddenly turned on, and then use the basic laws of physics to describe the development of the wind and temperature distributions. These laws may be expressed mathematically, and the resulting equations solved on fast electronic computers. Such computer "models" of the atmosphere (see Section 6.3 for a further discussion of numerical models) correctly describe many of the observed features of the real atmosphere. In particular, the mathematical models clearly show the tendency for the atmosphere to concentrate winds into narrow jets and produce large temperature changes over short distances. They not only explain the development of fronts and jet streams, but also correctly describe the tendency for air near the ground to move from the tropics toward the equator, rise at the equator, move poleward above the ground, and sink at about latitude 30° north or south.[3] This mean cir-

[3] This simplified circulation applies only over the oceans. Over the continents, monsoon circulations dominate (see Section 3.5.3).

culation, averaging less than a mile per hour and, therefore, being much slower than the west-east motions described above, has been known for a long time (since 1735); it is called the *Hadley cell,* after its discoverer, George Hadley (see Figure 3.7). This circulation cell, with its rising motion in equatorial regions, explains why there is so much rain in this area (rising motion leads to condensation and precipitation). Most of the rain forests of the world are located in the rising branch of the Hadley cell, where surface winds are light and variable. This region is called the *doldrums,* meaning "much rain and light winds." Actually, satellite pictures have shown recently that the precipitation avoids the equator itself and the maximum rainfall occurs just north and south of the equator. This peculiarity is probably a result of a narrow region of cold water in the oceans right at the equator, which inhibits cloud formation.

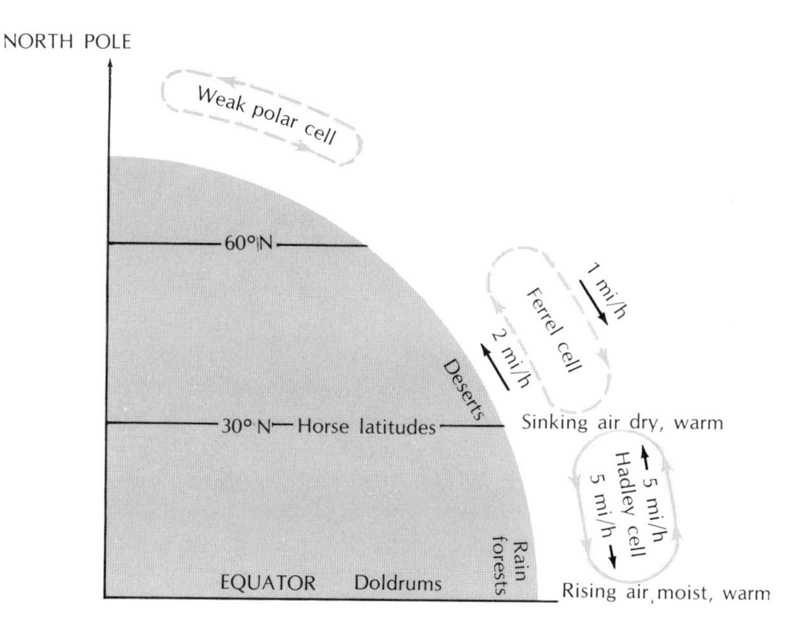

Figure 3.7 *Mean vertical circulation cells in the general circulation.*

At latitude 30° north or south, the air sinks and warms, and skies are clear most of the time. Again, there is little organized motion near the ground; sailors have often been becalmed in this area. Because of the mean downward motion, there is also almost no rain, and most of the world's deserts are found here. We call the latitude band near 30°N or S the *horse latitudes,* because sailors sometimes threw overboard (or perhaps ate) horses they could not feed. Between the horse latitudes and the doldrums, the air near the ground flows toward the equator where the surface pressure is low. The Coriolis force deflects the wind to the right in the Northern Hemisphere and to the left in the Southern Hemisphere (Figure 3.8). In both hemispheres, therefore, the wind blows toward the west, with a strong component toward the equator. These

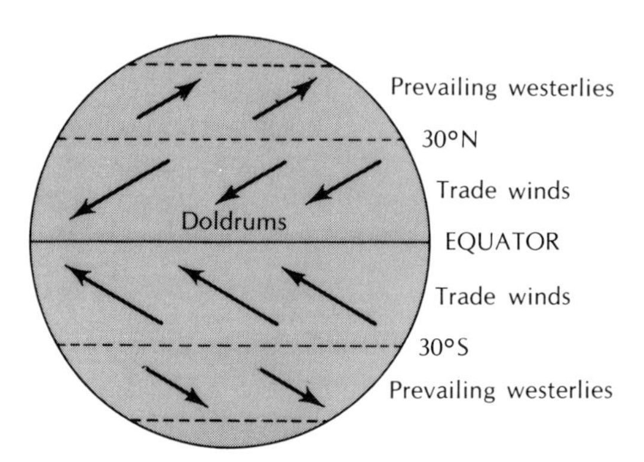

Figure 3.8 *Mean surface winds in the general circulation.*

winds (northeast in the Northern Hemisphere and southeast in the Southern Hemisphere) are very steady and reliable and are called *trade winds*. Sailors made good use of the trades to reach America from the Old World.

The Hadley cell emerges clearly from wind observations. But observations and general circulation models also suggest two other, much weaker cells (Figure 3.7). These are difficult to detect from observations, because the small average north-south motions (less than 1 mi/h) are masked by much larger motions associated with circulations around traveling high- and low-pressure centers. The more important of these two weak cells is the *Ferrel cell* (named after William Ferrel), where winds near the ground tend to blow from west to east. Thus, in regions between roughly 30° and 60°N (latitudes of Jacksonville, Florida and Anchorage, Alaska, respectively), most of the surface winds have westerly components; winds from the southwest, west, and northwest are much more common than easterly winds. But these westerly winds are not as reliable as the trade winds in the tropics, and are frequently disrupted by moving storm systems.

Fast and narrow jet streams, and sharp fronts that separate cold air and warm air, seem to be fundamental properties of the atmosphere, and can be duplicated well by computer models. Satellite pictures often show such fronts because they are associated with narrow regions of clouds and precipitation (Figure 2.6). These fronts and jet streams meander north and south, and, in an average picture of the atmosphere, are quite indistinct, since averages eliminate the extremes. Because averages smooth out characteristic features of great practical importance, it is inaccurate to consider climate as simply average weather.

3.5 Seasons

So far, we have discussed typical global distributions of wind, temperature, and precipitation without regard to season. We know, however, that it is colder in winter than summer, and that on the American west coast there is more precipitation in winter than summer. We find that many weather phenomena in the United States show a preference for certain seasons. Tornadoes occur most frequently in spring, hurricanes in the fall. Thunderstorms are common in spring, summer, and fall, but rarely disturb the cold peace of the northern winter. These seasonal changes, which are caused primarily by the motion of the earth around the sun and the tilt of the earth's axis of rotation, are also influenced by the type of terrain.

3.5.1 *Why the seasons?*

To begin with, the earth revolves about the sun in an ellipse which is nearly a perfectly round circle, as illustrated in Figure 3.9. Here, the ellipse is exaggerated to emphasize the characteristics of an elliptical orbit. The sun is at a focus of the ellipse, so that both the maximum distance and minimum distance between the earth and the sun occur when the earth is located on the major axis. The minimum distance, *perihelion* (Greek for "around the sun"),

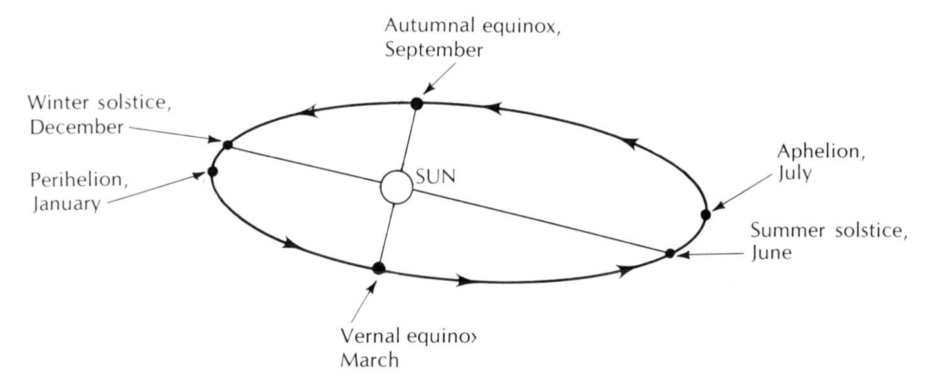

Figure 3.9 *Elliptical orbit of the earth (eccentricity exaggerated).*

occurs about January 3. Maximum distance, *aphelion* ("away from the sun"), occurs about six months later. The obvious result of this difference in distance is that July should be colder than January. Yet it is not, at least for people in the Northern Hemisphere. So, the changing distance between earth and sun cannot be the reason for seasonal change. Does this mean that we can forget about the difference in distance between the earth and the sun in July and January? For many purposes, we can. The difference is only about 3 percent, which, considering that the intensity of radiation varies as the inverse square of the distance from the source, means that the entire earth gets about 6 percent more radiation from the sun in January than in July. If we are to make accurate calculations, however, we should take this difference into account. For example, if summer in the Northern Hemisphere occurred at perihelion, the summer would be hotter. Also, in that case, winter would be at aphelion, and winters would be colder than they now are. Therefore, seasons in the Northern Hemisphere would be more severe than they are at present. Actually, this situation did occur 10,000 years ago. Not only that, but astronomers have computed the orbit of the earth for the more distant past and have found that several tens of thousands of years ago the orbit was less circular than it is now, and the radiation received at the closest approach was 20 percent greater than at aphelion. Certainly, this would produce a substantially different climate from what we have now.

In Figure 3.9, we note another consequence of the elliptical shape of the earth's orbit. The distance the earth has to travel from the beginning of spring (March 21) to the beginning of autumn (September 23) is longer than the distance from the beginning of fall to the beginning of spring. Not only that, but Johannes Kepler showed long ago that the earth travels more slowly when its distance from the sun is greater than when its distance is small. Therefore, in the Northern Hemisphere, the total length of spring and summer is substantially longer (about a week) than fall and winter. If you do not believe this, count the number of days from March 21 to September 23 and compare it with the number between September 23 and March 21. In the distant past, this difference in the length of the seasons has been as much as a month. So, we in the Northern Hemisphere have nice long springs and summers. We must remember, though, that the seasons are reversed in the Southern Hemisphere, which, therefore, has relatively short springs and summers.

Now we come to the real reason for the difference between seasons (Figure 3.10): The plane of the earth's equator is not parallel to the plane of the

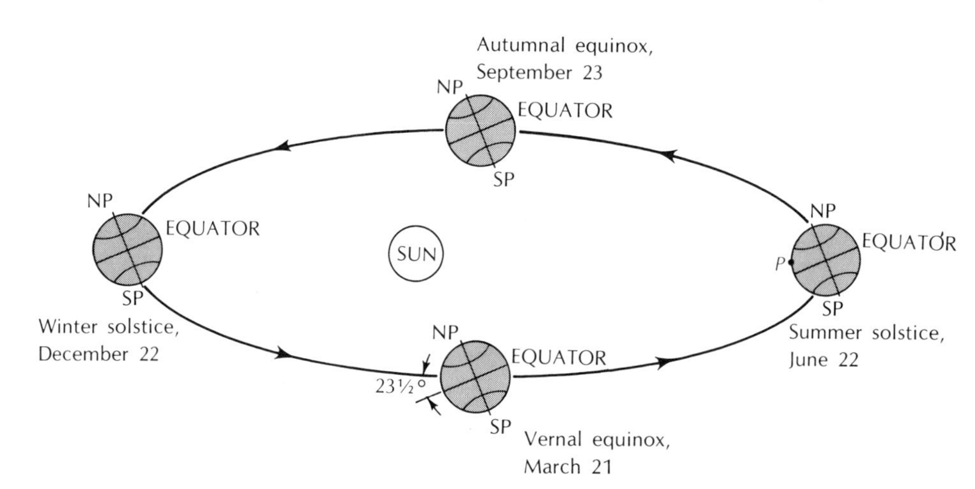

Figure 3.10 *Orbit of earth in perspective, showing seasons.* Note that the plane of the equator is not parallel to the plane of the earth's orbit.

earth's orbit. Instead, there is an angle between the plane of the equator and the plane of the earth's orbit (also called the *ecliptic*). This angle, which has the impressive name *obliquity of the ecliptic,* is now 23½ degrees. In the last 100,000 years, this angle has changed a bit, because the earth rocks back and forth a little. Roughly, it has varied between 22 and 25 degrees. We shall see that this angle has important effects on climate, and when it changes, the climate changes accordingly.

As the earth revolves about the sun, its axis points in the same direction in space. Figure 3.10 shows the earth in four positions: on March 21, June 22, September 23, and December 22. On June 22, the Northern Hemisphere is directed toward the sun. The North Pole is in sunlight all day, and the South Pole is shaded all day. Also, the point *P*, where the sun is directly overhead, is in the Northern Hemisphere, and daylight lasts a long time. In the Southern Hemisphere, days are short, and the sun is never far from the horizon. Therefore, all points in the Northern Hemisphere get more sunlight on June 22 than the corresponding points in the Southern Hemisphere. This date, when the sun is higher above the horizon than on any other day of the year, is called the *summer solstice,* which means "summer sun-stand-still." The sun "stands still" in the sense that it ceases its daily northward migration in the noon sky, hesitates, and then heads back toward the southern skies.

The satellite picture of the Western Hemisphere at 0430 EST July 5, 1974 (Figure 3.11), illustrates the effect of the tilt of the earth's axis on the length of night and day. Because this date is close to the summer solstice, the North Pole is bathed in sunlight, while the South Pole shivers through its polar night. The satellite is centered over the equator at 45°W longitude so that the line of sunrise on the earth makes an angle of nearly 23½ degrees with the 45° meridian. Thus, the comma-shaped mass of swirling clouds over the northern Atlantic is illuminated by the early morning sun. Farther west, dawn is breaking over eastern Canada and northern New England, while locations at the same longitude but farther to the south, such as Miami, have an hour or two of night left.

On September 23 and March 21, both hemispheres get exactly the same amount of sunshine. These dates are called *autumnal equinox* and *vernal equinox,* respectively. Equinox means "equal night"; every place has nights and days of equal length. Therefore, at the equinoxes, the time from sunrise to sunset should be 12 hours. It would be exactly 12 hours, too, if the atmosphere did not bend the rays of the sun. Because of this bending, the sun can be seen when it is below the horizon, so that the time between sunrise and sunset appears to be a little longer than 12 hours.

On December 22, the *winter solstice,* the Southern Hemisphere gets much more light than the Northern Hemisphere, where the sun rises only a little above the horizon and remains for only a short time.

These changes through the year would not exist if the obliquity of the ecliptic were zero, in which case, there would be no seasons. Over tens of thousands of years, this angle has changed, and, as a result, the severity of the seasons has also changed. When the angle is small, the seasons are less harsh than when the angle is large.

In most places in the United States, the date of the greatest amount of sunshine (around June 22) is not, on the average, the warmest (see Appendix 2 for examples). Since it takes time to heat the ground, the oceans, and the atmosphere, the period of maximum temperatures follows the period of maximum sunshine. Therefore, the warmest month over most continents outside the tropics

Figure 3.11 *Satellite view of the Western Hemisphere for 0430 EST July 5, 1974.*

is July in the Northern Hemisphere and January in the Southern Hemisphere. Similarly, it also takes time to cool the air, so that the coldest date in the Northern Hemisphere is not December 22, but is usually later in the season.

3.5.2 *Seasons in different parts of the world*

The warmest and coldest times of the year lag behind the times of the maximum and minimum radiation, for it takes time to warm and cool the surface of the earth and the air. How much time depends on the properties of the surface. The oceans take more time than land, for reasons given in the discussion of the diurnal variation of temperatures over water (Section 3.2.2). Land heats more rapidly in summer and cools more rapidly in winter than the sea. Thus, continents in the Northern Hemisphere in middle latitudes are warmest in July and coldest in January. The oceans, on the other hand, are coldest in March and warmest in September. Also, the annual range of temperature is much greater over the land than over water.

It is apparent that maritime and continental climates are quite different: maritime climates are mild and have their extreme temperatures late in the seasons, compared to continental climates. We have both maritime and continental climates in the United States. Because weather moves from west to east, California weather is brewed over the Pacific and the climate is maritime, with late seasons, and little temperature contrast from winter to summer. On the other hand, the central and eastern United States have continental climates, with early seasons and large seasonal contrasts. Thus, the annual range in temperature in San Francisco is only 6°C (11°F), compared to a range of 29°C (53°F) in Chicago (see Appendix 2).

Another difference between continental and maritime climates is the difference in annual distribution of precipitation. In continental climates, most precipitation occurs in summer. At this time, the ground gets so hot compared to the air aloft that strong convection currents are produced in the afternoon, leading to thunderstorms and large amounts of precipitation.

Thus far, we have contrasted continental and maritime seasons. The intensity of the seasons also changes with latitude, from the arctic to the tropics. In the tropics, the sun is always high at noon, and the days and nights always last about 12 hours. In high latitudes, the sun is low in the sky in winter, or does not shine at all. Therefore, Minnesota has much more severe seasonal contrasts than Texas, even though both have continental climates.

There is little annual variation in temperature at the equator, but much in the arctic, and so we find another important difference between summer and winter. In summer, the north-south temperature difference is small; in winter, it is much larger. For example, in winter there is a large temperature difference between New York and Miami, which is why so many New Yorkers go to Miami in winter. In summer, however, the temperatures at New York and Miami are about the same, so relatively few Miamians go to New York.

We noted earlier that temperature differences in the horizontal are related to vertical differences in wind speed, known as *vertical wind shear*. Therefore, vertical wind shears are much greater in winter than summer; west winds aloft are about twice as fast in winter. Weather patterns in middle latitudes also move faster from west to east in winter than summer, and airplane schedules have to be adjusted for this seasonal wind change. Strong wind shear also tends to stir up the air and make it turbulent. This turbulence, which is most common and severe in the winter months, shakes planes and passengers, and is quite common at the flight level of conventional jets.

A less pronounced, but still important, seasonal change is that most weather patterns, including cyclones and anticyclones, shift southward in winter and northward in summer (as do the sun, birds, and some people). These weather pattern migrations alter the position of the doldrums and other basic features of the general circulation. For example, the doldrums are about 8 degrees of latitude north of the equator in summer, but are 4 degrees south of the equator in winter. Similarly, the dry belt at about 30° latitude drifts about 5 degrees north of its winter position in summer. In California, a belt of relatively high precipitation moves southward in winter and is responsible for the fact that the heaviest precipitation occurs in that season.

Up to now, we have been mainly concerned with annual changes near the ground, but it turns out that some huge variations occur in the high atmosphere. For example, at the North and South poles, 30 kilometers up, the temperature decreases from a "warm" −35°C (−31°F) in summer to about −80°C (−112°F) in winter. Also, at high levels, most of the world's wind circulation outside of the tropics completely reverses in direction from summer to winter; 55 kilometers up, 200-mi/h west winds in winter are replaced by 100-mi/h east winds in summer.

A most peculiar thing happens in the stratosphere above the equator. Here, the wind direction does not change from summer to winter, but reverses, on the average, every year. Such a year-to-year change is called a *quasi-biennial oscillation*, that is, an oscillation with a period of about two years. A very complicated theory of this reversal has been given, having to do with atmospheric waves from the ground traveling upward and interfering with the stratospheric flow. Since the discovery of this phenomenon, meteorologists have tried to find such oscillations in weather near the ground. One two-year cycle is a tendency for warm, pleasant summers in northern Europe to occur every other year (good wine years), with cold, rainy summers (and poor grapes) in between.

We saw earlier that seasonal variations are smaller over the sea than land. This difference produces great differences in climate between the two hemispheres; the Southern Hemisphere, consisting of 80 percent water, is more maritime than the Northern Hemisphere, which is only 60 percent water. Thus, seasonal variations are generally smaller in the Southern than in the Northern Hemisphere, in spite of the slightly greater contrasts between the amount of radiation received by the Southern Hemisphere in its summer versus its winter season. Only the polar regions are exceptions. The North Pole is covered by water and ice (the Arctic Ocean). In contrast, the South Pole is surrounded by the antarctic continent, an ice-covered land mass of considerable height. Therefore, the coldest ground temperatures observed anywhere in the world (around −87°C, or −125°F) occur near the South Pole in winter.

3.5.3 Monsoons

The difference between land and sea produces seasonal wind patterns which interfere with the simple, regular patterns discussed earlier. Because land is hotter than the sea in summer, and colder in winter, these land-sea circulation patterns reverse from winter to summer. Such seasonally reversing patterns are called *monsoons*. To see how a typical monsoon works, consider Figure 3.12, which shows a monsoon circulation associated with an idealized square continent in the Northern Hemisphere in summer and in winter.

In summer, the continent is hot and the continental air rises. To replace this rising air, other air rushes in from the oceans toward the land. But the Coriolis force deflects the air to the right so that a counterclockwise (cyclonic) circulation results. In winter, the circulation reverses, as the land is then cold. Cold, heavy surface air flows off the continent over the seas, acquiring a clockwise (anticyclonic) circulation under the influence of the earth's rotation.

Real monsoons are not quite so simple, because continents are not squares, and mountains interfere. The bigger the continent, the better the idealized patterns work. Monsoons are therefore especially prominent in Asia.

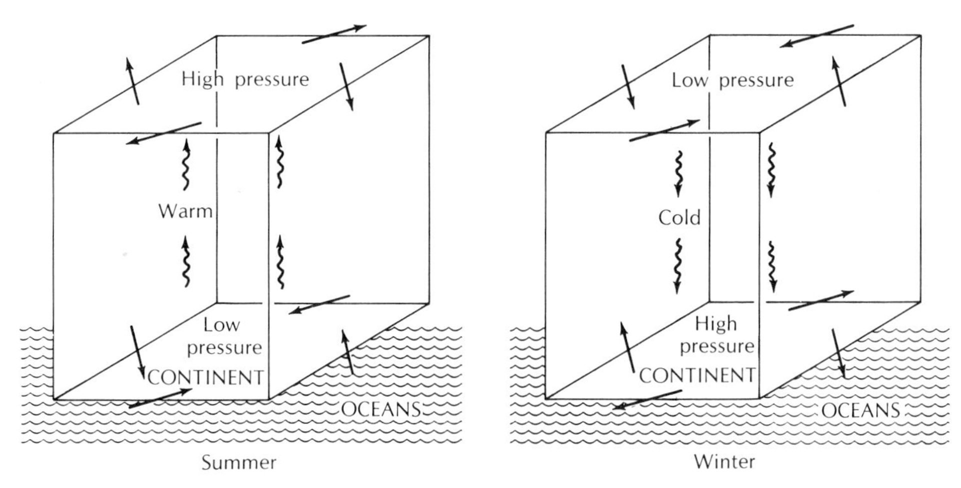

Figure 3.12 *Monsoon circulations around continents in the Northern Hemisphere.* Warm air rises over the continent in the summer, and air from the ocean moves in to take its place. In winter, cold air sinks over the continent and spreads outward over the ocean.

Consider, for example, China, which has a well-defined monsoon. China would be located on the right of the box-shaped continents of Figure 3.12. We see that the winds in China should be (and are) mostly northwest in winter and southeast in summer. In winter, the northwest winds bring cold air out of Siberia, making winters much colder than they would be if the monsoon were not present. Now consider India, which is located at the bottom of the box. Summer winds are mostly southwest. The rising humid air from the Indian Ocean flows up over the Himalaya Mountains and releases huge amounts of moisture in the form of rain. Some of the largest rainfalls in the world have been measured for summer at Cherrapunji in northeast India, which holds the world's record for rainfall in one year (2647 centimeters, or 1042 inches) and in one month (930 centimeters, or 366 inches). No wonder that to the Indians, and many other people, the term *monsoon* has come to mean large amounts of rainfall.

The effect of land-sea differences is, of course, most obvious in connection with large land and water masses. Even near lakes, pronounced differences occur. For example, the east shore of Lake Michigan is milder and more humid in winter than the west shore; peaches can be grown in Michigan, but not in eastern Wisconsin.

3.6 Effect of Mountains on Climate

A final, major influence on climate is the presence of mountain ranges. First, air ascending mountains cools and deposits precipitation on the upwind side. Thus, where the winds are westerly, most rain falls on the western slopes. A conspicuous example is the west side of the Olympic Mountains in Washington State, where the annual precipitation is over 100 inches, allowing rain forests to grow. On the east slopes, some towns get only about 15 inches a year.

Mountains which are oriented along an east-west direction tend to restrict the exchange of air from south to north, so that cold arctic air does not mix with tropical air. For example, the Himalayas make possible enormous differences between temperatures in Siberia (–60°C) and India (+30°C) in winter. In America, the changes from north to south are gradual because the mountain chains are oriented north-south.

Finally, mountains tend to channel the low-level winds. For example, on the east coast of the United States, the prevailing wind in winter is generally northwest. But since the Appalachians run southwest-northeast, the wind in the Appalachian valleys is most commonly from the southwest.

3.7 Climate Close to the Ground (Microclimate)

Because most of our activities are carried out in the lowest ten feet of the atmosphere, there is special interest in the climate next to the ground. This lowest layer is a rather complicated region which shows tremendous variations from one height to another and from one horizontal location to another. For example, we have all experienced surfaces so hot that we cannot walk on them barefoot, e.g., a macadam road on a sunny July day, or the sand on a beach. In these cases, the surface temperature may be 60°C (140°F) or more. Yet, merely five feet above the surface, where our noses are, the temperature may be only 30°C (86°F). Along with such large temperature contrasts immediately above the ground, we find correspondingly vast variations of wind, moisture, and other variables. We might wonder how such great changes are possible over these short distances. For an explanation, we will start with the temperature.

3.7.1 Vertical temperature distribution at low levels

The atmosphere is largely transparent to incoming solar radiation. Hence, most of the sunlight heats the ground, not the air. How hot the ground gets depends on how much sunlight is reflected and on how much heat is conducted into the ground. If the ground is dark (a poor reflector) and conducts heat poorly, the temperature can be extremely high. Sand is a poor heat conductor, mainly because air is trapped among the grains. Air is an especially poor conductor, which is why we use double-paned windows with an air space between to keep heat in buildings.

After the ground gets hot, the air in contact with the ground begins to warm by conduction. However, since air is such a poor conductor, it would take a very long time to heat the air a few feet above the ground if conduction were the only mechanism. Conduction does heat the air very close to the ground, perhaps 0.04 centimeter above it. This heated air expands and becomes so light that the air above the next layer is actually denser. This heavy air begins to sink in

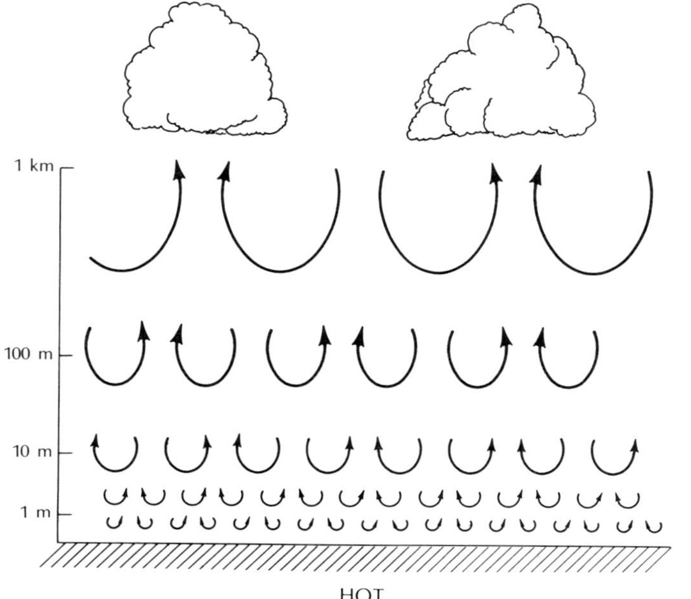

Figure 3.13 *Vertical eddies and convection currents.*

certain places, while the warm, light air rises in others. These upward- and downward-moving bubbles of warm and cool air are called *convection currents* (Figure 3.13). These convection currents heat the air higher up by carrying hot air upward in the thermals and mixing with the air aloft. The convection currents can actually be seen, because the associated variations in density cause shimmering of light.

As the ground continues to warm, convection currents rise higher and higher, eventually reaching about one kilometer on a typical day. If there is enough moisture in the air, the rising air becomes visible in the form of cumulus clouds. The updrafts and downdrafts of the thermals are also felt by airplane passengers. Birds and sailplanes use the updrafts to stay aloft without effort.

From our point of view, the function of convection or thermals is to mix the hot air in contact with the ground with the cooler air higher up. The result is that the air at 5, 10, and 100 meters, and eventually at thousands of meters, is being heated. At the same time, if the sun continues to shine, the ground is heated more and more. The question is, what will the temperature distribution be eventually? To answer this, consider again Figure 3.13. Thermals close to the ground are small because they have little vertical space in which to grow. Higher up, away from the constraining ground, the eddies are larger, and the larger the eddies, the more efficiently they mix air of different properties. It is because the thermals close to the surface do not mix air efficiently that large vertical temperature gradients can exist. Thus, we get temperature distributions such as those shown in Figure 3.14, with huge temperature differences near the ground and smaller differences higher up. A 10°C temperature difference in the lowest meter is quite a common occurrence on a hot day.

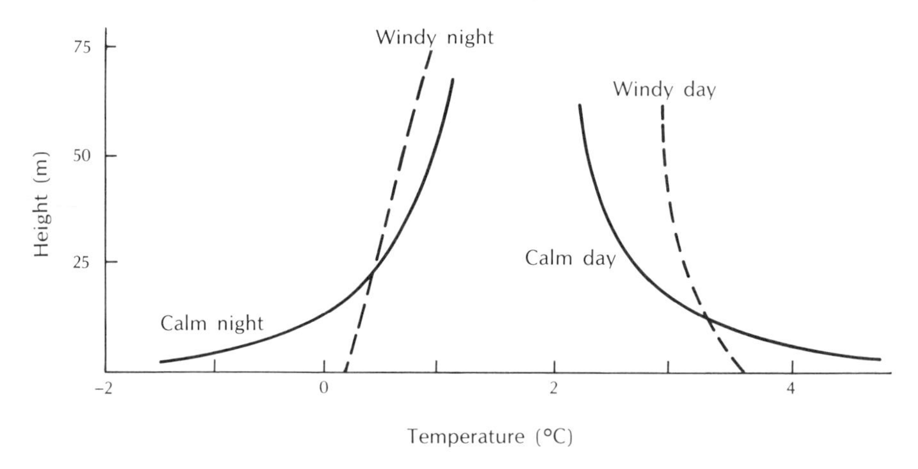

Figure 3.14 *Vertical temperature distributions near the ground on calm and windy days and nights.*

Thermals produced by heating are not the only eddies able to mix air with different properties. Eddies can also be produced mechanically, by rapid changes in wind speed or direction with height (wind shear). We have already met this type of eddy in connection with jet streams at high levels. Such mechanical turbulence also occurs near the ground when the wind blows. At the ground itself, the speed is zero, but just a few meters above the ground, the wind speed can be considerable, which means that the wind shear near the ground is usually strong and will produce eddies. Again, because of space limitations, these eddies are small close to the ground and larger and more efficient higher up. Such mechanical eddies can help the thermals carry the heat upward. Therefore, on windy, sunny days, the air above the ground is heated more rapidly than on calm days. At the same time, the surface will not get quite as hot, since some of the heat is carried aloft. Figure 3.14 compares temperature distribution on calm and windy days.

So far, we have considered air motions near the ground during the daytime. At sunset, the ground cools rapidly and eventually gets quite cold, as anyone who has slept outside on the ground knows. This is because the ground is an excellent radiator (it radiates heat away easily). It gets especially cold if the air is dry and cloudless. How does this cold ground cool the air at higher levels? No thermals are generated, because the air just in contact with the surface is colder and denser than the air above. If there is enough wind, though, mechanical mixing can cool the air aloft and simultaneously prevent the surface from cooling as strongly, producing the "windy night" distribution shown in Figure 3.14.

If there is not enough wind to cause turbulence, the air above the ground slowly cools by radiation. Some of the radiation from the air is returned to the ground. On clear, dry nights, this returned radiation is relatively small, so that the ground cools rapidly. The radiation from the air to the ground increases as the relative humidity increases, so that on humid nights the ground cools less than on dry nights. The maximum return of radiation from the nocturnal sky occurs when low clouds are present. These clouds may return nearly as much radiation to the ground as the ground is losing to the air, and so prevent any significant cooling. Thus, the coldest mornings (which hold the greatest possibility of frost near the beginning and end of the growing season) are calm, dry, and clear.

3.7.2 *Vertical distribution of wind at low levels*

In some respects, the vertical distribution of wind near the ground is similar to that of temperature. There is much more wind shear from the ground to a height of one meter than from one meter to two meters. You can observe this characteristic on a windy day at the beach; by lying down, you will feel much less wind than when standing. A typical wind distribution is shown in Figure 3.15. The theory of such a distribution is well understood, and a logarithmic equation fits it well. The reason for this behavior is, again, that eddies near the surface are much smaller than they are higher up; hence, large differences of wind can occur only close to the ground, where the eddies are small and mixing is inefficient. If the surface is rough, e.g., if it is the top of a forest or a city, bigger eddies are possible right at the surface, so that large gradients in wind speed cannot develop without being destroyed by mixing. Typical wind distributions over rough and smooth terrain are contrasted in Figure 3.15.

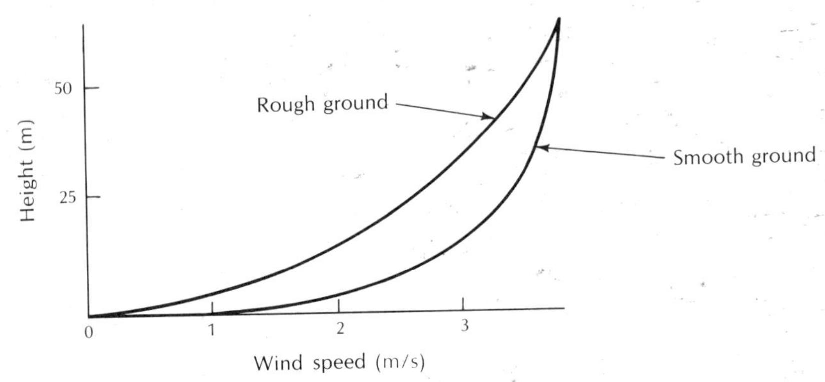

Figure 3.15 *Vertical wind distributions near the ground over rough and smooth surfaces.*

3.7.3 *Horizontal variations of temperature near the ground*

So far, we have emphasized the enormous vertical variations of meteorological variables close to the ground. Huge horizontal variations are also possible. For example, we have seen that the surface temperature depends on the characteristics of the surface. A dark, poorly conducting surface becomes very hot, while lighter surfaces remain cool. Different types of surfaces are frequently located very close to each other, with the result that large changes in temperature can occur over very short distances. Large differences can also be produced by even small changes in topography. Cold air at night collects in low spots, which, on calm nights, may be 10°C or more colder than higher elevations. Sometimes, the air cools below the dew point, so that such low spots often fill with fog, which is absent at higher elevations.

Proximity to bodies of water may also produce large horizontal temperature variations near the ground. On cold nights, oceans, lakes, and even ponds, retain enough heat to keep the nearby shore areas warmer than inland areas. Conversely, on hot days, the water heats slowly, and the temperatures are frequently 10°C cooler on the beach than inland. This latter effect is more pronounced near large bodies of water in the spring, when water temperatures may be close to 5°C (41°F), while air temperatures away from the water may stand at 20°C (68°F) or higher.

There are many aspects of microclimate which are important to specialists, for example, the special character of the climate of cities. Cities are warm compared to the country, and this temperature difference (recall the importance of differential heating) produces peculiar wind-flow patterns around cities. Because these small-scale circulations play an important role in understanding phenomena such as the dispersion and chemical reactions of air pollution, we will describe them here.

3.7.4 *Urban climate*

Man's activities tend to be concentrated in cities, and the result of a large number of people living in a small area has been a significant change in the urban climate. The reasons for these climatic changes

Plate 3 *Hurricane Gladys, taken by Apollo 7 astronauts in October, 1968 (photo supplied by Dr. Robert Sheets, National Hurricane Research Laboratory).*

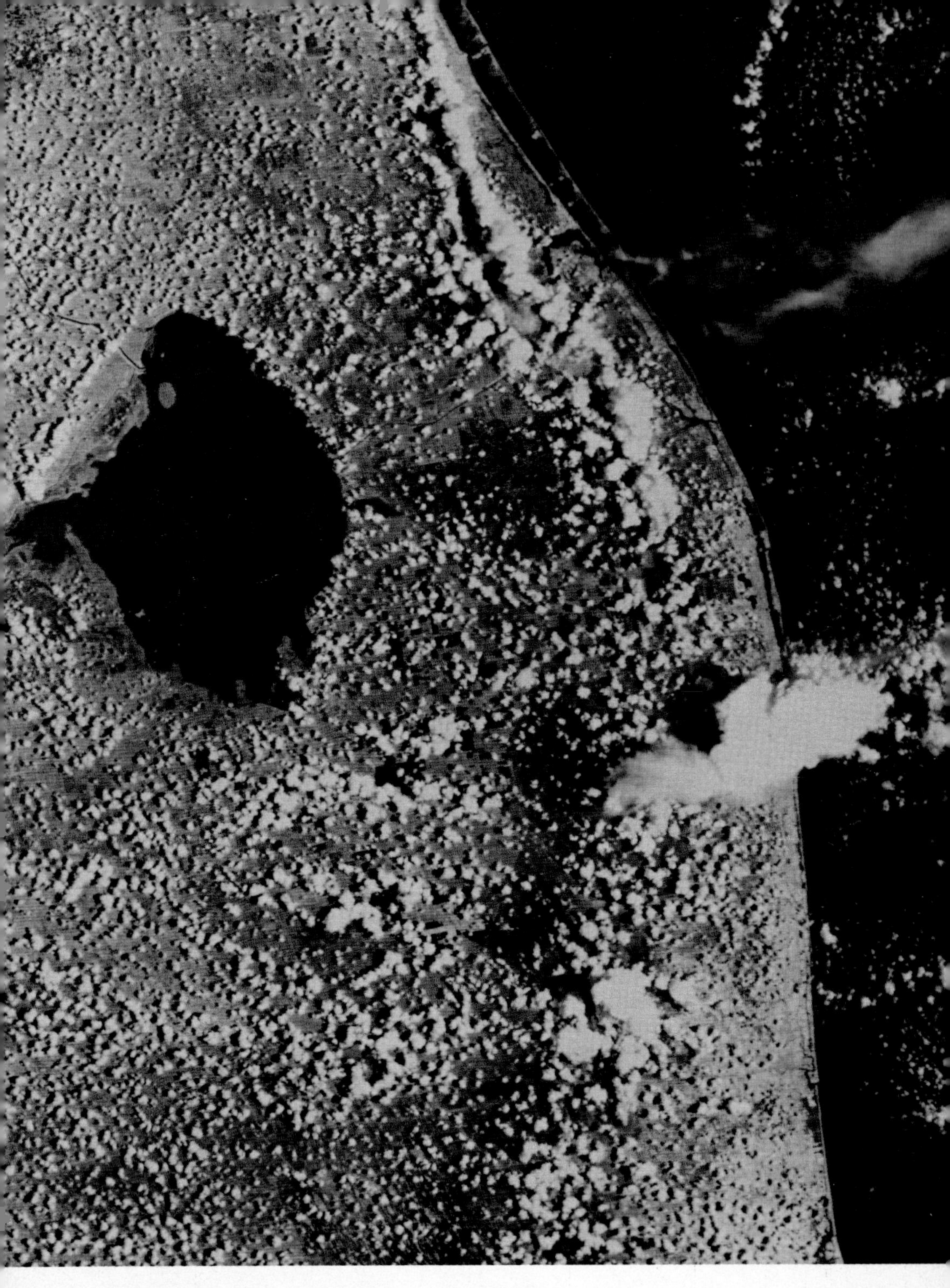

Plate 4 *ERTS-1 photograph of cumulus clouds over Florida, 1000 EST August 18, 1972.* The line of towering cumulus clouds along the east coast marks the "sea-breeze front."

can best be understood by examining the physical differences between the city and the surrounding countryside. These differences become more pronounced as the city grows.

Cities are usually made up of granitelike materials with large thermal capacities (ability to store heat) that act something like a body of water, absorbing heat during the day and releasing it at night. Thus, the city cools more slowly in the evening than the country. Additionally, the three-dimensional nature of the city with its tall buildings presents a complex geometery for heat exchange. The vertical sides of city surfaces such as apartment houses or office buildings do not allow thermal radiation to be lost to space as readily as in the country, where surfaces are flatter. The retained heat is in part reradiated between buildings, rather than toward the sky, and is therefore slowly dissipated. The city also has major artificial heat sources from industrial, domestic, and automotive consumptions of energy, which make cities warmer than rural areas.

The humidity and air quality also vary from city to country. The reduction of transpiration in the city (as a result of a lack of vegetation) and the efficient removal of precipitation from relatively impervious surfaces cause urban areas to be deficient in moisture when compared to the rural surroundings. The wastes of industry, domestic activity, and modern transportation contribute to major changes in the type and amount of materials suspended in the air, thus decreasing urban air quality.

What are some of the results of these physical differences? One of the important changes is in temperature. The warmth of the city has been recognized for over 100 years. Figure 3.16 shows the average minimum temperatures in the Washington, D. C. metropolitan area for the winter seasons (December–February) of 1946–1950. The warmest minimum temperatures (31°F) occurred near the Potomac River in the heart of the city. In the suburbs and surrounding countryside of Maryland and Virginia, the *average* minimum temperature was 6°F lower than in the city. A difference of 6°F in average minimum temperature is very significant, because on many nights (cloudy, windy nights) the differences are much less. In fact, on nights which are clear, dry, and calm, the temperature difference between city and country may be 20°F or more.

Because of the warmer temperatures (*heat-island effect*), the growing season in cities is three to eight weeks longer than in the country. For example, Figure 3.17 shows the average date of the last freezing temperature for different sections of the Washington, D. C. area. The last freeze of the winter occurs nearly a month earlier in the central part of the city than in the surrounding countryside. Because of the extended growing season, during World War II, "victory gardens" were promoted with the hope that the increased growing season would produce greater yields from plants. However, the quality of produce was somewhat decreased because of air pollution effects.

For a city with clean air, measurements have shown that the urban maximum temperature exceeds the rural maximum by only a small amount. When a significant pollution dome lies over the city, the daytime temperatures can actually be lower than the rural surroundings because of the reduction in the incoming solar radiation.

Under both bad and good air quality conditions, the relative warmth of the city increases at night. This increase exists even under moderate wind conditions, and is due primarily to the increased fuel consumption and the

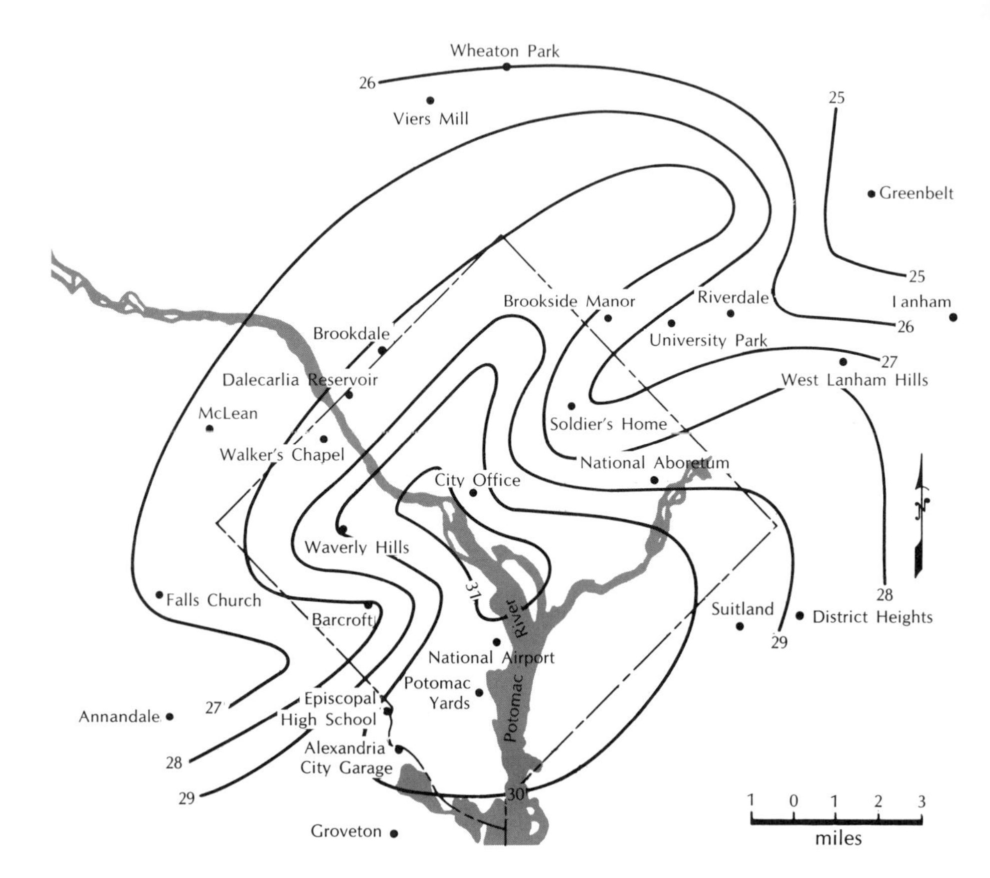

Figure 3.16 *Average minimum temperatures for the Washington, D.C. area during the winter season (after Clarence A. Woolum, "Notes from a Study of the Microclimatology of the Washington, D.C. Area for the Winter and Spring Seasons," Weatherwise, 1964, 17:6).*

recovery of heat stored during the day in streets and buildings. Poor air quality enhances the accumulation of heat in the city because of a partial closing of the radiative "window" (see Section 2.1.5) to the exit of long-wave energy from the city surfaces. Even small towns can exhibit temperature differences of 4–10°F in the evenings. Although large pockets of heat exist in urban areas, there are marked contrasts with relatively cold air. The coldest areas in the city tend to be found in the lowest areas, where the cold air accumulates.

The heat island effect, combined with the many air pollution particles present in urban air, probably cause a general increase in urban precipitation, although this effect is somewhat controversial. The urban heat island promotes thermal convection currents, which cause clouds to form and showers to occur. At the same time, the abundance of artificial *freezing* and *condensation*

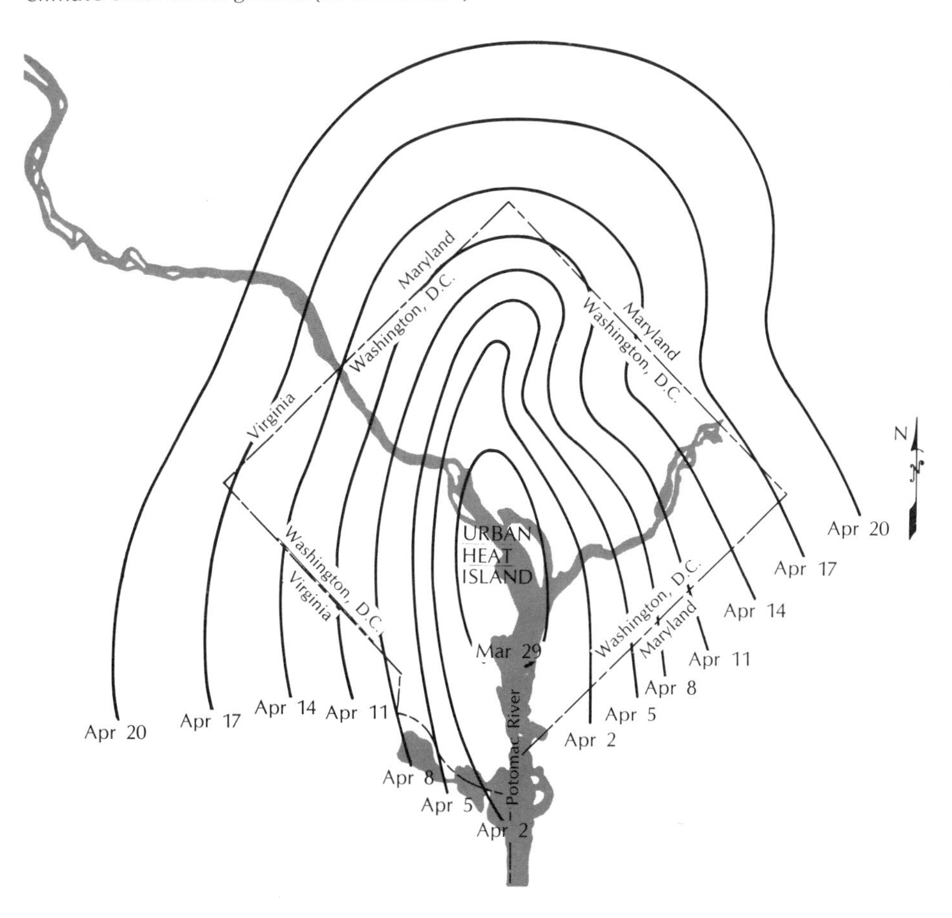

Figure 3.17 *Average dates of the last freezing temperatures during spring for the Washington, D.C. area (after Clarence A. Woolum, "Notes from a Study of the Microclimatology of the Washington, D.C. Area for the Winter and Spring Seasons," Weatherwise, 1964, 17:6).*

nuclei (small particles which facilitate the formation of precipitation, as discussed in Section 8.3) in pollution emissions actually seed the clouds, increasing the probability that additional precipitation will occur. The total annual precipitation may be increased by 10 percent in the city as compared to the country. Associated increases in the amount and frequency of clouds are also noted. It appears that the majority of increases in precipitation and cloudiness come as a result of small increments on relatively calm, moist days, when the convection over the city is well developed. Additionally, rural areas downwind from cities are affected by the *urban plume,* which is the stream of pollution generated in the city and carried by the wind. For example, LaPorte, Indiana, downwind from the Chicago-Gary complex, has experienced annual precipitation increases as great as 30 percent, although the relationship to the upstream pollution is very controversial.

In addition to the increase in total precipitation, a number of other climatic anomalies result in the city. There tends to be about a 15 percent decrease in the number of days with snow in the city as a result of the increased temperature in the winter. Cities also tend to experience considerably more fog than the country because of an excess of condensation nuclei (30 percent more in summer and 100 percent more in winter). Finally, the wind is also modified. The city streets tend to channel air flow. Because of increased surface roughness in the city, velocities are lighter. On generally calm, clear nights, heating in the city, and subsequent upward air motion, cause a light wind to blow into the city from the surrounding countryside.

3.8 Change of Climate

The climate of the world is not constant over long periods of time. A meteorology textbook written 50,000 years ago would have described very different weather patterns from those of today. At that distant time, the mean global temperature was lower by several degrees Celsius, a very significant difference. The large-scale wind patterns, including the jet stream, were shifted toward the equator, as were the precipitation patterns. Virginia's climate then was more like Pennsylvania's climate today. Any such changes in the climate may be classified as natural or man-made. Although the greatest changes have been natural, recently there has been concern that man's activities may also be contributing to important changes.

3.8.1 Natural climate changes

Climate has changed on various time scales. There are changes over relatively short time periods, with fluctuations over 100 years or so. Much larger changes occur over tens and hundreds of thousands of years. We do not know the causes of these changes, but there are many theories. In general, the problem with explaining climatic change is that we have too many theories, all quite plausible. Still more theories are hatched all the time, with no conclusive proof for any of them. It is clear, however, that quite different mechanisms may be responsible for the changes in climate on the different time scales.

Throughout geological history, there have been long periods, lasting about 100 million years, with very quiet and warm conditions, interrupted by relatively short periods of rapid change. We are now in one of the epochs of change. In the last million years (the Pleistocene epoch), there have been several ice ages interrupted by warm periods (*interglacials*). For 60 million years prior to that, there were no ice ages at all. The ice ages of the last million years have occurred in different locations from those of earlier ice ages. This difference suggests that the long, calm periods alternating with the shorter periods of climatic contrasts are caused by changes on the earth's surface.

It is now well established that the continents migrate slowly, floating on the liquid core of the earth. One hypothesis states that ice ages can occur only if major land masses are distributed around one of the poles. This situation exists now, with the major continents in the Northern Hemisphere all surrounding the North Pole. Another theory says that ice ages occur during times of volcanic activity and mountain-building. For one thing, mountains limit the exchange of air between cold and warm regions, permitting strong cooling of the arctic. Furthermore, volcanoes throw particles of dust into the atmosphere, which cut off some of the sunlight reaching the earth. In fact, there is considerable evidence that recent volcanoes have temporarily depressed the mean world temperature. The period of time between major episodes of volcanoes, or between changes in the location of continents, is about right to account for the 100 million years or so between the unsettled periods.

During the last million years, there have been at least four equatorward advances of ice sheets, which have coincided with wet, cold climates. The glaciers at these times advanced into the central United States. Between the advances, there were relatively dry interglacial periods, when the glaciers retreated toward the poles. Right now, we are between the extreme conditions on this time scale.

What are the explanations of changes on the shorter scale of 100,000 years? Again, we have numerous theories. One theory holds that the sun, like most stars, emits variable amounts of radiation with time. However, we have no evidence of variation in amounts of sunshine. To measure variations in solar radiation intensity, measurements would have to be made above the earth's atmosphere, an impossible task until just recently, when satellites became available. Before the advent of satellites, the intensity of the sun was measured on mountains. To estimate the effect of the atmosphere on the measured radiation, the measurements were made at different times. It was found that the lower the elevation of the sun above the horizon, the more radiation absorbed by the atmosphere and the less received at the ground. A graph was then drawn of the intensity of the sun as a function of the amount of air, and this curve was extrapolated to the point of no air at all. In this way, the solar constant (see Section 2.1.2) was determined to be about 2 cal/cm^2/min. Early observers found considerable changes in this number from time to time, but we now believe that these changes were a result of variations in the transparency of the atmosphere rather than real changes in sunlight emitted by the sun.

It would appear that observations of the sun from artificial satellites which circle the earth above the atmosphere would provide the definitive answer on whether the solar constant is indeed a constant. The trouble is that, in order to see whether the brightness of the sun varies, we must compare it with a constant light source. It turns out that the sun is as constant as any artificial source. In summary, we are quite certain that the intensity of the sun has varied less than 1 percent, if at all, in the last 70 years.

When we say that the solar radiation has been constant, we really mean that the visible radiation, which accounts for most of the total radiative energy, has not varied. However, radiation with wavelengths outside the visible range, such as X rays, vary a great deal. For example, X rays are more powerful when the sun is full of spots than otherwise. Such rays have little total energy, however, and are not likely to change the weather.

Even though there has been no appreciable change of sunlight in the last 70 years, many scientists have suggested that sunlight may have varied thousands of years ago. Of course, we have no proof of such a variation. Even if it could be shown that solar radiation varied, meteorologists do not even agree whether increasing or decreasing the radiation would produce an ice age. The paradox is that increased sunshine would raise temperatures, but would also produce more evaporation from the oceans, adding to the snowfall of the world. So, in the absence of a quantitative theory, it is not clear whether the intensity of the sun would have to increase or decrease to produce an ice age.

We have already mentioned that there is some evidence that major volcanic eruptions have been followed by cooler global temperatures. As recently as 1963, Mt. Agung, a volcano on Bali, erupted and threw huge numbers of particles into the atmosphere. The increase of dust lowered the transparency of the air for the next few years. In this case, no temperature effect at the ground was obvious, but there were substantial changes in the temperature of the equatorial stratosphere. The trouble with this *volcano theory,* as well as with so many other hypothetical causes of climatic change, is that we do not know whether the causal mechanisms acted on the correct time scale. For example, why have ice sheets advanced about every 200,000 years?

There is one theory which does give an explanation of the time scale—the *orbital change theory* of Milankovich. Astronomers have calculated from Newton's law of gravitation how the earth's motions have changed over the last million years. We already mentioned some of these changes earlier in this chapter: the obliquity of the ecliptic has been larger at times, and smaller at others; the orbit has at times been more eccentric, that is, less circular; and there have been periods when the closest approach of the earth to the sun occurred in the Northern Hemisphere summer instead of the winter.

In explaining climatic change by all these orbital changes, it is important to note that they have no significant effect on the total amount of sunlight reaching the ground. They, instead, change the contrast between seasons. For example, if the orbit were very eccentric, and the closest approach of the earth to the sun occurred in winter, winters would be mild. If the obliquity were small at the same time, winters would be even warmer, and summers cooler. In other words, the seasons would be very much milder than they are now. Milankovich postulated that ice ages are produced under just such conditions. If winters are warm, relatively more snow falls because the atmosphere holds more water vapor. The heavy winter snowfall will not melt as much in summer if the summers are cool. Thus, ice sheets can grow every year.

The orbital change theory has been accepted in the past, then rejected, and now seems to be quite popular again. The time scale fits the known data, and there is some confirmation from geological evidence, including ocean floor deposits. Objections against the theory are mostly that the orbital changes are too small and that the theory predicts different effects for the Northern and Southern hemispheres. For example, if the sun and earth are close in January, northern seasons are mild, while those in the Southern Hemisphere are relatively severe.

Although there are many other theories about climatic changes on the time scale of a few hundred thousand years, one more, the *carbon dioxide theory*, merits attention. This theory plays an important part in speculation about future climate, which may be influenced by carbon dioxide being added to the atmosphere. Carbon dioxide is added when power and heat are produced by burning coal, oil, and other fossil fuels.

The importance of carbon dioxide (CO_2) lies in the fact that it is transparent to sunshine, but not to invisible infrared (heat) radiation leaving the ground. Part of this infrared radiation is absorbed by the CO_2 in the air and returned to the ground, keeping the air near the surface warmer than it would be without the CO_2. In a sense, CO_2 acts like a blanket, though one full of holes. Most recent estimates suggest that doubling the CO_2 content would raise the surface temperature about 2–3°C.

A quantitative theory to explain past climatic changes by variations in CO_2 is quite complex. If, for some reason, the CO_2 content were raised in the atmosphere, there would be warming. But, then, some of the CO_2 would be dissolved in the ocean, reducing the amount in the atmosphere and producing a subsequent cooling. Eventually, the oceans would return CO_2 to the atmosphere, which would then begin to warm again. According to some scientists, the time scale of this cycle is similar to the time scales of recent ice ages.

Finally, we come to climatic changes with the shortest time scale—those with periods of a few hundred years. These fluctuations are smaller in magnitude than those over longer periods and represent temperature changes of just a few degrees. However, even such slight differences are responsible for important shifts in the weather patterns. For example, navigation in the far northern Atlantic was possible about 1000 years ago. Today, such navigation is not possible.

There are again many theories for the causes of climate fluctuations on the short time scale, some involving hypothetical changes in the sun, and some involving volcanoes. However, it is quite possible that climate would fluctuate this way naturally without any changes in outside factors. Unfortunately, we are not yet able to investigate this possibility by reproducing the behavior of the atmosphere over such long periods with mathematical models. In any case, we must realize that changes in temperature on the order of 1°C per century have always occurred and are occurring now. For example, the world's average temperature has increased about 1°F (0.6°C) from 1900 to 1950 and has cooled at about the same rate since.

3.8.2 *Man's impact on climate*

We have seen that man has produced a significant change in the climate of urban areas. Some scientists have also argued that the recent changes in global climate have been produced artificially by human activities. However, since these large-scale changes are no larger than those in the past, there is absolutely no proof that man has had anything to do with recent global changes. This is not to say that man's actions will not eventually have an important effect on climate. The need for power, heat, and transportation is increasing so rapidly that our output of CO_2, smoke, and other contaminants is growing at a faster rate. Some of the potential consequences of man's activities on global climate will be discussed in Chapter 7.

4 The changing weather outside the tropics

Weather, wind, women, and fortune change like the moon.

[French proverb]

There are many weathers in five days, and more in a month.

[Norwegian proverb]

4.1 Rhythms in the Weather

We are accustomed to certain rhythms in the weather: the daily cycle of the earth's rotation that warms the air from morning to afternoon, stirring the wind and drying the morning fogs; fluffy daytime cumulus clouds; afternoon showers in the mountains; clearing skies at sundown; and the nighttime thunderstorms of the Midwestern summer. Annual rhythms mark seasonal temperature swings at the high latitudes of Leningrad and Montreal, the wet winters and dry summers of San Francisco and Rome, winter clouds and skies brightening toward summer at New Orleans and London, and clean, refreshing summer lake breezes at Cleveland and Buffalo which accelerate to a howling blizzard in late fall.

Rhythms aside, the weather also changes from day to day or after a few days, albeit irregularly. Why is it sunny today, with rain tomorrow? The answer has little to do with seasonal or daily cycles; instead, it is related to traveling storm systems that cover hundreds of miles and have typical time scales of about four days. Spatially, the weather that we observe is highly variable, even over very short distances. Who has not noticed rain at one end of town and not at another, deeper snow lying a few miles distant, or cooler temperatures literally within walking distance? Nevertheless, consistent weather patterns do occur on larger scales: beneficial rains water the entire Corn Belt, snow may blanket several states, and cold waves engulf half of the United States.

Thus, it is appropriate to consider the larger-size patterns of weather, recognizing that there may be considerable variation of actual weather within them. Implicit in this approach is the idea that the larger scale usually controls the framework or range of the smaller-scale events that are "permitted." For example, over 60 percent of all tornadoes in the United States move from southwest to northeast; these tornadoes are very small, but the larger-scale winds steer them preferentially northeastward. Plainly, the condition for tornadoes must involve southwesterly flow on a large scale.

The characteristic time between cloudy or rainy periods at most locations in the United States, Canada, or Europe is a few days; the cause lies in sporadic weather disturbances, now lifting the air to form clouds, now compelling it to sink, wiping the clouds out.

In what follows, we examine these limited regions of disturbance, where clouds and precipitation are favored, leaving the more extensive volumes of fair weather as the normal background. We shall find that the disturbances, called cyclones, move about; only irregularly do they affect any one location. Interestingly enough, they move faster in winter than in summer, but are farther apart in the winter, so that the characteristic time between them is comparable throughout the year. But it is never so regular as to allow us to say that it will rain every three days, or four, but only that the weather will change, sooner or later.

4.2 Migrating Cloud Systems as Viewed from Satellites

None of this rhythm would greatly surprise an observer perched on a satellite and viewing much of the hemisphere, for he would notice collections of clouds, not necessarily uniform, but at least regions of enhanced cloudiness, moving about the face of the earth.

Figures 4.1–4.5 illustrate the migrations of clouds for the period of May 1–May 5, 1973. These are composite daytime satellite pictures produced by the NOAA-2 satellite, which orbits over an approximately north-south path. The picture is built up by strips as the earth rotates eastward. Thus, it is always mid-morning underneath the satellite. Although the pictures are not taken at one instant, they do depict the general cloud conditions on a given day. The clouds, being good reflectors, appear white from sunlight reflecting back up to the satellite. By contrast, in the dark regions of the pictures, the skies are more nearly clear. One exception is the strong reflection from snow, notable over Greenland and the polar regions. We can differentiate the deep clouds from snow by looking at an *infrared* picture. Figure 4.1b shows the infrared view for the same time. Infrared sensors read the *temperature* of objects viewed rather than their reflective properties, so this picture is white for only the high clouds, whose tops are very much colder than the snow (perhaps as low as −40°F). Throughout early May, there is snow at high latitudes and broken bands of tropical clouds north of the equator, but we will concentrate on the changing middle latitudes.

On May 1 (Figure 4.1), extensive clouds cover Japan with a distinct band extending southwestward to China. Much of the Pacific Ocean is clear, except for a sprawling disturbance near the dateline (180th meridian) and broken cloud sheets off lower California. A band of clouds is invading British Columbia, and swirls off southwestward to nothingness north of Hawaii. The western states of the United States are clear; then heavy clouds run from east Texas to the Great Lakes, curling back to northeastern Colorado. Eastward from a clear east coast, cyclones can be seen in the western Atlantic and along the coast of Norway.

By May 2 (Figure 4.2), partial clearing has occurred over Japan, but we can see a great cloud mass over China. Thus, the clearing over Japan may not persist for long. The mid-ocean storm is approaching the Aleutians now, and clouds have penetrated the Rockies somewhere in British Columbia. The storm in the central United States has spread clouds northeastward to Labrador, but clearing is very slow in the Plains. To the east, two cyclones over the Atlantic are spewing clouds into regions that were clear the previous day, including southern Ireland.

On May 3 (Figure 4.3), a vigorous cyclone northeast of Japan produces the characteristic cloud pattern that looks something like a toboggan, a walking stick, or a broken wishbone. This one resembles a toboggan, with the "sled" giving another cloudy day in southern Japan. In the Pacific, a "broken wishbone" cyclone has overrun the Aleutians. Clouds infest the northern Rockies, while the Plains have finally cleared as the midwestern storm heads for the east coast. The mid-Atlantic storm lacks coherence, but the toboggan shape rewards the imagination. Farther east, the slow-moving storm south of Ireland has spread its clouds further north and east, so that now only Scotland is clear.

One day later, on May 4 (Figure 4.4), Japan is finally clear, and two "toboggans" ride the Pacific waves, moving inexorably eastward. Clouds in the northern Rockies and near the Great Lakes stand out in contrast to an otherwise clear United States. The mid-Atlantic "toboggan" is easier to see and is also moving eastward, but the cyclone near the British Isles and Ireland looks complicated and is very slow moving. In fact, look carefully—those clouds are moving westward. The western edge of the clouds was at about 17°W longitude on May 3, but now appears to be backed up to 20°W.[1] So much for the old forecasters' rule, "If it's west, bring it east," although most weather systems outside the tropics do move eastward. They do not simply move, however; they spread northward or southward, expand, shrink, intensify, or weaken.

By May 5 (Figure 4.5), the cyclone that originated near Japan has moved far off to the northeast, and is now approaching the Aleutian Islands. But notice that a bump, as though the "toboggan" hit a rock, has appeared on the long

---

[1] The zero longitude passes through Greenwich, England. It is the longitude of Greenwich mean time; 1200 GMT (noon of Greenwich) is the mean time at which the zero meridian passes under the sun. It is often abbreviated Z; for example, 1200Z.

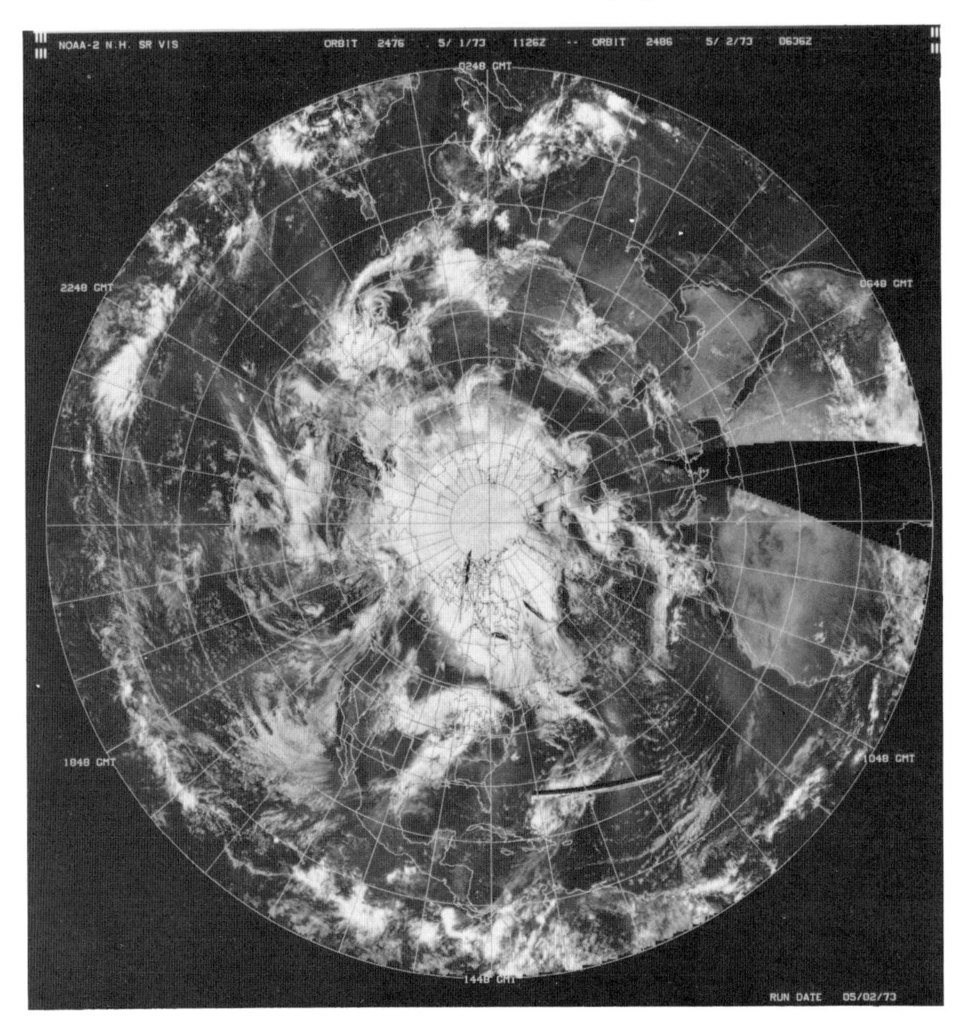

(a)

Figure 4.1 *(a) NOAA-2 satellite view of the Northern Hemisphere for May 1, 1973. (b) Infrared view of the Northern Hemisphere for May 1, 1973.*

band of clouds that trail behind the cyclone. A new storm is being born, as the increased cloudiness suggests. Here, in one part of the world, we see the life cycle of cyclones. The old storm that formed on May 2 near Japan is, a few days later, already past its prime, heading for extinction in the Gulf of Alaska. But a new cyclone, called a *wave cyclone* (doesn't it look like the new storm is the crest of a wave rippling along the cloud band?) is just beginning to form. In a few days, it will probably dissipate off the coast of Washington State, but not before it stirs up the Pacific. It is amazing that a group of Norwegian meteorologists, during the period just after World War I, described the basic cyclone structure and this "family" behavior of cyclones from surface observations taken from only a few

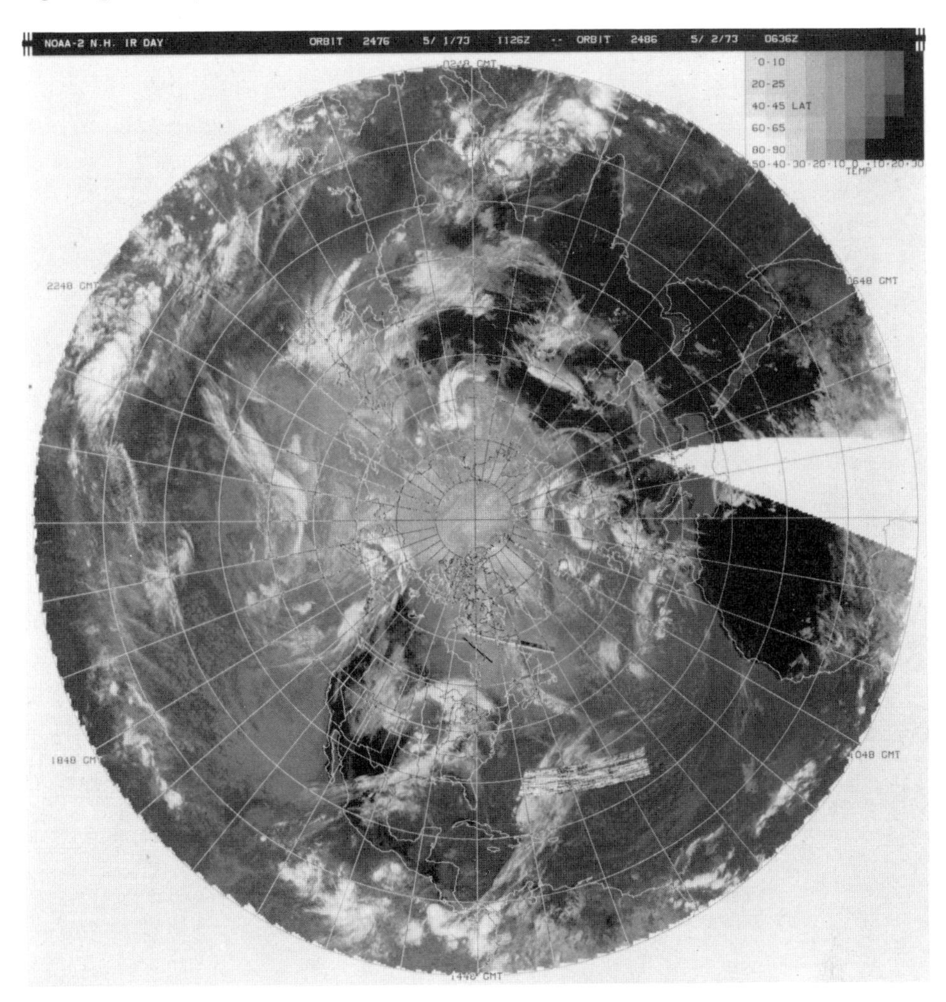

(b)

ship and land stations. They even predicted that each successive track would be farther southeast of its parent.

Figure 4.6 depicts the surface wind flows associated with the "toboggan" and its offspring, the wave. Note that the cloud bands tend to occur along fronts, or boundaries between warm and typically moist air to the south, and cold, dry air to the north. Note also that the second member of the family is traveling along a path south and east of its parent. The winds depict a counterclockwise rotation about the cyclone center; indeed, the westerly flow to the rear of the parent has moved the cold front southeastward, ordaining the future track of the new cyclone.

The other Pacific storm is plowing into the northern Rockies, and extensive cloudiness covers the West, in contrast to three days earlier. Even so, there are plenty of clear places. Dry weather regions are distributed uniformly

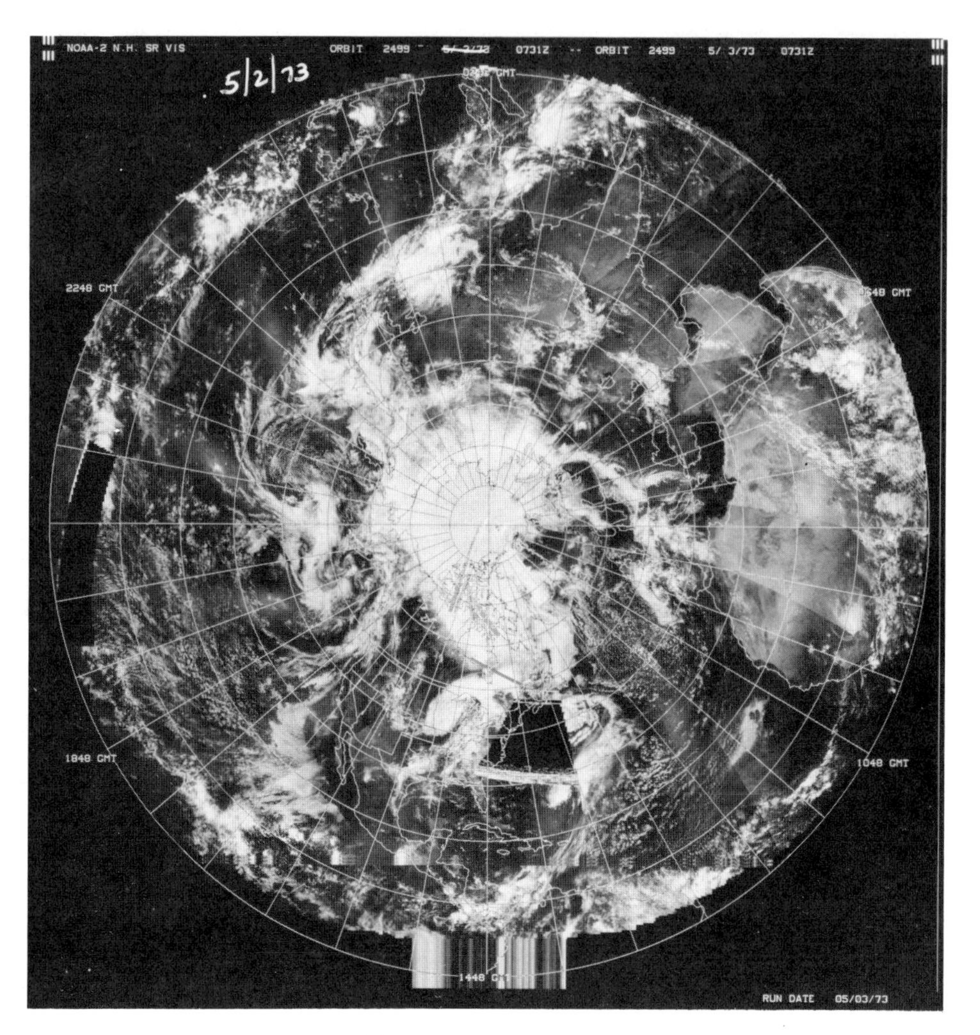

Figure 4.2 *NOAA-2 satellite view of the Northern Hemisphere for May 2, 1973.*

over large areas, with the result that the atmosphere has a strong bias toward good (dry) weather. Thus, forecasts of good weather are easy to make, and frequently correct, while forecasts of bad (wet) weather are more challenging and less likely to be perfectly accurate. The pictures tell the story: The dimensions of the good weather systems, which are anticyclones and dry flows, are much larger than those of the cloud bands within cyclones. Forecasters can often afford to make errors in locating anticyclones, because they are so big; the same error in locating a cyclone can mean a completely wrong forecast.

Finally, over the Atlantic, a wishbone near Newfoundland, and a small toboggan near the Azores, mar an otherwise quiescent ocean scene. Some observations need to be made here, however. First, the quiescent weather scene is not necessarily clear; pebbly sheets of clouds appear over the cold Canaries Current, as well as over the California Current in the eastern Pacific. These stratus

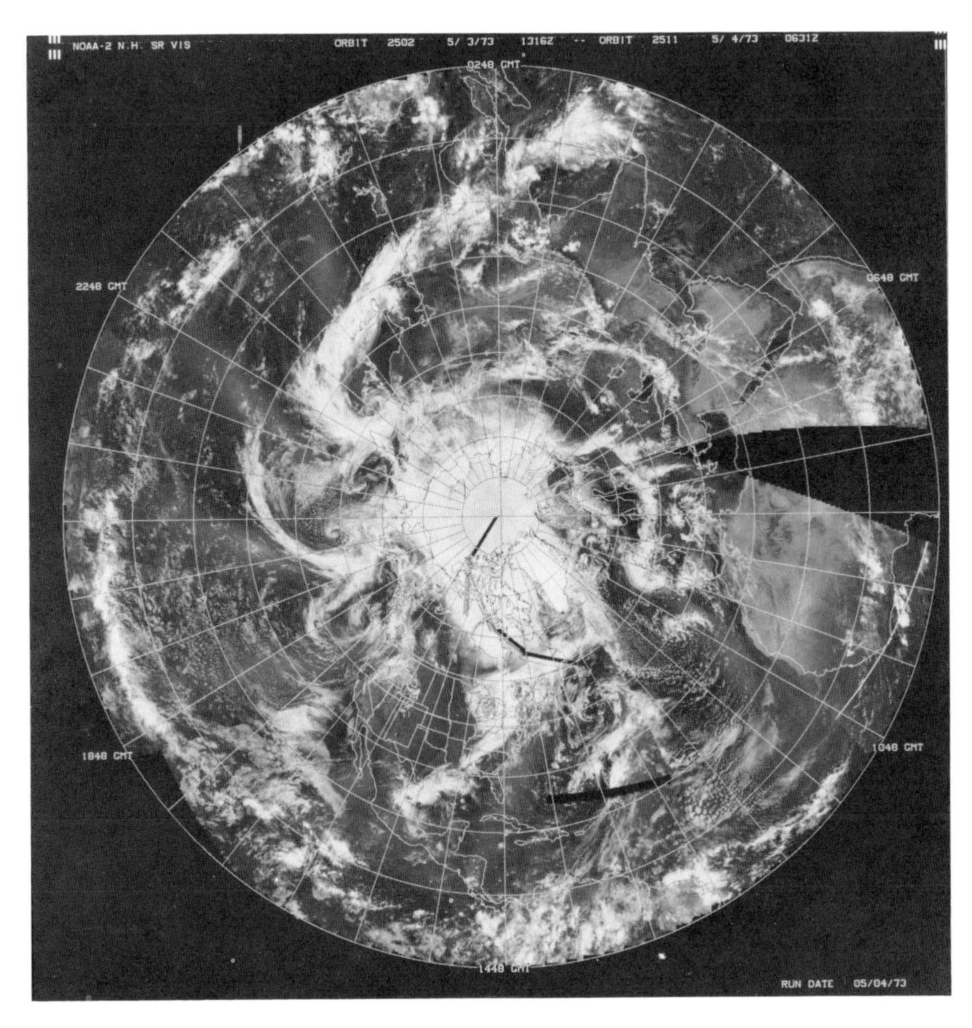

Figure 4.3 *NOAA-2 satellite view of the Northern Hemisphere for May 3, 1973.*

and stratocumulus cloud sheets are the result of mixing cool, moist air up from the earth's surface, causing it to expand and cool to its dew-point temperature. Though shallow and inert, these clouds are quite persistent and extensive. A steamer trip to Hawaii can be depressingly cloudy. The bright spot north of Haiti is a cumulonimbus cloud, so there may be random showers as well as organized bands of rain. We infer that cumulus convection, that is, showers, may break out at any point where the lapse rate is sufficiently large to permit unstable vertical motions due to buoyancy. Such unstable conditions are especially likely to occur over the continents during daytime, during the summer, and in lower latitudes.

But although apparently random showers do occur, there are favored spots for convection, and these are within the portions of cyclones that get heated. Not atypically, the morning clear bands, under the influence of strong

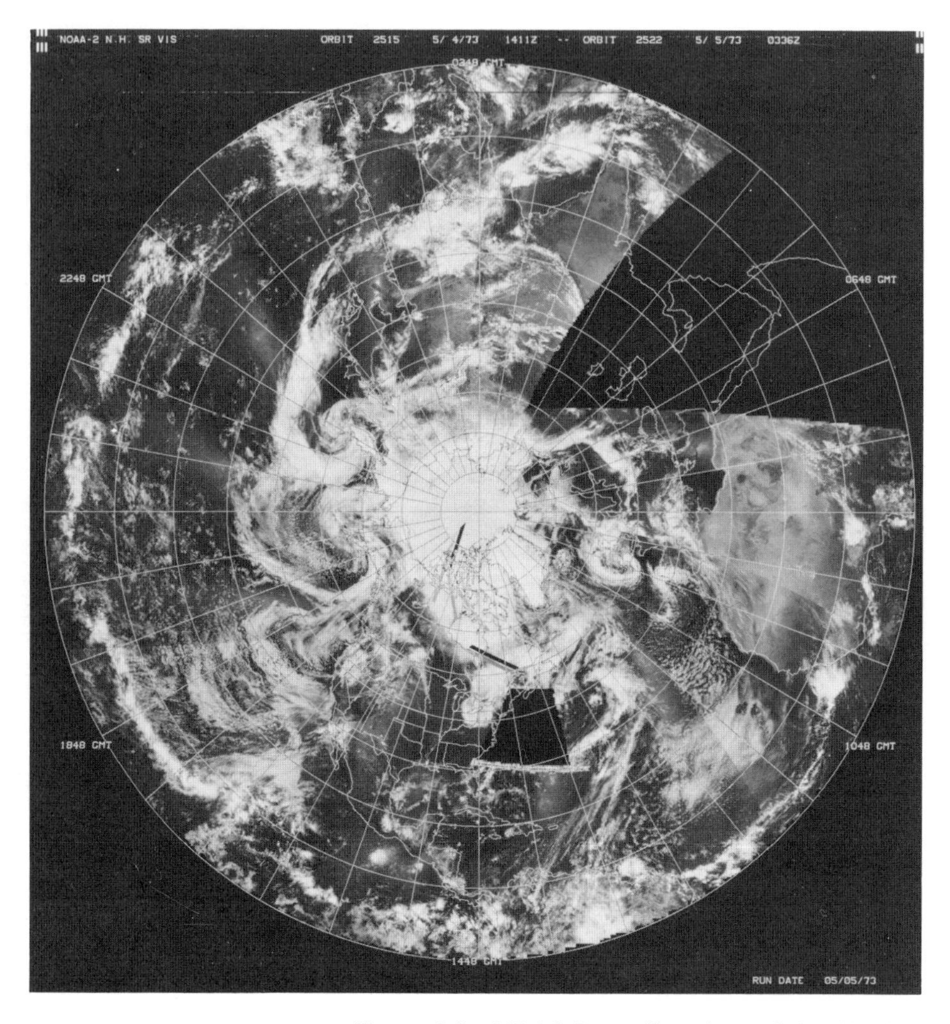

Figure 4.4 *NOAA-2 satellite view of the Northern Hemisphere for May 4, 1973.*

solar heating, become the location of afternoon showers within the large storms. On the ground, we sense the buildup of swelling cumulus in the humid, sticky air, the wind freshens, and we watch for the almost inevitable showers or thunderstorms.

4.3 Dead and Alive Clouds

Mist is the residue of the condensation of air into water, and is therefore a sign of fine weather rather than of rain; for mist is as it were unproductive cloud. [Aristotle, *Meteorologica,* with an English translation by H.D.P. Lee (Cambridge, Mass.: Harvard University Press, 1952), p. 71]

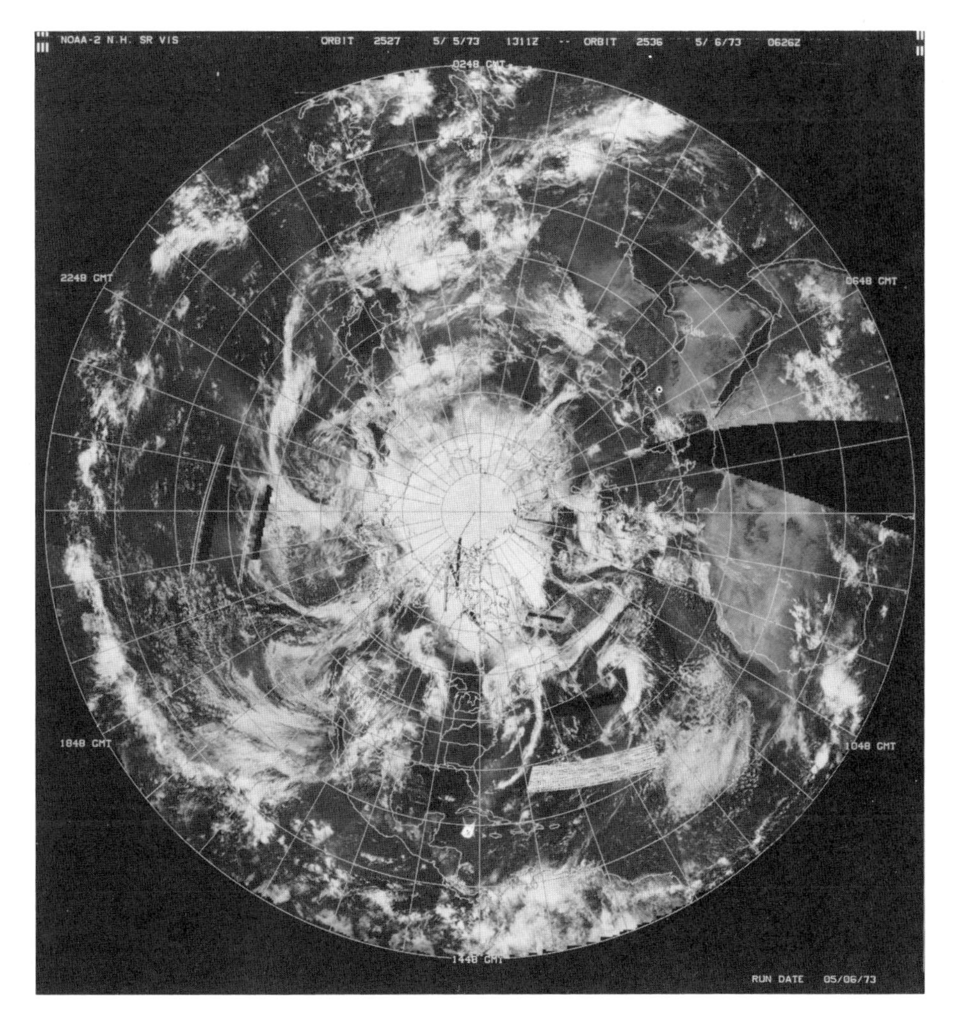

Figure 4.5 *NOAA-2 satellite view of the Northern Hemisphere for May 5, 1973.*

Of the many clouds we have seen in the previous pictures, only a few were actually producing precipitation. Some clouds are building in upward-moving currents of air; these are the active clouds that are likely to drop snow or rain. Other clouds are dead; they did not necessarily form where they are seen, but may be debris from old upwind showers, or simply shallow clouds that never bore their aqueous fruit. Indeed, most clouds are barren. Even rain clouds may be raining themselves to extinction, especially if they are of the cumulonimbus variety, which has lifetimes on the order of an hour. Nevertheless, within the masses of old, dead clouds in a cyclone, there are usually spots where new ones are forming and generating precipitation. Occasionally, a large cloud mass might produce rain or snow for hours, or even days. In these cases, the cloud system must be constantly regenerated by inflow of moist air near its base, upward motion within the clouds, and condensation

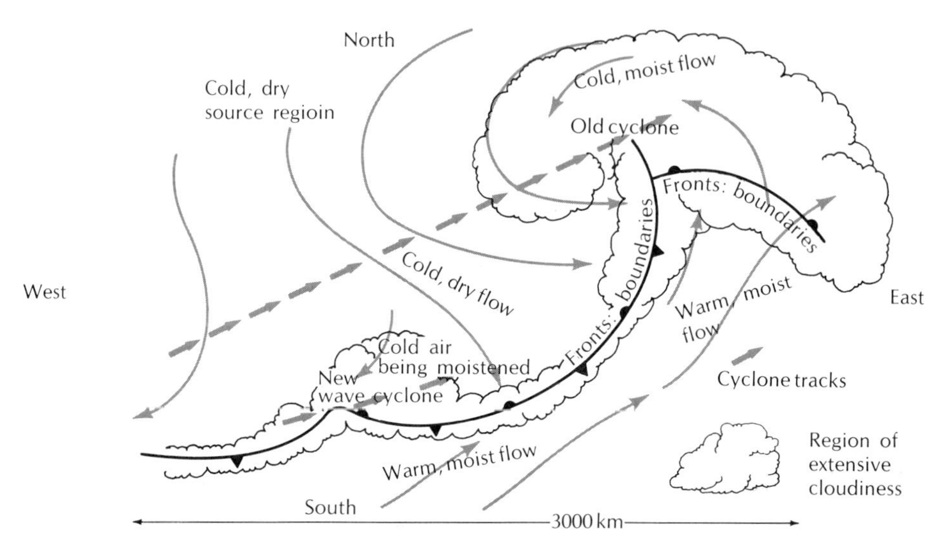

Figure 4.6 *Schematic diagram showing relationship of clouds, wind, and fronts in an extratropical cyclone.*

of vapor to rain, with the dessicated air flowing out aloft. Figure 4.7 illustrates this process, which is valid for any size of cloud system, whether an individual cumulonimbus or a vast rain shield of a great winter storm.

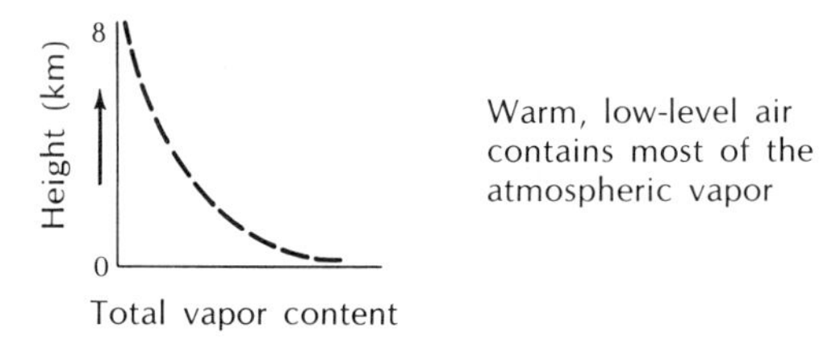

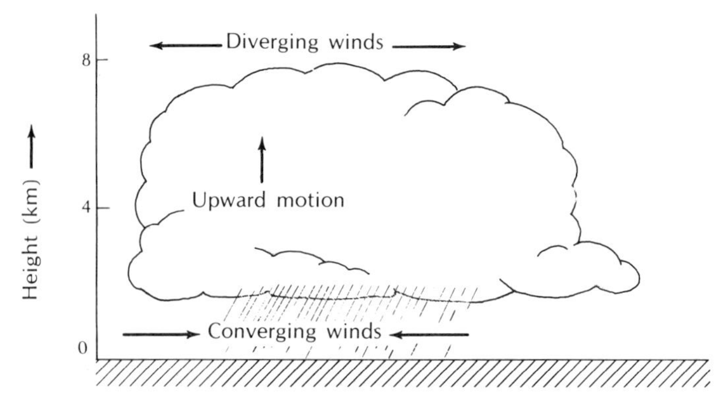

Figure 4.7 *Vertical cross section of a precipitating cloud system.*

Because so many clouds are dead, the change to a cloudy tomorrow may be much more certain than to a rainy one, for the active clouds occupy much smaller volumes than the entire cloud system. But rain or snow is certainly favored where the cloud collections are observed. So, we will study the parents of clouds, the cyclones, for they have the environments that are favorable for some of the more interesting weather events, such as thunderstorms, blizzards, tornadoes, and wild snowstorms. Nevertheless, we must recognize that not all cyclones will actually have these severe phenomena; many will pass with no more than a sprinkle or a subtle shift in wind direction.

4.4 Cyclones and Anticyclones—Big Eddies in the Westerlies

We have seen that the earth, being heated strongly near the equator and cooled near the poles, sets up circulatory motions within its atmosphere that redistribute the heat. In the atmosphere on the tropical side of latitude 30°, this process is accomplished rather simply by northeasterly winds near the surface and southwesterly winds aloft. But poleward of latitude 30°, the flow pattern becomes complicated by the rather rapid rotation of the local horizon about a vertical axis. To visualize the rotation of the local horizon about its vertical axis, imagine a plane that is tangent to the rotating earth at some middle latitude (Figure 4.8). As this plane rotates with the earth, a point on the plane rotates about the local vertical axis. Outside the tropics, this rotation causes the flow near the earth's surface to be characterized by rotating eddies rather than by simple straight flow. The result is that the winds in middle and high latitudes are quite variable, as the eddies, which are the cyclones and anticyclones, form and move about. The cyclones rotate in the same sense as the earth, counterclockwise as viewed from above the Northern Hemisphere and clockwise as viewed from above the Southern Hemisphere. Anticyclones, as their name implies, rotate in the opposite sense, i.e., clockwise in the Northern Hemisphere. Both cyclones and anticyclones have typical lifetimes on the order of a few days, or at most, a couple of weeks. Their horizontal dimensions are from a few hundred to a few thousand kilometers across. These vortices rotate about their centers at the same time that they move horizontally above the face of the earth. The situation is similar to what one sees when a paddle is drawn through a stream: The paddle causes whirls in the water, which then move away horizontally in response to the flow of the stream, all the while rotating about their own centers. This analogy can be carried one step further by noting that the direction of movement of the eddies in the stream is usually guided by the larger-scale current of the stream itself; that is to say, downstream. Similarly, in the atmosphere, the cyclones and anticyclones rotate about their own centers, but also tend to be carried downstream by the large-scale flows in which they are embedded. As a result, cyclones and anticyclones are observed to move toward the east on the average, because the broad-scale flows aloft are usually from the west. However, we hasten to remind you that individual cyclones

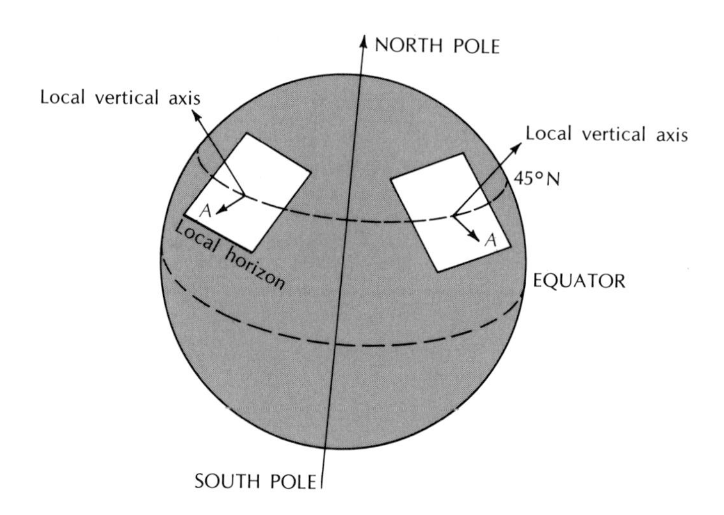

Figure 4.8 *Rotation of the horizon about the local vertical axis.*

and anticyclones may travel very differently. They may meander, remain stationary for a while, travel westward, execute loops, speed up, slow down, and so forth. Besides their general eastward motion, anticyclones usually have a component of motion toward the equator. On the other hand, there is a strong tendency for cyclones to move with a component toward the pole, so that a typical one moves northeastward over the Northern Hemisphere.

There is much more of interest than wind changes in all these cyclones and anticyclones drifting about. Cyclonic rotations occur about points of minimum pressure, and anticyclonic rotations occur about pressure maxima. A cyclone is just another name for a low-pressure system, affectionately called a *low* by meteorologists. Similarly, the anticyclone is a *high*. It is the pressure difference between the highs and lows that causes the winds. We will later see just how the wind and pressure differences are related. For the present, we are simply stating that the pressure distribution not only causes the wind and controls its sense of rotation, but also plays a role in determining the relative amount of cloudiness, as we have seen in the pictures. We also know that both cyclones and anticyclones are the offspring of the temperature difference between the equator and the poles, and that both tend to mitigate that difference by transporting heat poleward.

Although there is a certain amount of inevitability to cyclone formation, it is not obvious that all cyclones should have a common structure. But it is a remarkable fact that virtually all cyclones outside the tropics have many similar characteristics; they even bear notable resemblance to tropical storms. This is not to say that all cyclones cause the same weather, for some cyclones are stronger than others. In addition, available moisture supplies, hills and mountains, lakes and seas, the time of year, prevailing temperatures, and countless small-scale effects intervene to present the bewildering variety of actual weather that is both the glory and the curse of weather forecasting. But these complications may also thrill the astute observer, for he can learn to assess regional and local topographic effects, sense or measure the moisture and the local winds, watch the clouds, and thus anticipate, often as well as the professional meteorologist, the changing weather. These are the reasons that many farmers and fishermen are such good short-range weather forecasters; better, indeed, than meteorologists who believe that repeating weather patterns cause repeating weather. They are good observers, and accurate observation is at the heart of all good science.

But for all the local "tuning," the weather does march to the sound of a much larger-scale drum, and when the weather is bad, that drum is the cyclone.[2] Knowledge of cyclone structure makes it possible to go a step beyond the observation of local effects, extending our vision beyond the horizon. With satellites, radar, and television, the task is getting somewhat easier for laymen and meteorologists alike; however, we can apply the ideas discussed in the following pages without the use of any modern communications. Cyclones are cyclones, however we observe them, and the local weather they produce can be understood and anticipated. But this task is not easy, and we have to watch carefully. Even so, we will be disappointed at times, for there are many aspects of the weather that remain mysteries.

4.5 The Norwegian Cyclone Model

Throughout the twentieth century, the Scandinavian countries have produced a number of remarkably gifted meteorologists, all in the tradition of the founder of a noted school at Bergen, Norway—Vilhelm Bjerknes. Shortly after World War I, Norwegian meteorologists proposed a model of a typical cyclone, which has achieved wide acceptance. While it is true that we have learned to be skeptical of its overenthusiastic use, and to apply it flexibly, it is also true that the Norwegian cyclone model is the best single concept for the interested layman or the vacationing meteorologist to know. Satellite pictures, radar, and weather maps help us go beyond the typical, or idealized, cyclone; but it is a rare satellite picture that does not depict some part of a cloud structure that reminds us of the Norwegians' great insight. And, they did it without satellites!

The Norwegian wave cyclone model links the cloud pattern, the precipitation duration, type, and intensity, winds, action of the barometer, temperature, visibility, and air quality to the sea level pressure distribution, fronts, and the wind direction and speed aloft.[3] The basis of relating all these aspects of the weather is the idea that pressure differences at the same elevation

---

[2] Snow, rain, thunderstorms, etc. are bad weather to the average person, but, of course, are "good" weather for the meteorologist, who would otherwise be without a job or a hobby.

[3] You can correct a barometer to sea level pressure simply by calling a friend, neighbor, or nearby weather station which knows the sea level pressure at the moment. Set your barometer to the correct value. If your home, school, or other observation point is not the same level as theirs, it does not matter. Your correction is then different, but you don't care, or need to know, by how much.

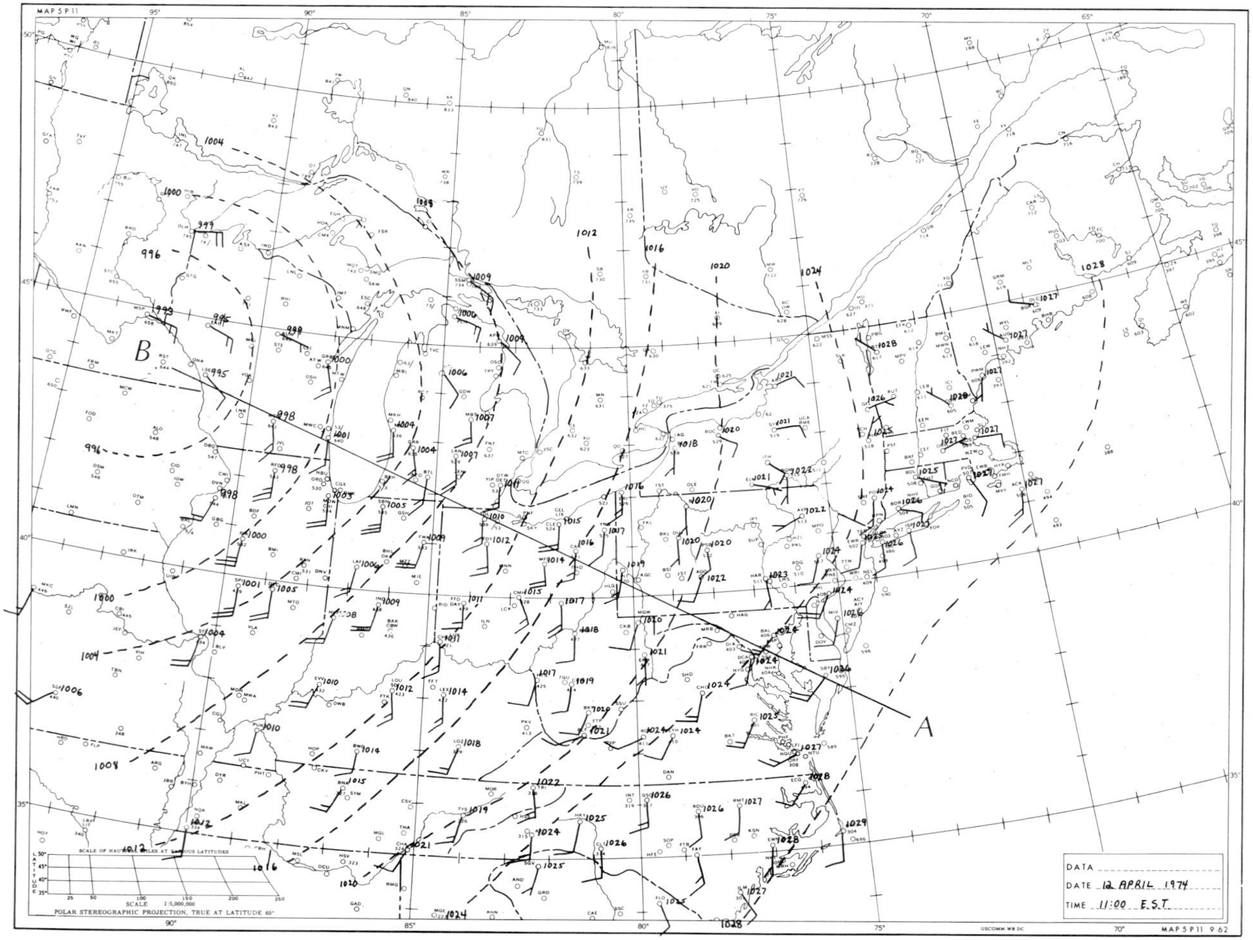

Figure 4.9 *Surface weather map for the northeastern United States on April 12, 1974, showing winds and sea level pressures.*

in the atmosphere (say, sea level) cause the air to move; that is, horizontal pressure differences produce wind. The winds then carry moisture, lift and cause clouds, or sink and evaporate them, move the fronts, change the temperature, and so forth. This is why weathermen are so preoccupied with highs and lows. The lows are regions of low sea level pressure, or cyclones. High pressure cells are anticyclones. Pressure differences cause the winds, and indirectly, the weather. Let us see how the wind and pressure are related to each other when we look at flows much larger than those across a beach or down the side of a mountain.

4.5.1 The Coriolis force

Figure 4.9 is a plot of the winds and sea level pressures over the northeastern United States on April 12, 1974, at noon EDST. The pressures are expressed in millibars (1013.2 millibars is the mean sea level pressure), and the winds are plotted as arrows which fly with the wind. Barbs on the arrows give wind speeds; a full barb is 10 knots, a half barb 5 knots. Thus, at Richmond, Virginia (RIC on the map), the wind is from a little west of due south, or what is called a south-southwest wind, at 15 knots.

We immediately notice that there is a low-pressure system, a cyclone, somewhere off toward the northwest, beyond Minneapolis, Minnesota, where the pressure is lowest of all those reported (993 millibars). The highest pressure is apparently off the East Coast, for Cape Hatteras reports 1029 millibars. We have dashed in a few isobars (lines of equal sea level pressure) which show the pressure increase toward the coast. The difference in pressure along a horizontal surface (in this case, sea level) is called the horizontal pressure gradient. The closer the isobars, the greater the pressure gradient.

Note that the wind, which is mainly from the south, certainly does not flow directly from the region of highest pressure toward the region of lowest pressure, that is, from *A* to *B*. There is some tendency for the air to flow that way, but most of the air looks as if it is missing to the right. In fact, it looks as though the air flows as much across the line *AB* as along it. (Recall the rotary wind controversy in Chapter 1.) Actually, it is doing a good bit of both, spiraling in toward the center of low pressure. Thus, we observe that the large-scale winds around cyclones and anticyclones do not blow simply from high to low pressure like the sea breeze does at noon, or like canyon winds do at night, or the way water flows when we open a faucet. Their movement is complicated by a marked deflection toward the right of the line connecting the high and low. We will show that the earth's rotation produces this deflection toward the right of the direction of motion (in the Northern Hemisphere). First, let us consider Table 4.1, which summarizes what we see here, and a few other things we do not see.

Now, before explaining the Coriolis force, we need to be clear about one thing: The wind is air moving relative to the earth. We often forget that the earth, the trees, our rooms, and we ourselves are moving at all times. Wind is air movement relative to the fixed earth, but the earth, together with its atmosphere, is traveling eastward at high speeds at low latitudes, and more slowly at higher latitudes, causing the local horizon to rotate about a

Table 4.1 *Relations of observed winds to the distribution of pressure*

(1) We note that the wind blows mainly along the isobars, with low pressure to the left and high pressure to the right of the wind. If a similar chart were prepared for any region in the Southern Hemisphere, we should find the reverse, namely, that the wind blows mainly along the isobars with low pressure to the right of the wind. Because rotation of the earth's surface in the Southern Hemisphere is opposite that in the Northern Hemisphere, we suspect that the different behavior of the winds is due to the rotation of the earth.

(2) We find that the wind is not altogether parallel to the isobars, but tends to stream toward the side where the pressure is low. However, if we prepared charts for any level above 2000 or 3000 feet, we should find no systematic drift toward lower pressure. We are thus led to believe that the drift toward lower pressure is caused by friction along the earth's surface, and that its effect is not noticeable above about 2000 to 3000 feet above the ground. This is true in both hemispheres.

(3) We find that the wind is strong where the isobars are crowded, and weak where they are wide apart. If we ignore the drift toward lower pressure, we gain the impression that the wind blows in isobaric channels in such a manner that the speed stands in inverse proportion to the width of the channel. This is true in both hemispheres.

(4) If we extended our chart from pole to pole, we should find that the relation between isobar orientation and wind direction is rather firm in high and middle latitudes, but weakens as we approach the equator. Between about 10°N and 10°S, we should have considerable difficulty in relating the winds to the pressure distribution.

(5) If we were in a position to follow the motion of an individual parcel of air and measure the rate at which its speed changes (that is, its acceleration), we should find that the accelerations are very small. In fact, if we consider the large-scale currents and ignore the fluctuations associated with short-period gusts and lulls, we should find accelerations which are about 0.0002 m/s^2 (0.0006 ft/s^2). (If this rate continued for an hour, the wind speed would change by less than 2 mi/h.) In the large wind systems, the air is a slow starter, but when it has worked up speed, it will carry on for a long time.

(6) If we could measure the vertical component of the air's motion, we should find that it is large in thunderstorms, tornadoes, etc., and also in the very small eddies which we call turbulence. However, if we consider the large-scale currents of the atmosphere, we shall find that the average vertical motion is small and that the wind is overwhelmingly horizontal.

vertical axis (picture a flagpole) once per day at the poles, and somewhat more slowly away from the poles. This effect, called the *Coriolis effect,* vanishes at the equator, where the north and the south horizons move eastward at the same speed. A flagpole at the equator tumbles over once per day as it races eastward at about 1000 mi/h, but neither it nor the horizon rotates about a vertical axis. Hence, for the horizontal winds, Coriolis deflective effects due to variable eastward rotation speeds are significant at middle and high latitudes, but not at the equator.

Consider a cylindrical planet rotating on its long axis (Figure 4.10). There would be Coriolis effects on such a planet only at the top and bottom; all points on the sides would be turning eastward at the same speed, so there would be no Coriolis effects there. This would, in effect, be like making the earth flatter

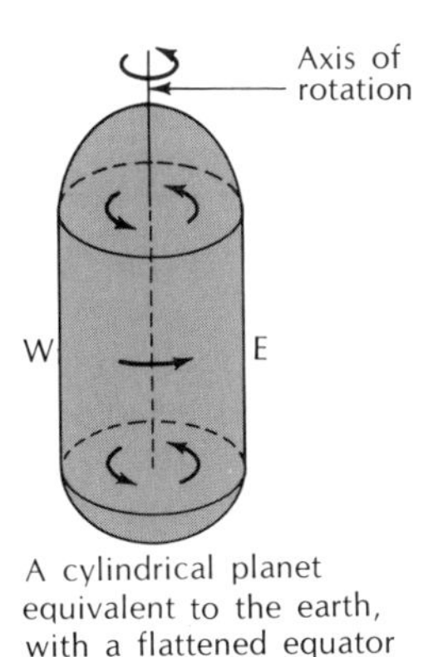

Figure 4.10 *Cylindrical planet rotating on its long axis.*

at the equator; it shows that differences in eastward speed are small or absent near the equator.

Figure 4.11 shows that the apparent deflection of the wind or any object moving relative to a horizontal plane fixed to the earth is rightward on the Northern Hemisphere, which is appropriate, because the faster eastward speed on the southern end results in the horizontal plane rotating counterclockwise. What is more, it can be seen that the greater the relative speed of the wind, the greater the apparent deflection. Because of these effects on the winds as we observe them, it is necessary to invent a deflective Coriolis force having the properties of acting 90 degrees to the right of the flow in the Northern Hemisphere, increasing with latitude from zero at the equator, and being proportional to the relative speed of the flow. We also note that the Coriolis force changes the direction of the flow, and not its speed.

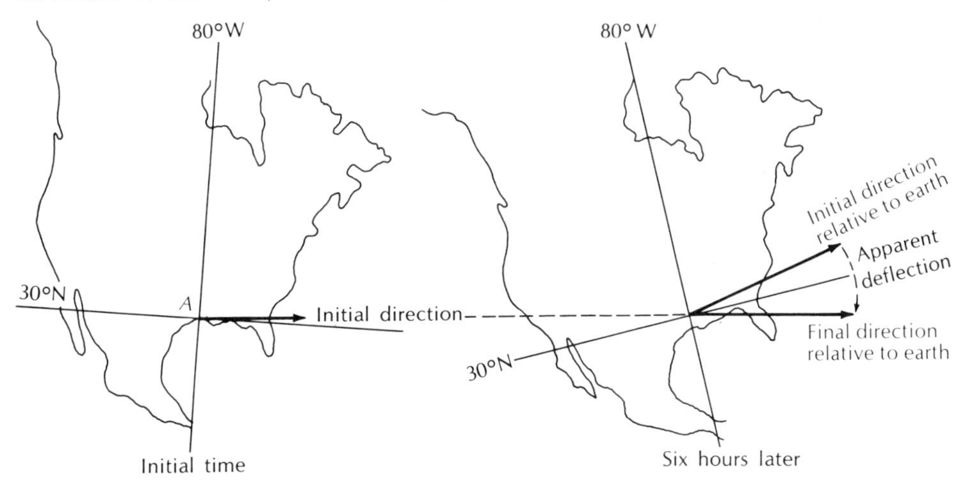

Figure 4.11 *Apparent deflection of the wind or any object on a rotating plane.*

These deflections of the winds are not limited to air flowing from west to east. Test your own understanding by picturing the apparent deflection caused by differences in the eastward rotation speed of the underlying ground when the wind is from the south. The ground is executing a counterclockwise rotation, so the air appears to turn off to the opposite direction, that is, to the right.

Coriolis deflections usually need to be considered when objects or fluids move substantial distances relative to the earth, either by traveling rapidly or by traveling for long times. Some cases for which Coriolis forces are too important to neglect include ballistics problems, flow in rivers, ocean currents, and the case of interest to us, large-scale flows of air.

4.5.2 Geostrophic flow

Many forces can accelerate airflow, speeding up or slowing down the winds and turning them into new paths. In general, forces arise from differences in pressure, differences in electrical potential, friction, magnetic fields, gravitation, and as consequences of rotation (Coriolis force). We know that electrical and magnetic forces are unimportant in affecting tropospheric air movements. It is the pressure difference force, or what is the same thing, the pressure gradient force, that is the principal cause of air motion (wind). We also know that a net unbalanced force would continually accelerate the air. But the winds do not seem to be speeding up, at least not noticeably since we have been measuring. They are about the same speed today as they were centuries ago. Therefore, we conclude that some other force, or combination of forces, is acting opposite the pressure gradient force. What is more, we have already reasoned that Coriolis forces are important. If the pressure gradient and Coriolis forces were the only ones operating, would they exactly balance each other? It turns out that they do balance each other, at least approximately, and we call the resulting balanced motion *geostrophic* ("earth-turning") *flow.*

We hasten to admit that the real wind rarely behaves as a true geostrophic wind should, but it is often a close enough approximation that we can neglect all the other forces and consider what happens when the pressure gradient force is exactly balanced by the Coriolis force. We should not expect the approximation to be especially accurate near the ground, where friction is important, nor for strongly curved flows where centripetal forces (forces directed toward the center of a circle) cannot be neglected. But where flow is straight and more than a kilometer above the ground, where friction is weak, the following conditions of geostrophic motion apply (refer to Figure 4.12):

(1) The pressure gradient force and Coriolis force are equal in strength and opposite in direction.
(2) The wind is strong (Coriolis force proportional to wind speed) when the pressure gradient is large, that is, when equal-interval isobars are crowded together. Similarly, the wind is weak when the pressure gradient is small (small Coriolis force balancing a small pressure gradient force).

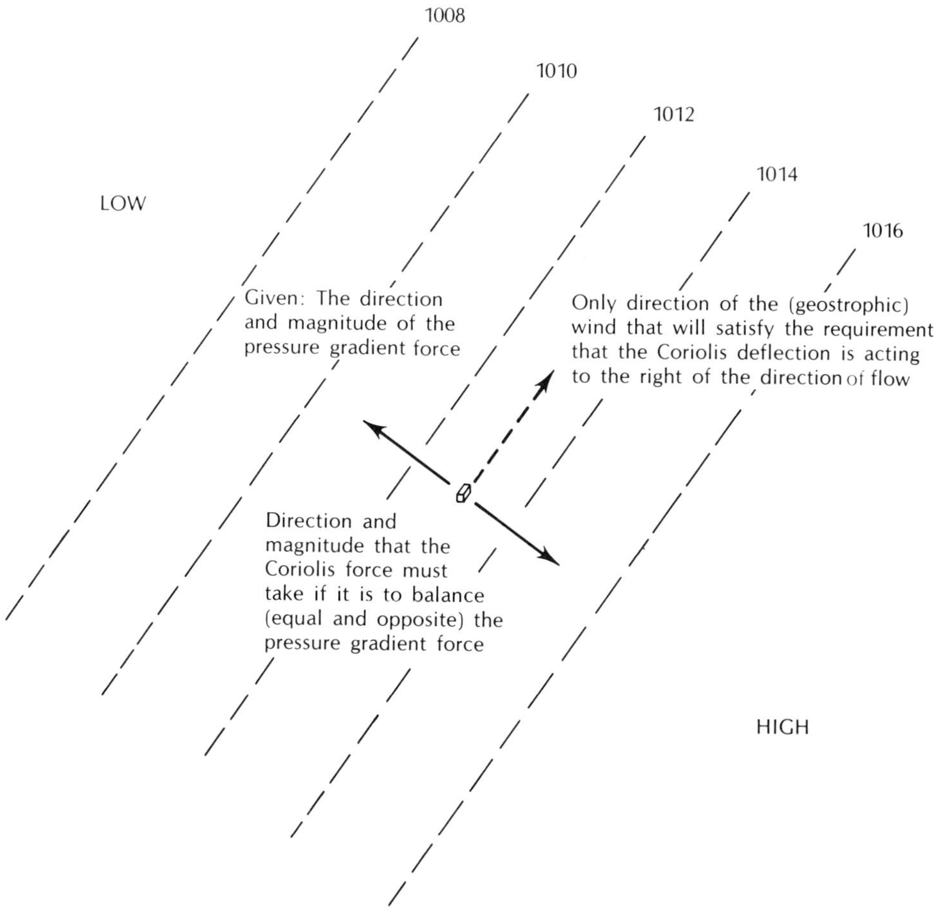

Figure 4.12 *Schematic diagram of forces in geostrophic flow showing balance between pressure gradient force and Coriolis force.* Isobars are labeled in millibars.

(3) When there is no pressure gradient, there is no wind and no Coriolis force. This is true in pressure maxima and minima, the centers of highs and lows, respectively. The pressure gradient is weakest in highs, which spread out more.
(4) Isobars are by definition perpendicular to the pressure gradient force. The geostrophic wind is also perpendicular to the pressure gradient force and, thus, blows parallel to the isobars, because the pressure gradient force and Coriolis force are oriented along the same line, and the Coriolis force is always at right angles to the wind flow.
(5) Toward the equator, Coriolis forces are weak and are balanced by weak pressure gradient forces; therefore, the isobar spac-

ing for balanced winds of given speed increases toward the equator. Stated another way, for a given isobar spacing, the geostrophic winds are stronger near the equator.

(6) Inasmuch as the Coriolis force acts to the right of the wind flow in the Northern Hemisphere, the pressure gradient force must act to its left. The pressure gradient force acts from high pressure toward low; therefore, an observer looking downwind in the Northern Hemisphere finds low pressure on his left, high pressure on his right. This rule was formulated in 1857 by a Dutch meteorologist, Buys Ballot, into Buys Ballot's law: On the Northern Hemisphere, with the wind at your back, high pressure is on the right, low pressure on the left; on the Southern Hemisphere, with the wind at your back, high pressure is on the left, low on the right.

The last part of Buys Ballot's law is true because the Coriolis force acts to the left in the Southern Hemisphere. However, in either hemisphere, application of Buys Ballot's law is sometimes chancy at ground level, where frictional effects or local perturbations to the winds may obscure the geostrophic wind direction. Notwithstanding, the rule is often of value, even at the surface.

4.5.3 *How friction modifies geostrophic flow*

Frictional effects on the wind speed are fairly obvious; they tend to slow the wind. Not so, however, is the frictional influence on the direction. Looking back on Figure 4.9, we can now see that the wind direction at the surface along the line *AB* is mainly from the south, but the direction of the geostrophic wind at sea level is about southwest. Picture what happens when friction is introduced to flow that was previously geostrophic. The brakes are applied, and the speed diminishes. As a result, the Coriolis deflection, which is proportional to the wind speed, is reduced. Accordingly, the pressure gradient force dominates, pulls the air to the left, and we see a component of flow from high pressure to low. And, indeed, we do see it in Figure 4.9; in fact, we can see that the approximate angle at which the flow cuts across the isobars toward low pressure is 30–40 degrees. We thus infer that surface friction turns the wind in toward low pressure by this amount, while it also slows the wind. Both of these effects increase with the roughness of the terrain, so that we cannot be rigid about the exact angle of inflow.

4.5.4 *Extension of the geostrophic wind to circular flow around cyclones and anticyclones*

When we consider the wind above the friction layer, from typically a few hundred meters upward, there is very little flow across the isobars, and geostrophic conditions are much more likely. Accordingly, we rely on geostrophic wind ideas to interpret flow patterns through much of the atmosphere. The beauty of the geostrophic wind is that it is simple, reasonably faithful to nature, and involves the pressure field, which is easy to measure accurately. Hence, if we know the spacing of the isobars, and the latitude, we can evaluate the geostrophic wind speed and direction and be confident that they approximate the real winds without measuring the wind at all. Hence, it is possible to plot a finite number of pressure observations on a map, obtain the pressure gradients, and then calculate the wind velocity at an infinite set of points on the map. This is the genius of good analysis—it extends the utility of limited resources. Current practice increasingly involves going the other way, observing the winds and determining pressures that are consistent with them. But the basic idea is the same: We use dynamic relationships between the wind and pressure to extend the information provided by a limited number of observations.

We now want to extend the relationship between pressure and wind to the pressure patterns commonly encountered on weather maps. In particular, circular cells of high and low pressure, elongated pressure maxima called ridges, and elongated pressure minima called troughs usually display flow parallel to the isobars even though centripetal effects must be present where the isobars curve. Careful examination will show that the geostrophic wind speeds are erroneous by 50 percent or more with curved flows; however, the geostrophic wind direction is still reasonably accurate. Because the geostrophic direction rule states that high pressure is on the right of the flow in the Northern Hemisphere, then the wind will blow clockwise around anticyclones. Similarly, the geostrophic wind direction and the actual wind direction are such that low pressure is to the left of the flow; thus, the wind blows counterclockwise about cyclones in the Northern Hemisphere. However, in the Southern Hemisphere, the wind blows counterclockwise about highs and clockwise about lows. Low-pressure centers are still called cyclones there, so it is not correct to say that the wind always blows counterclockwise about cyclones.

Although the actual wind direction does not depart significantly from the geostrophic direction around circular highs and lows, the speed does. For the same pressure gradient, the flow is faster around highs than lows. This may seem surprising, because our experience tells us that strong winds are most likely to be associated with storms (low-pressure systems) than with fair weather (anticyclones). The paradox is explained by noting that the pressure gradients associated with lows are normally much greater than those associated with highs.

Let us now see what effect curvature has on the wind speeds for isobars of equal spacing around highs and lows (Figure 4.13). We first note that to make the air turn in a circle, a net force (called the *centripetal,* or "center-seeking" force) on the air must be directed inward toward the center of the circle. Otherwise, the air would continue to flow in a straight line. The only two forces that can produce this net centripetal force are the pressure gradient force, which acts from high to low pressure, and the Coriolis force, which acts to the right of the wind direction. By assumption, the magnitude of the pressure gradient forces is the same in both cases shown in Figure 4.13. However, the pressure gradient associated with the low is directed inward (toward the center of rotation), while around the high, the pressure gradient force is directed out-

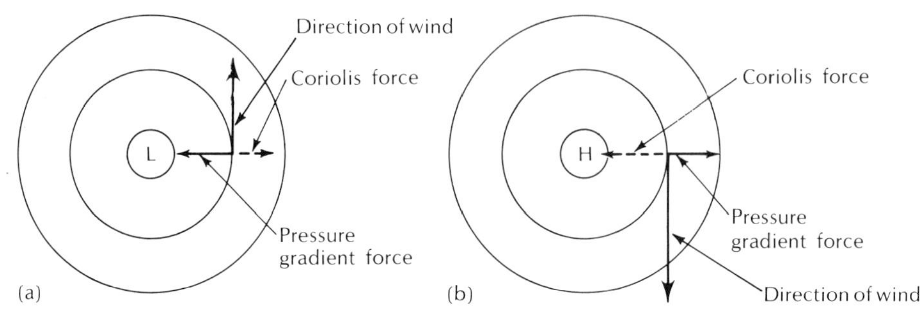

Figure 4.13 *Balanced flow around circular highs and lows.* (a) Cyclonic flow and (b) anticyclonic flow around lows and highs when the pressure gradient force is the same. To make the air turn in a circle, the air must have a net force directed toward the center of the circle.

ward. Because the sum of the Coriolis and pressure gradient forces must be directed inward, a greater Coriolis force is required in the case of flow around the high, because it must overcome the outward-directed pressure gradient force. At a given latitude, the only way to increase the Coriolis force is to increase the wind speed. Thus, we have shown that for a given pressure gradient force, the wind speed around anticyclones must be greater than that around cyclones. Note also that in contrast to straight geostrophic flow, in curved flow, the pressure gradient and Coriolis forces are unequal in magnitude. The resultant force is simply the centripetal force that is required to make the air turn.

If we refer back to Figure 4.6 and to the composite satellite pictures for May 1–5, 1973 (Figures 4.1–4.5), we can now appreciate why the wind is swirling in a counterclockwise direction of rotation about those cloudy cyclones. The pressure is low there, and the pressure gradient force is trying to accelerate air toward the centers to fill them. But the Coriolis force intervenes and deflects the air paths to the right, so that the air ends up blowing around the center, not perfectly symmetrically, but still in the direction of counterclockwise rotation. Furthermore, we can see that what inflow there is occurs in the lowest few thousand feet, as a result of friction.

If we can now see why the wind blows as it does about high- and low-pressure centers, ridges of high pressure, and troughs of low pressure, we have still to deal with the reason we can see the cyclones so markedly in the pictures. Low-pressure systems are cloudy because upward motion of air is favored near cyclones. Let us talk about that.

4.5.5 Vertical motions

The association between the pressure and wind fields is so strong that even when the wind departs strongly from geostrophic conditions, we find useful relationships between the winds and ridges, troughs, anticyclones, or cyclones. What is more, we can construct the pressure distribution from temperature data. Thus, three principal atmospheric variables, namely, pressure, temperature, and wind, are intimately related to each other. But all that would be nothing if we could not relate winds, cyclones, and temperature fronts to the upward motions that produce clouds and precipitation, for, in the last analysis, it is the upward and downward motions of air that govern the weather. Unfortunately, these vertical motions are often weak, erratic, prone to operate on small scales, and variable with height, and therefore are notoriously difficult to measure and forecast accurately.

We do know, both from observations and theory, that the vertical motion depends heavily on the wind, pressure, and temperature fields. Therefore, when we infer vertical motions indirectly, it is from these fields that we do our inferring. Let us now see how wind, temperature, and pressure cause upward motions and clouds.

Upward and downward motions are associated with four primary processes, each of which tends to have a characteristic magnitude and affects an atmospheric volume of a characteristic scale. In many situations, two or more operate simultaneously. The first of these processes is buoyant convection, which is the rising of volumes of air that are lighter than their surroundings, or, conversely, the sinking of heavy air. We know that buoyancy operates only when the lapse rate is steep (temperature decreases rapidly with height) and that it can then produce very strong upward and downward motions of from 1 to 20 or 30 m/s (40–60 mi/h). The horizontal scale of buoyancy is always quite small (a few kilometers or so), and any clouds will invariably be of the cumulus variety. *Buoyancy* is often represented by the shorthand statement "warm air rises," but it really means "light air rises." Water vapor weighs less than nitrogen or oxygen, which comprise 99 percent of the air, so the optimum condition for buoyant exchange in the vertical occurs when the lapse rate is steep and the low-level air is humid.

On larger scales, three other processes lift the air. In two of these three cases, the upward motion is related to the upglide of light air over something that is more dense, either the ground or cold, dense air. Air rising over sloping terrain is called *orographic lifting*. When it rises over cooler air, it is called *overrunning*. Overrunning refers to warm air advancing toward and over colder air. It is associated with warm advection aloft, which means that warmer air is replacing cooler air. Either upgliding process is likely to generate weak vertical motions on the order of 5 cm/s (0.1 mi/h) over a relatively large area, although lifting by terrain can be much stronger and on a smaller scale with strong winds in the vicinity of steep hills. Nevertheless, these weak motions frequently persist for a time long enough to produce significant amounts of precipitation.

Conversely, downslope motion or cold, dense air advancing toward warm air tends to produce sinking motions. Air flowing down mountain slopes leads to familiar rain-shadow (precipitation minimum) effects on the lee of mountains. Advancing cold air, called *cold advection,* is just the opposite of overrunning. The advancing cold air sinks, warms somewhat as it is compressed, and dries out relative to saturation. Thus, it usually turns colder after a storm. The cold advection is one of the factors that causes the storm to end, for the turn to colder produces sinking motions.

Finally, the fourth, and most difficult to understand, cause of upward motion is attributed to the atmosphere's fluid properties. When heating, strong upper winds, or any other cause transports air out of a region (*divergence*), vertical motions arise to compensate for the changed pressure. These vertical currents then induce horizontal *convergence* (flowing together of air) at some other level, as shown in Figure 4.14. The compensating horizontal motions always occur at different levels of the atmosphere, and the vertical circulations that couple them are important weather-producers, even though they may be very weak (5 cm/s). That is because these inflow/outflow patterns not only produce vertical circulations; to the extent that they are unbalanced, they produce pressure changes as well. Thus, net divergence can reduce the surface pressure and form lows or destroy highs, while net convergence destroys lows and forms highs. The changing highs and lows then can modify the geostrophic winds and feed back through overrunning or orographic effects to other vertical circulations. In summary, dynamic vertical motions arise when disturbances aloft cause pressure changes through their associated divergence patterns.

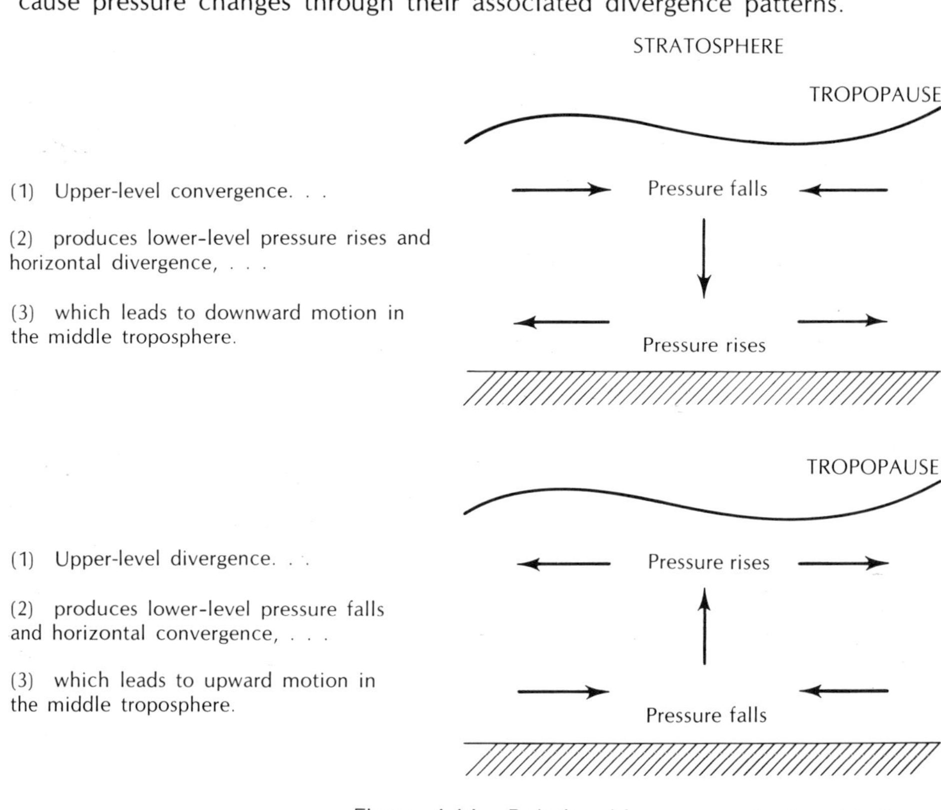

Figure 4.14 *Relationships among horizontal divergence, horizontal convergence, and vertical motion.*

4.5.6 *Formation of cyclones and anticyclones*

The real key to the formation of cyclones and anticyclones lies in disturbances in the upper westerlies. Figure 4.15 shows a vertical cross sectional view of the relationships among upper-level divergence/con-

Plate 5a (above) *Mature thunderstorm showing well-developed anvil (photo by Ronald Holle).*

Plate 5b (below) *Photograph from space showing anvil (courtesy of NASA).*

Plate 6 *Gust front near Las Cruces, New Mexico (photo by Richard Anthes).*

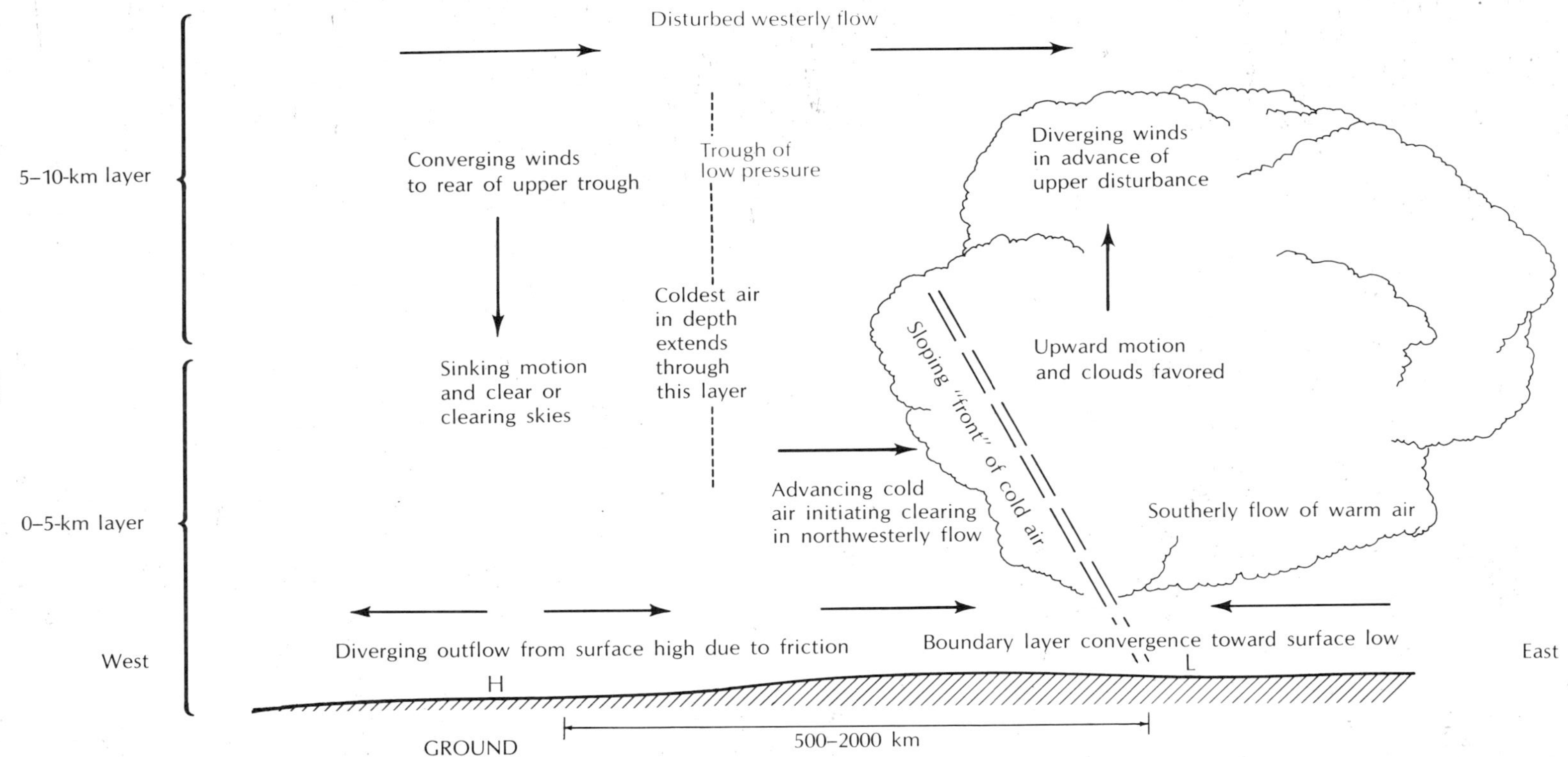

Figure 4.15 *Vertical cross section showing relationships among upper-level divergence/convergence patterns, surface highs and lows, and vertical motions in a frontal system.*

vergence patterns, surface highs and lows, and the vertical motion pattern associated with a frontal system. The cyclone (low) normally slopes upward from its surface position toward cold air. If the diverging winds aloft are stronger than the converging winds near the surface (net divergence), the low intensifies, and the central pressure falls. Then, the inflow gradually strengthens and ultimately overcomes the upper divergence, killing the cyclone. For this example, we can see that the upper disturbance is the driving force behind the formation and intensification of surface cyclones. In fact, because the intensification of cyclones must be characterized by net divergence in an atmospheric column to reduce the surface pressure, and because low pressure at the surface results in frictional inflow or convergence, divergence must occur aloft. Therefore, storms are created as well as steered by the upper-level winds.

Once an organization of the winds about an intensifying storm occurs, overrunning is favored where warm air flows toward cold air. This occurs ahead of (or east of) a counterclockwise rotating circulation in the Northern Hemisphere, which explains why the cloud and precipitation pattern associated with a typical cyclone is not symmetric about the low center. Most cyclones are "right-handed," with more clouds and precipitation on their east sides, where the wind is from the south (see Figure 4.16). Figure 4.17 shows the typical precipitation probabilities in various regions of the model cyclone. Because most cyclones travel from southeast to northeast at any given place, rain mainly falls when the cyclone is approaching, that is, when we are on its northeast side. Thus, rain occurs when the barometer is falling as a result of a cyclone's motion toward us. Once the center of the low arrives, at the time of the lowest barometric pressure, the rain is usually nearly over. Of course, if the cyclone is intensifying, the barometric falls will be stronger, and so will the upward motion and rain. If the cyclone is old and dissipating, the barometer will hint at its weakened state by not falling so fast. All this information is summarized in Figure 4.16, in which the Norwegian cyclone model is modified by what we have learned from satellite pictures. Indeed, compare this to the satellite photograph of the storm over the midwestern United States in Figure 4.2.

4.5.7 *Use of the cyclone model in interpreting local weather*

The principal advantage of the cyclone model is that it is a good tool for understanding the weather we observe at any one point. In other words, the observer who knows the model can usually diagnose the weather situation without weather maps, satellite pictures, or anything else. A barometer is some help, but two good eyes are all that is essential. First, you can determine the winds both at the surface and aloft (watch the clouds that are straight overhead). Applying Buys Ballot's law at both levels, you can determine where the pressure systems lie, and by applying the steering principle using the winds aloft, where they will go. For example, if the wind at the surface is from the southeast, and clear skies have recently been replaced by cirrus or cirrostratus, you can surmise that a cyclone is somewhere to the west or northwest. By watching the cirrus cloud motion, you can then predict whether the cyclone is approaching, or passing off to the north. If there is little cloud

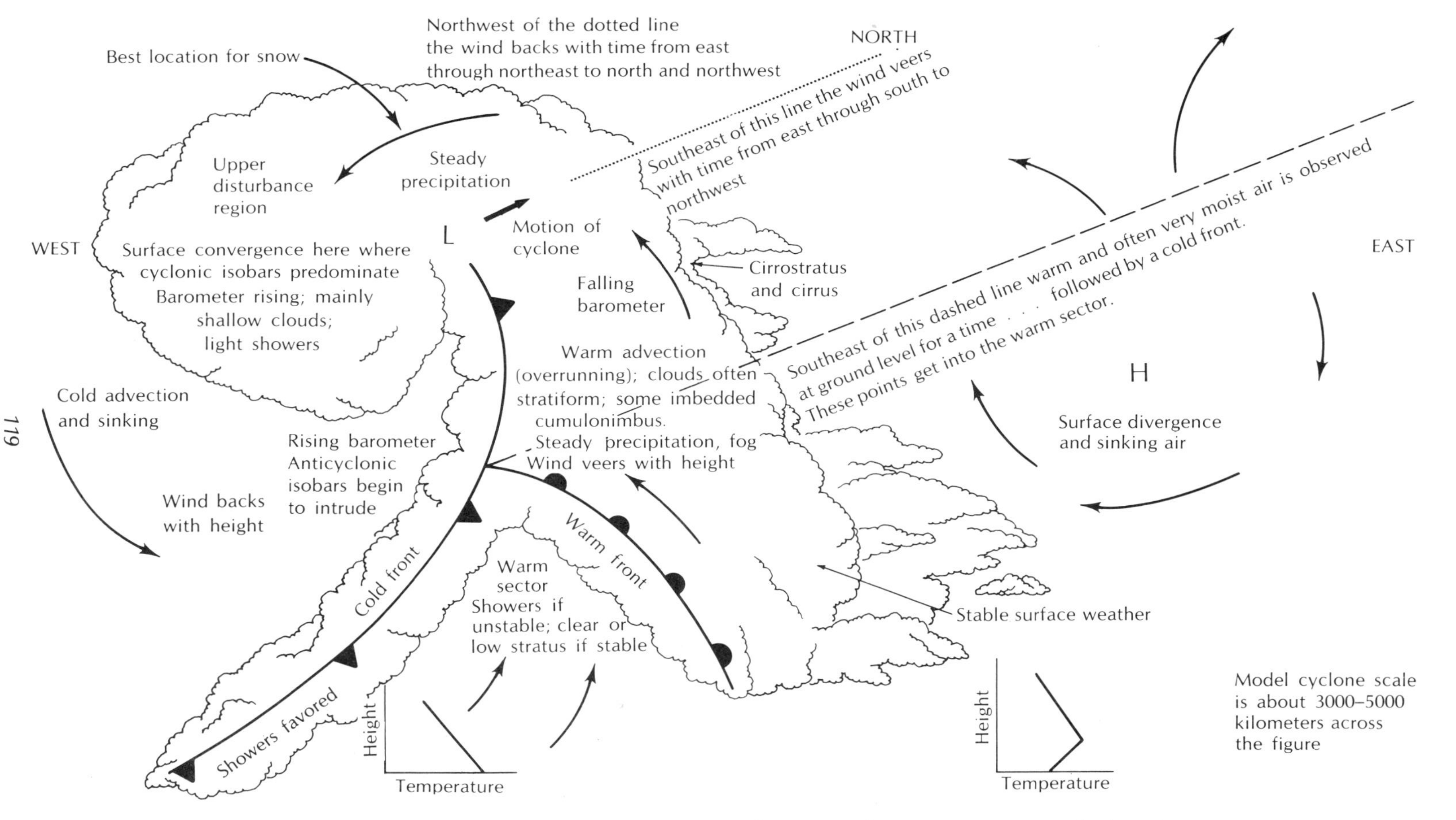

Figure 4.16 *Distribution of clouds and weather around a mature extratropical cyclone.*

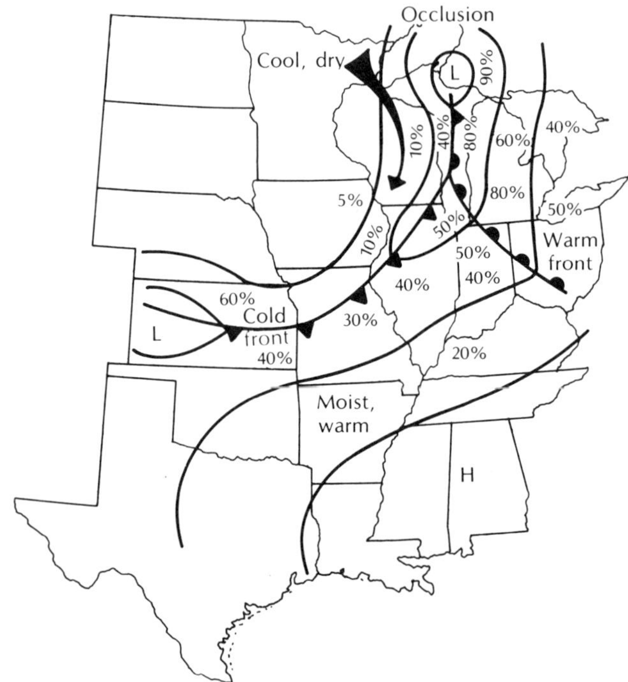

Figure 4.17 *Probability of precipitation in various regions with respect to the extratropical cyclone. (after Cooperative Extension Service, Purdue University).*

motion aloft, you could guess that the cyclones and anticyclones are moving very slowly, implying that any local changes in weather could only come about as a result of sea breezes, valley breezes, or the like. This principle of *steering* is worth some emphasis. If there were a swarm of bees or a cloud of dust suspended through a great depth of atmosphere, you would not be shocked to find it moving in the direction of the winds aloft, which occupy a deep layer and usually blow from a consistent direction that may be different from the low-level winds. So, too, with clouds and fronts and cyclones. They tend to be steered downwind by the winds aloft. All these signs are available for the looking.

4.5.8 Warm frontal precipitation

There is much more in the cyclone model. Notice that in regions of overrunning (Figure 4.18), the warm air over cold air stabilizes the atmosphere with respect to buoyancy forces, suppressing cumulus clouds in favor of deep and extensive sheets of stratus clouds. The schematic temperature soundings (measurements) in Figure 4.18 show the stable conditions that are favored (but not guaranteed) in the region ahead of the warm front. Notice that the upper-level front is associated with the top of an inversion in the soundings. Thus, to the extent that the atmosphere ever produces steady rain over flat terrain, regions of overrunning on the cold side of warm fronts or stationary

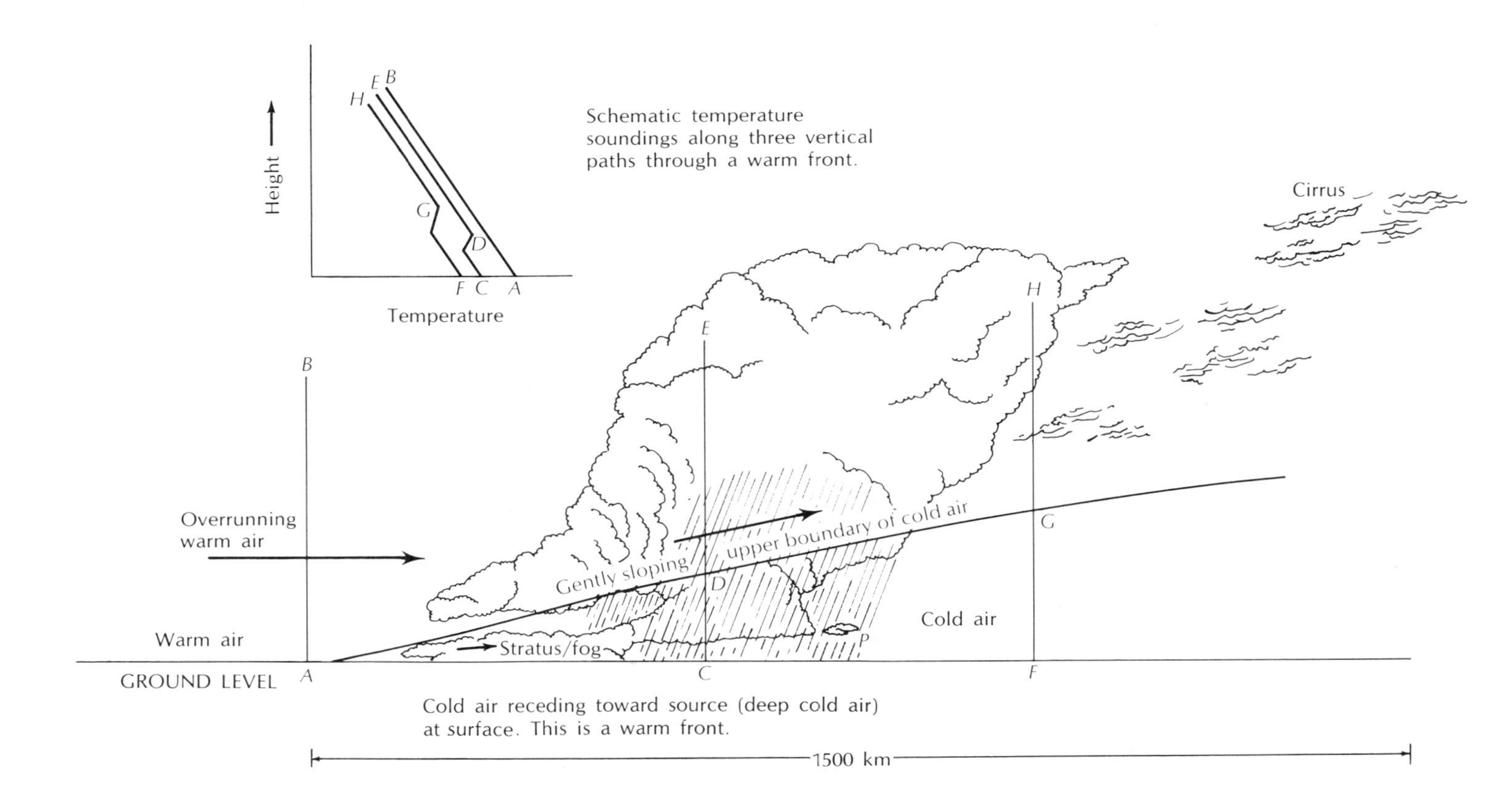

Figure 4.18 *Clouds and precipitation associated with warm air overrunning cold air.*

fronts are the places to find it. Notice that we do not insist that the frontal boundary at the surface, that is, the edge of the cold air mass, be moving. If it is moving in the sense of cold air receding back toward its source regions, usually to the north, the front is called a *warm front*. If it is not moving at all, we call it a *stationary front*. In any case, the behavior of the surface front does not really matter; as long as there is significant overrunning aloft, stratiform (layered) clouds and steady precipitation will occur on the cold side of the front.

The actual frontal type depends on the surface winds just on the cold side of the front, with northerly winds indicating a cold front, southerly, a warm front. The main overrunning will be hundreds of kilometers back over the cold air. Significant overrunning is usually associated with warm, or at least stationary, fronts, because cold advection, to persist, would have to infect a large enough volume to cause sinking and drying in the cold air. Then the structure would be a *cold front,* with cold air advancing and rapidly clearing skies on the cold side of the front, which is the situation to the rear of typical cyclones.

Note also in Figure 4.18 that the steadiness of the overrunning precipitation depends intimately on the initial stability of the overrunning warm air. If the warm air is stable, layered clouds such as altostratus (gray, layered clouds occurring in the middle troposphere) will prevail in the warm air, and the precipitation will be steady. If the warm air is unstable, however, thunderstorms may erupt as the warm air is lifted. But underneath, in the cold air, it will always be stable, so stratus-type clouds will almost certainly predominate when viewed from below.

One process of low cloud formation that is of particular interest for pilots frequently occurs in an overrunning situation. The clouds that are created by this process are marked *P* in Figure 4.18. Initially, the low-level air in the cold air mass is dry, because cold air masses originate in the arid and often ice-covered portions of the globe. The overrunning rain, however, humidifies the dry, low-level air as the first drops come down and evaporate. Because the surface air usually has been prehumidified to some extent by surface vapor sources, there frequently follows the rapid formation of low stratus clouds when rain starts falling from higher clouds. The unwary amateur pilot who needs to maintain visual contact with the ground may be flying at 2 kilometers (7000 feet) with an overcast far above, say at 4 kilometers (13,000 feet). But suddenly, an undercast appears down near 1 kilometer (3000 feet) or so, and with it, real problems. The tipoff, of course, is spatters of rain on the windshield.

4.5.9 *Winds around the cyclone*

Additionally, the cyclone model can be used to forecast temperature changes by watching the variations with time of the wind. We use the term *veering* when the wind changes in such a way that our wind arrow turns in the clockwise direction when looking down on it. Conversely, if the wind arrow turns counterclockwise when viewed from above, we say that the wind is *backing* (see Figure 4.19). Notice that at all points south and east of the dotted line (the storm track) on the schematic cyclone model (Figure 4.16) the wind veers with respect to time through the south, and the weather generally

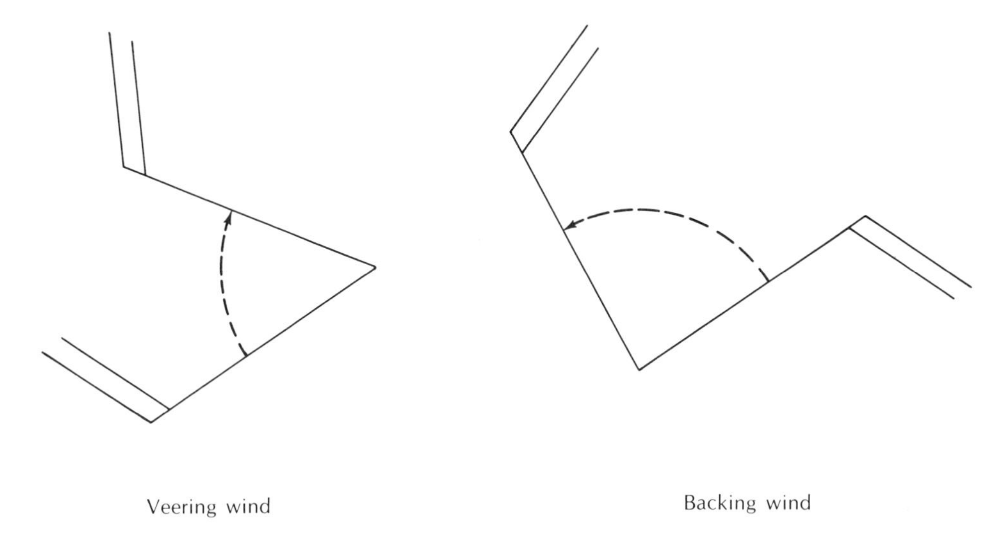

Figure 4.19 *Veering and backing winds.*

improves. This is the history of the warm (south wind) side of the storm track. On the other hand, points to the north and west of the track experience wind backing with time. These locations are on the cold side of the track, and thus much more likely to observe snow, sleet, or freezing rain during the cold seasons. This model thus explains the following proverbs:

A veering wind, fair weather;
A backing wind, foul weather.

If wind follows sun's course, expect fair weather.

So, if you want to forecast temperature, and especially snow, when a cyclone is headed in your general direction, watch whether the wind veers or backs with time; you'll become the local groundhog!

4.5.10 *Temperatures and precipitation around the cyclone*

Points south and east of the dashed line in Figure 4.16 will have a period of high temperatures, because they will get into the warm sector for a time.[4] Then, they will be approached from the west by the cold front, which is typically associated with a band of showers, or, during warm seasons, thundershowers. The small scale of these showers means that not everyone in the path of the cold front will get one. However, the cold frontal region and the zone a couple of hundred kilometers wide just ahead of it are the favored regions (Figure 4.20). The schematic sounding labeled *G-H* in Figure 4.20 shows that in these regions which have been in the warm sector, and especially at the times of the day and year when the sun is strong, the lapse rate

[4]Cold fronts often "pinch off" (see Section 4.5.11) warm sectors at the surface, because they travel more rapidly than warm fronts. Thus, this line could shift southward with time.

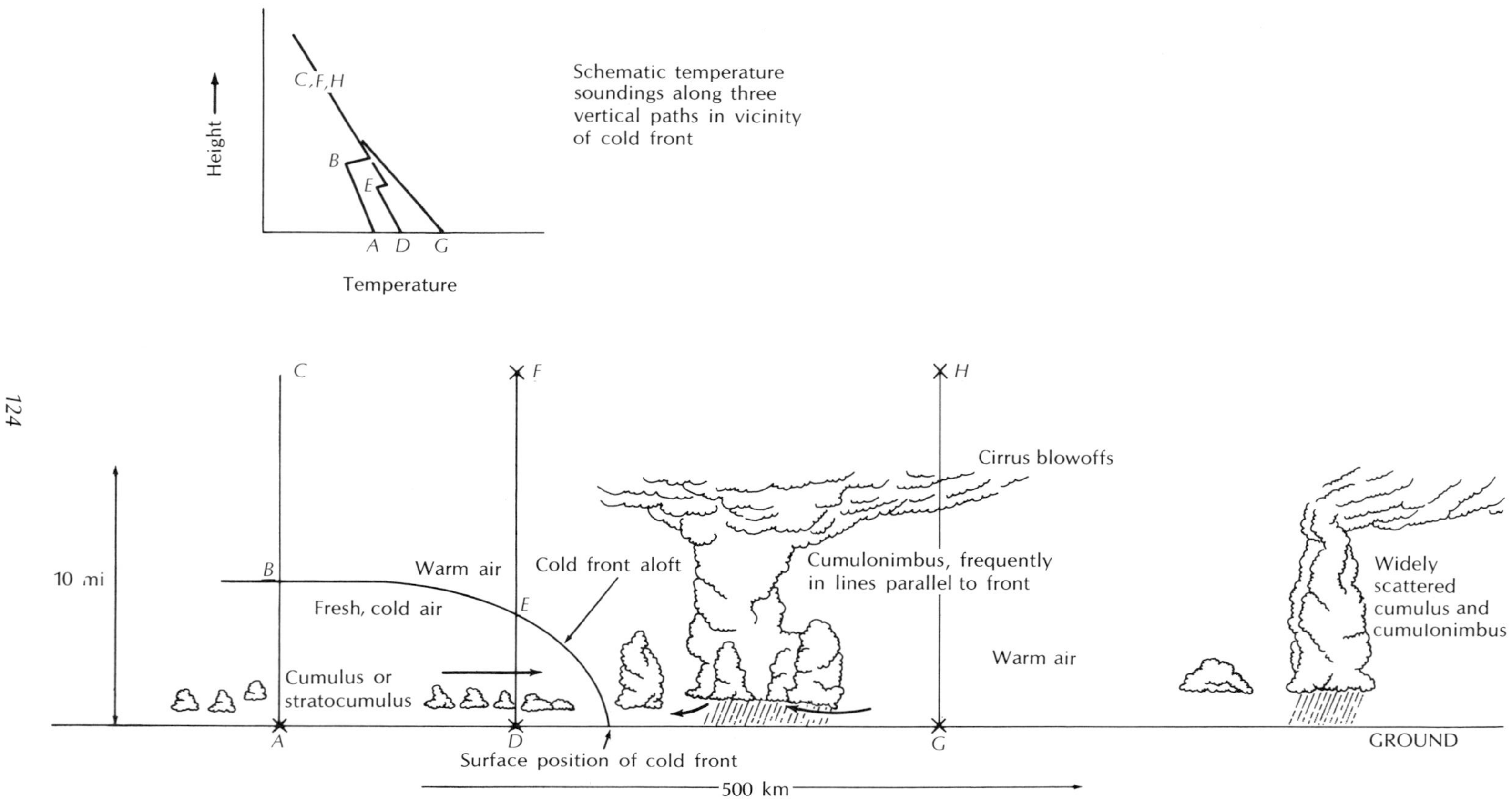

Figure 4.20 *Vertical cross section showing clouds and precipitation in warm sector ahead of cold front and vertical temperature profiles in warm sector and behind cold front.*

is quite steep, and buoyancy production of deep cumulus and cumulonimbus clouds is quite likely. The intensity of the convection, of course, depends on the details of moisture availability, strength of the low-level convergence, frontal intensity, and steepness of the lapse rate. Some of our most devastating thunderstorms, and even tornadoes, occur along squall lines that are just ahead of cold fronts.

In contrast to the unstable low-level soundings ahead of the cold front, the air is generally more stable behind the front (note soundings *G–H* and *A-B-C* in Figure 4.20). Therefore, the clouds, if any, tend to have little vertical development. As in the case of the warm front, the cold front aloft appears as an inversion in the soundings.

In discussing the composite satellite pictures, we called the cyclones "toboggans" and "wishbones." Now we can see that the main features of these cloud patterns resulted from the extensions of cloudiness along fronts. We have not, however, focused on the causes of fronts. No one would challenge the idea that temperature variations should exist between low and high latitudes and along coastlines, mountains, etc. When the flows are such that the horizontal variation of temperature gets crowded into a small space, fronts are produced. Recognizing that low-level winds that are convergent can pack the variation into a small space, we should not be surprised to find cloudiness maxima along fronts, for low-level convergence results in upward motion and, hence, clouds. Further, we should thus expect fronts to be associated with lows rather than highs, and, indeed, fronts always occur in troughs of low pressure where the low-level air is convergent.

4.5.11 *The occluded front*

So far, we have discussed the typical structures of cold and warm fronts, but have not yet discussed what happens when the surface cold front overtakes the surface warm front, a phenomenon which occurs in the latter stages of a cyclone's life. Let us first consider the case when the overtaking cold air is colder than the cool air ahead of (on the cold side of) the warm front (Figure 4.21a). The coldest air pushes under the cool air, just as it does with the warm air, but the warm air mass is still present aloft even though the warm front disappears from the surface. The boundary between the coldest air and the cool air is called a *cold-type occluded front*. Because the advancing air is colder than the air being displaced, the sequence of weather events that accompanies a cold-type occlusion is similar to that which accompanies a cold front. Indeed, the distinction between these most common occlusions and cold fronts is quite small. A cold-type occluded front is depicted in the model of the extratropical cyclone (Figure 4.16). The occluded portion of the front extends from the low-pressure center to the point where the cold and warm fronts intersect.

A second, less common, type of occlusion occurs when the retreating cold air ahead of the warm front is colder than the overtaking cool air mass behind the cold front (Figure 4.21b). In this case, the advancing cool air rises over the colder air mass, producing weather that is similar to the weather associated with warm fronts. This *warm-type occlusion* occurs occasionally in

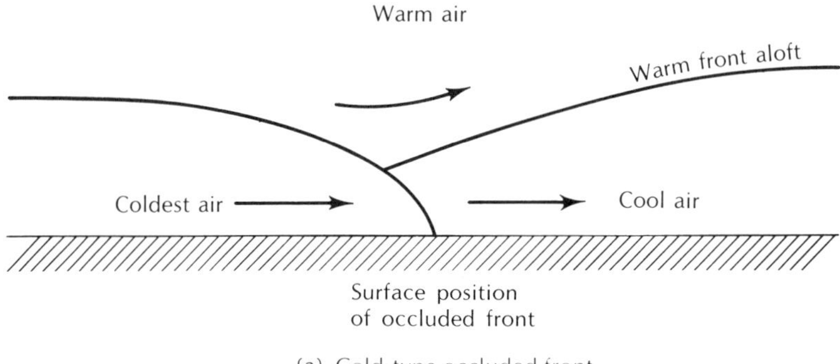

(a) Cold-type occluded front

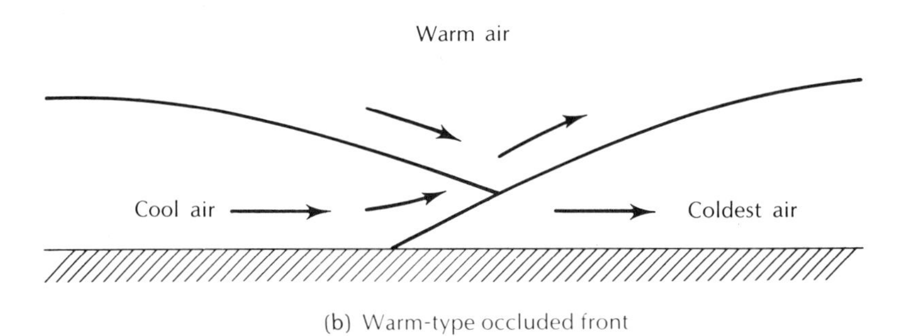

(b) Warm-type occluded front

Figure 4.21 *Vertical cross sections through cold-type and warm-type occluded fronts.*

winter west of the Rockies when westerly flow behind Pacific cold fronts is warmer than the cold air over the continent.

A third type of occlusion occurs somewhat spontaneously. Here, no air overtakes other air; a front simply forms within air of rather uniform temperature. The front can be observed with its shift in wind direction, cloud bands, and precipitation, but little temperature change occurs, except that related to cloudiness changes and precipitation.

4.5.12 Warm and cold advection indicated by veering and backing winds

One last concept for use with the cyclone model may be a bit difficult to apply, but it is worth the effort. Because the temperature and pressure aloft (not at the surface) are quite well correlated, we can usually tell where cold air and warm air are and how the temperature is likely to change by observing the winds aloft. You can do this in your backyard (see Figure 4.22). Just watch the middle or high clouds, preferably those straight overhead. Then remember Buys Ballot's law and "cold—low, warm—high."

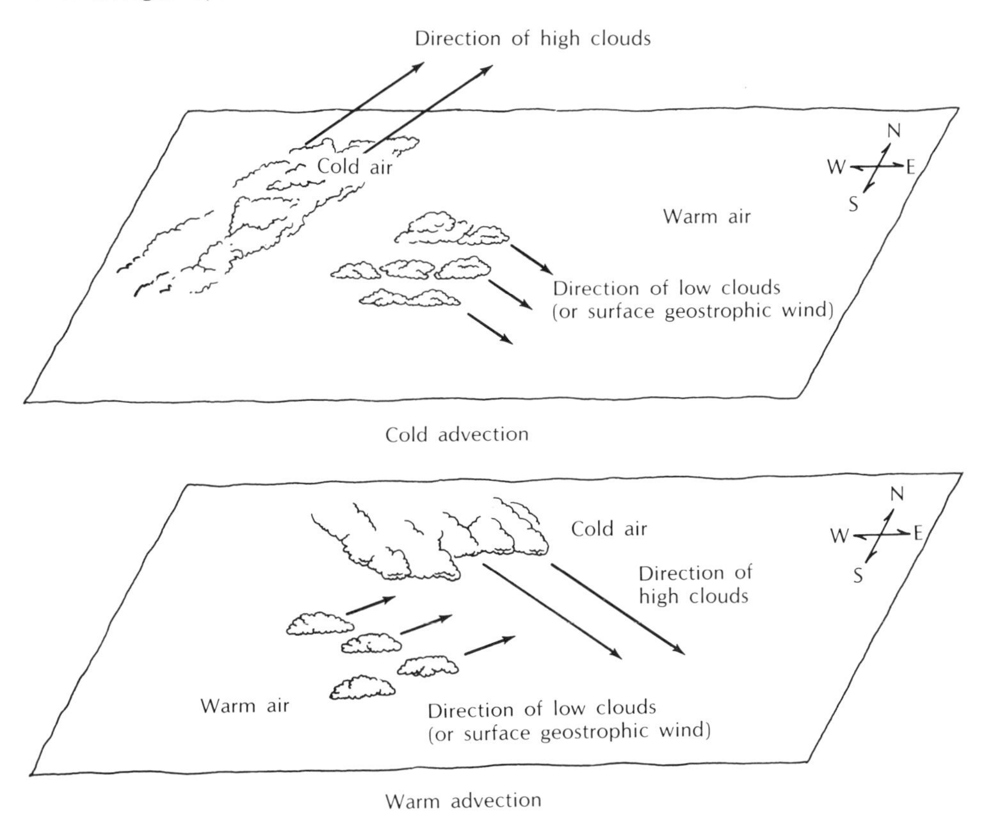

Figure 4.22 *Relationship of change in wind direction with height to horizontal advection of temperature.*

Now, pretend that you are standing on a very high tower with the upper-level wind at your back. High pressure and warm air are on your right, and low pressure and cold air are on your left. Check whether the low-level wind is bringing in that cold air, the warm air, or neither, and you are ready to make a temperature forecast that will be highly reliable until the next front comes along to change things. If the winds are bringing in air of different temperatures, you can diagnose the situation more precisely by using the following rule, which is just a generalization of the ideas noted above: If the geostrophic wind direction veers upward through the atmosphere, it is getting warmer (warm advection); if it backs, the atmosphere is getting colder (cold advection). Notice that you have to determine the geostrophic wind direction for at least two different levels. The direction at the higher level can be obtained by the motion of the middle or high clouds. If lower-level clouds are also visible, these cloud motions may be used to estimate the lower-level geostrophic wind direction. However, if low clouds are not present, the surface winds may be "corrected" for the effects of friction to yield an estimate of the geostrophic wind. Because friction causes the surface wind to deflect about 30 degrees to the left of the geostrophic wind, we need to make a 30-degree correction to the actual wind to obtain the surface geostrophic wind direction. To do this, face into the surface wind direction, then turn 30 degrees to your right to allow for frictional inflow across the isobars. Now, you are facing into the low-level geostrophic wind, and you are ready to determine whether the geostrophic wind is backing with height and the atmosphere getting colder, or the geostrophic wind is veering with height and the atmosphere getting warmer. Of course, if the geostrophic winds at the two levels are from the same direction, no large-scale temperature change is taking place.

5 Watching the weather

All the pictures, diagrams, maps, and descriptions of Chapter 4 would mean little or nothing if we did not put the ideas behind them to everyday use. To be sure, we can learn from weather maps; they are included in this book for that purpose. But if we apply the ideas and concepts that emerge from an understanding of weather maps by using our eyes and brains, we will not need weather maps, except the one in our mind's eye.

So, let us set out in our imaginations on a cross-country trip. We will drive across the Great Plains, from the Rockies to the Appalachians, using Interstate-70 most of the way, from Denver to Pittsburgh. We choose for our trip the period in early May, 1973 that we have already "watched" from the satellites. We saw in Figures 4.1–4.3 that between May 1 and May 3 a slow-moving cloud system spread across the United States into eastern Canada, finally abandoning the Plains states on May 3.

5.1 The Synoptic View of the Weather across the United States on April 30, 1973

Figure 5.1 shows the weather map for the morning of April 30, 1973. Because this synopsis sets the scene for the events that follow, it is important to understand what is pic-

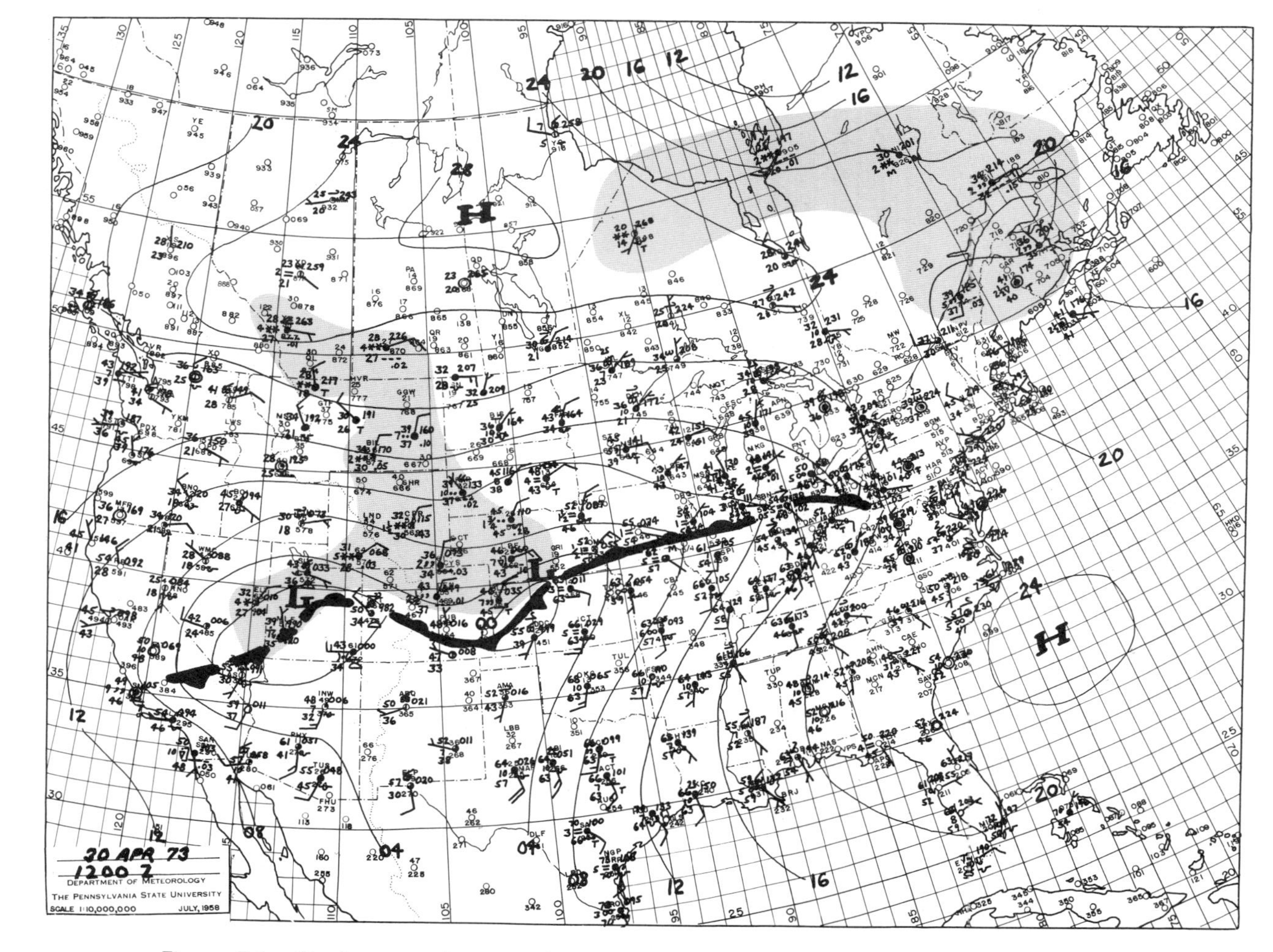

Figure 5.1 *Surface weather map for United States at 0700 EST (1200Z) April 30, 1973.*

tured. At each reporting station, the weather elements are plotted according to the scheme given in the corner of the figure. For example, the temperature is plotted to the upper left of the station; it is 46°F at Goodland, Kansas, but 66°F at Wichita, just to the east. The heavy line lying through Kansas denotes a front, a fact that the temperature contrast between Goodland and Wichita fairly screams at us. Indeed, everywhere to the north of that frontal line, the temperatures tend to be cool or cold; Churchill, Manitoba is coldest of all at 7 degrees above zero. South of the front, temperatures are warmer; most stations report temperatures above 50°F, with Corpus Christi and Brownsville, Texas at 73°F. A few spots are cooler, the result of local effects such as elevation or shallow pockets of cool nighttime air settling in regions of light wind and relatively clear skies.

The sea level pressure in millibars is plotted at the upper right-hand side (for convenience, the leading 9 or 10 and the decimal point are not plotted; thus, the number 981 means 998.1 milliabrs; the number 120 means 1012.0 millibars, etc.). The large-scale pressure pattern is depicted by the sea level isobars. Note that the high, or anticyclone, which lies to the north of the front is cold; another high, off the East Coast, is warmer. Between the highs, low pressure is observed along the front, as we have come to expect. Finally, we note that the lowest pressures of all are found in the southern Rockies. Recalling what we have learned about the wind circulations, it is reassuring to see that the surface wind reports, plotted as before, confirm the tendency for counterclockwise rotation about the lows and clockwise rotation about the highs. There is also a notable tendency for inflow and surface convergence near the lows and the low-pressure trough along the front, and a concomitant outflow from the highs. We are not confused by "oddball" wind reports, such as those from Boise, Idaho or Prince George, British Columbia, which do not fit the large-scale pattern; we have seen that local effects can dominate large-scale effects, especially in mountainous terrain. At present, however, we are interested in the broad picture painted by all the reports.

We can now see the reason for the abundant cloudiness over the Rockies and northern Plains that is visible on the satellite photographs during the first days of May (Figures 4.1–4.3). On the last day of April, the atmosphere was generating clouds there, with surface convergence pushing the air upward, forming clouds. What is more, the shading on the map and the symbols at stations (∇ = shower, •• = rain, ** = snow) show us that these clouds are producing precipitation at many locations. And notice that the precipitation is occurring mainly (but not exclusively) on the cold side of the front. It makes sense to attribute at least some of this precipitation to overrunning, for we have seen that overrunning of cold air by warm air always puts the surface precipitation into the cold air (recall Figure 4.18). And where the air is overrunning the east slopes of the Rockies, there is even more widespread precipitation. Easterly winds blowing clockwise around the cold high to the north and counterclockwise around the cyclonic system in the southern Rockies are climbing upstairs as they flow westward. And look at the weather—a cold, rainy morning in Denver, and it gets worse farther north along the east of the Continental Divide, where some snow is falling. What little precipitation that falls along the east slopes of the Rockies usually comes with upslope east winds. For us travelers, the day does not look too propitious for embarking on a long journey, but let us start out anyway.

5.2 The First Morning (April 30)—Low Clouds and Rain in Colorado

We head eastward along I-70, planning to spend the night of April 30 just outside Kansas City. All morning, and into the early afternoon, from Denver to Hays, Kansas, low clouds, rain, drizzle, and fog make the driving miserable. Notice Denver's 2½-mile visibility on the 0700 EST (1200Z) chart. The visibility improves only slightly across eastern Colorado, where warm air, gliding upward and westward over higher and higher ground, as well as over the cold air west of the Kansas front, is saturated almost at ground level. Clouds in such situations are widespread stratus-type with ragged, low bases. Patchy scud clouds, even lower, race southwestward on the low-level winds from the northeast. It is a raw morning for late April in this cool northeasterly flow, and the temperature is still only 49°F when we stop for lunch at Oakley, just east of Goodland, Kansas. Goodland was 46°F at 6 A.M., and as is normal for a cloudy, drizzly day with northeasterly wind, the temperature has risen very little during the morning hours. When this day is over, Goodland will have experienced a range of only 9 degrees between the warmest and the coldest readings.

We can study the clouds a bit now while we are stopped. The steady rain, which was falling over Colorado's higher ground, has diminished. Low stratus clouds still cover the sky, but now they have the peculiar rolled shape that is common in cold, low-level flows behind cold fronts. Meteorologists call these *stratocumulus* clouds. The rolled shape does give some cumuliform appearance, but these are really stratus clouds, formed mainly by the mixing of cool, moist air near the ground upward to a level where the air is cold enough to produce saturation.

5.3 The First Afternoon—Driving through the Front in Kansas

After a welcome hot lunch, we approach Hays, in west central Kansas, where a remarkable transformation occurs. We see the skies brightening to the east, and the stratocumulus layer begins to break up. Now it is easy to see some higher, broken clouds, mainly globules of altocumulus, with some cirrus; our spirits are also lifted by patches of blue, which widen to the east.

We are approaching the narrow band of clearing, sometimes called *false clearing,* that often lies just behind cold fronts. Figure 2.7 shows an example of this band, which is the dark streak behind the front where the satellite can "see" down to the ground. We call it false clearing, because, with a cyclone's normal motion toward the east or northeast, the clear band, being only 50–100 miles wide, is at most over any one point only a few hours. Today, we are driving eastward at 55 mi/h and overtaking this slow-moving (10-mi/h) cold front, and those bright skies ahead tell us that we are getting close to the front.

As we drive eastward across Kansas, we notice that the temperature is rising rapidly; we know at least two reasons why this is so. First, we are escaping the cold air by driving east. As we continue, we will cross the front into the warm air and escape the cold air altogether; even now, we have already driven close to the edge, and it is not as cold here. Second, daytime solar heating of the ground, which was suppressed in western Kansas by the deep, reflecting clouds, is strong in this brighter part of the state, so afternoon temperatures are higher.

We left the steady rain in Colorado, but now there are some showers scattered about, and we can see the bulging cumulus and cumulonimbus clouds which are strung out along the front. Here is where the surface winds that comprise two great currents of air, one warm, the other cold, are converging along a great arc. We saw the boundary of these two currents (the front) on the morning map and in the satellite picture. Now, we see it from the ground—warm and humid surface air along the very edge of the cold flow being lifted by the converging surface winds. It is obviously unstable air; a look at the towering cumulus clouds tells us that. And suddenly, we catch up to and disappear under some hard showers and thundershowers.

The rain beats on the windshield; we are not making 55 mi/h now. It is dark enough to see a lightning flash or two under these huge cumulonimbus clouds that are seven or eight miles deep. The raindrops pelting the windshield have grown to monstrous sizes; it is raining at the rate of one inch per hour. A partial evaporation of these raindrops has chilled the air to 64°F. We almost wish that we had stopped, but we reason that if we caught up with this thunderstorm, we ought to be able to overtake and break through it. And suddenly we do, for it soon brightens to the east, and we drive out of the rain and under the cirrus clouds that are stretched out ahead of the thunderstorm cell. It is warm again, and rather humid, with a south wind. The dark clouds obliterate the western horizon behind us. We are in the warm air, east of the front, and moving toward a marvelous expanse of blue skies dotted with patchy cumulus clouds.

The weather that is observed in warm flows such as this one may be quite varied. Often there is abundant sunshine, for warm air can sustain much water vapor and, therefore, frequently has a low *relative humidity,* which is the fraction of the maximum possible amount of water vapor actually present. However, other factors may cause even this warm air to become saturated. If the earth's surface is cold, the low-level air may be cooled to its dew point, causing fog and stratus clouds. Warm, humid winds in winter often produce this depressing situation on the southern Plains. Also, if there is a cyclone close by, surface frictional convergence into the low center may lift the warm air to saturation. If the warm air is unstable, bands of showers may be produced. Since warm air can be rich in moisture, even when unsaturated, these warm-air thunderstorms can deliver copious rains, which are blessings for American farmers and for all in the world who use the huge crops grown on these plains.

But just occasionally, when the winds aloft are very strong, the thunderstorms become too vigorous, generating hail, destructive winds, or even tornadoes. The signs to watch for reflect the factors that produce vigorous convection (see Section 6.4)—warm, humid surface air, altocumulus clouds that grow vertically, indicating unstable conditions aloft, strong surface winds that help produce strong convergence, a low barometer (more convergence), and cirrostratus clouds blowing off the tops of thunderstorms which are located upwind. These *anvil* clouds are often present with ordinary thunderstorms, but if the anvils indicate strong jet-stream winds aloft when the other factors listed are present, some of the thunderstorms are likely to be severe. On the Plains, where the ground is smooth, large rotating thunderstorms may organize their spin into the dreaded tornado (see Section 6.5).

Thus, we can see why tornadoes are favored on the U.S. Plains in springtime. They must occur in cyclonic circulations (low barometer and surface convergence), when the low-level wind is southerly (warm and moist), and when the jet stream aloft is strong and from the southwest (wind veering with height, indicating warm advection and rising air). So, we are not surprised to learn that five out of eight tornadoes in the United States move northeastward, and we remember now to watch the direction of these streaking anvils. Maybe we should have stayed in the cold air, where tornadoes are very rare.

5.4 The First Night—A Thunderstorm in Kansas City

We reflect uneasily on these matters on the muggy night of April 30 as lightning flickers and thunder rumbles over Kansas City. There are a large number of nighttime thunderstorms over the Midwest on warm nights, when it is difficult to see the sky. We turn our motel television to Channel 13, turn down the brightness until the screen darkens, and then switch to Channel 2. (These channel numbers are the same anywhere in the United States). If the screen glows steadily again at that setting, it is quite possible that a severe thunderstorm or tornado is within 20 miles. This technique is based on the extraordinary electrical activity of severe thunderstorms.[1] They are terrific lightning producers, and the emission of high-frequency energy by the nearly continuous lightning strokes can be monitored in this way.

Fortunately, Channel 2 remains dark and we do not hear the terrifying roar of a tornado this evening. There is considerable moderate thunderstorm activity, however, and on the morning weather map for 0700 EST (1200Z) May 1 (Figure 5.2), we can see a line of thunderstorms (squall line) from Cape Girardeau, Missouri to Shreveport, Louisiana. Thunder during the past six hours is reported at Des Moines, Iowa, Moline, Illinois, Wichita, and Springfield, Missouri, as well as at Kansas City. Cape Girardeau received almost an inch of rain in its thunderstorm, and we can infer that plenty of rain fell this night

[1]This technique should not be used to supplant the National Weather Service warning system.

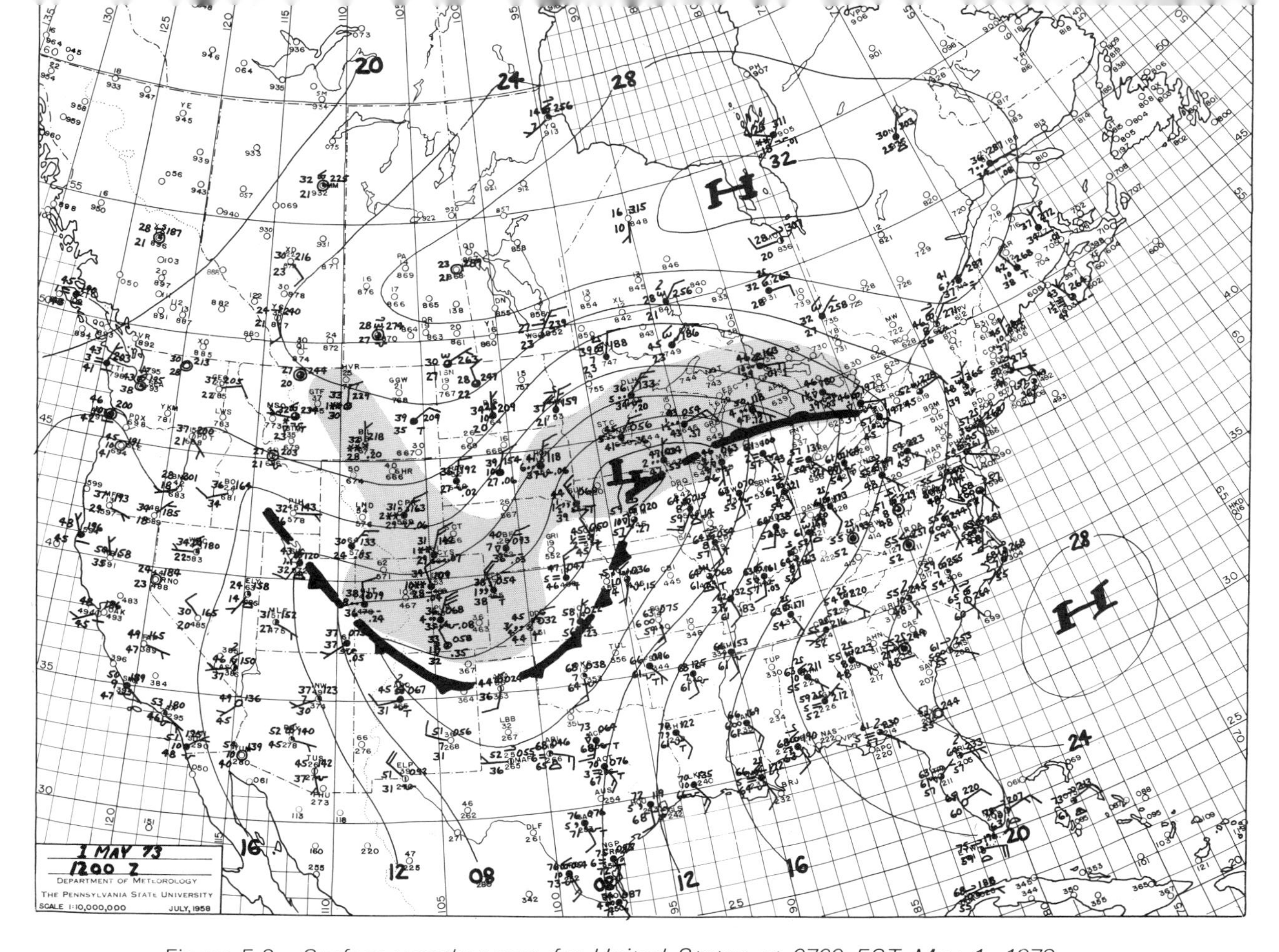

Figure 5.2 *Surface weather map for United States at 0700 EST May 1, 1973.*

between the reporting stations. The winds aloft charted on the 500-mb map (Figure 5.3) show that there is a strong jet stream overhead. We should not be surprised to hear of tornadoes today.

5.5 The Second Day—Racing the Cold Front across the Mississippi Valley

As we prepare to depart on the morning of the second day (May 1), the cold front is lurking to the northwest and threatens to overtake us again. The winds across eastern Nebraska have turned into the northwest as the first of the Rocky Mountain lows has raced northeastward to Wisconsin. Those winds will swing the front across Kansas City today, but we will drive eastward and stay ahead of it. This 57°F air in the morning is certainly more comfortable than yesterday's chilly weather; looking back at Denver and Cheyenne, Wyoming, we find snow in May!

This morning's weather at Kansas City gives us many clues as to what the weather map looks like. It is warm for the date, and humid, as indicated by the 54° *dew point*. (The dew point is the temperature at which saturation and the appearance of visible water droplets occur.) We observe patches of fog in the low places along the highway. A south wind and the thunder of the previous night confirm the fact that there is high pressure to the east, low pressure to the west, and a warm, unstable air mass south and east of us. But there is more. Altocumulus clouds in the morning sky are moving northeastward (see the winds aloft chart in Figure 5.3). The wind at the surface is southerly, so the wind is veering with height. When we correct the surface winds by 30 degrees or so to account for the effects of surface friction, the veering is reduced, but in any case, the wind is certainly not backing. Therefore, the atmosphere may be getting somewhat warmer. With abundant blue sky, the May sun will drive the surface temperature up quite rapidly today. The little turrets on the altocumulus clouds also tell us that the middle troposphere (about 3 kilometers, or 10,000 feet) is unstable, so residents of Kansas City may expect more thunderstorms to form as the surface heating progresses.

We can observe one other thing with our eyes. The rapid northeastward movement of the altocumulus clouds tells us there is a strong jet stream, and strong jet streams are found over fronts. The front is close, but the southwest winds aloft are nearly parallel to it, so that it is moving very slowly. In fact, we can see on the May 1 weather maps (Figures 5.2 and 5.3) that these southwest winds aloft are carrying another cyclone toward Kansas City. It is the second member of this cyclone family, following a track southeast of its parent. This morning, it is over the Texas Panhandle, but Missourians will sense its approach today as barometers, thundery rains, and hopes for the spring planting fall together. Falling weather![2] Once this second cyclone goes by, temperatures

[2] This expressive term is reportedly of Pennsylvania Dutch origin. Modern barometer watchers have learned that the pressure tends to rise every morning and fall every afternoon, so they usually concentrate on large or persistent trends in the barometer reading rather than on the instantaneous change. In the present example, barometers in Missouri would fall all day, but especially in the afternoon.

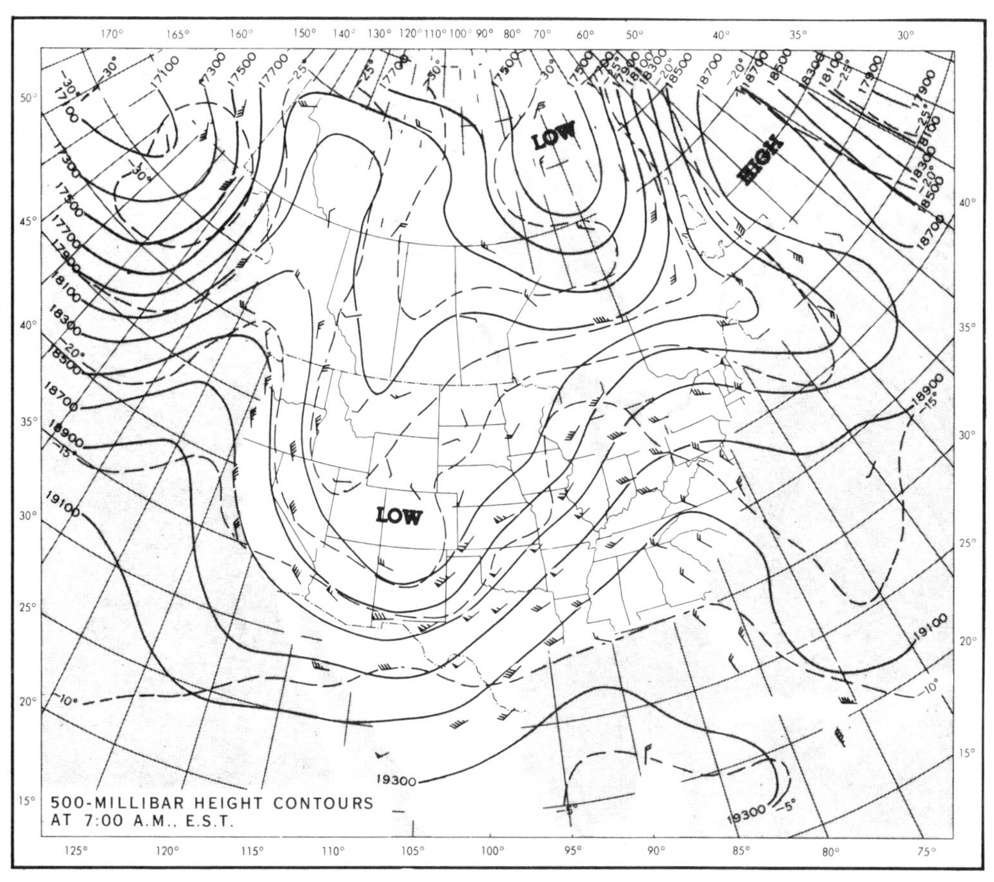

Figure 5.3 *Upper-air (500-mb) map for United States at 0700 EST May 1, 1973.*

are going to fall, too, as northerly winds follow the passage of the cold front.

As we drive on eastward across the Mississippi Valley toward St. Louis, enjoying this warm sector, we know that a cold front is nipping at our heels. However, at our average speed of 50 mi/h, we gradually pull ahead of it, since it is moving only at about 20 mi/h. We also know that we may see some scattered showers or thundershowers today in this unstable air. However, the chances are not nearly as high as they would have been had we stayed in Kansas City, close to the cyclone path.

By driving toward higher pressure, we reduce the odds of our meeting bad weather, and so we have a mostly sunny, warm day with afternoon temperatures in the 70s and numerous swelling cumulus clouds reminding us of the high humidity. Late in the day, a few of the towering cumulus grow to cumulonimbus clouds, but they are widely scattered, and we do not drive under any of them. But the telltale cirrus anvils visible on the afternoon horizon remind us of the persistent cold front and its associated squall line following us.

5.6 The Second Night—More Thunderstorms in Indiana

During the twilight of our second day, we stop in western Indiana. We have built up a good lead on the front, but fronts do not sleep or tarry for meals, and after we stop, the front begins closing the gap between us. Although the scattered daytime cumulus clouds are dying with the sunset, the western sky reveals fresh bands of altocumulus merging with thickening altostratus, indicating the approach of the front. The night will be cloudy, and because this warm air is rather unstable, we expect showers and thunderstorms to be embedded in the layers of altocumulus and altostratus. And sure enough, later that night, we are partially awakened by rumbles of thunder and flashes of lightning as thundershowers visit Indiana.

5.7 The Third Day—Overtaken by the Front in Pittsburgh

By Wednesday morning, May 2, the surface map (Figure 5.4) shows us that we are practically in the same position relative to the front as we were on the previous morning. Yesterday's drive kept us ahead of the front, but now the second cyclone has moved northeastward to the western shores of Lake Michigan, and the front has continued its relentless march eastward. Note that the cyclone track shows that the center of this second cyclone passed right over Kansas City yesterday after we left. The Kansas City airport received well over an inch of rain.

On the morning of May 2, the 7 A.M. Indianapolis weather report shows overcast skies and a temperature which is 5 degrees warmer than Kansas City was yesterday at the same time. We are still in the warm sector with a southerly wind at the surface. However, through breaks in the low clouds, we can see a more westerly movement in the upper-level clouds, and, as expected, Figure 5.5 shows that the flow aloft is definitely more westerly than it was over Kansas City yesterday. (Compare the two upper-air charts in Figures 5.3 and 5.5.) The difference is that the low-pressure center in the upper atmosphere has moved from western Colorado to western Wisconsin. Thus, the steering winds aloft have shifted, and now, instead of being parallel to a slow-moving front, the flow aloft is blowing perpendicular to the front, sweeping it eastward. This front may catch us today.

We can see from the importance of the steering principle why weather forecasters want to be able to anticipate changes in the steering winds. But we do not need to know all of the details of the upper-air chart. The strong flow from the west aloft, as indicated by the rapidly moving upper-level clouds, tells us that our pursuing cold front should approach quite rapidly now. And the Norwegian cyclone model tells us that colder, drier air will flow behind it from a northerly quarter. We know all that, standing by our wet car in western Indiana on a warm, humid morning in early May, just by using our eyes and remembering the cyclone model.

Looking again at the surface map for the morning of May 2 (Figure 5.4), we can see that there is no third member of this cyclone family. The surface

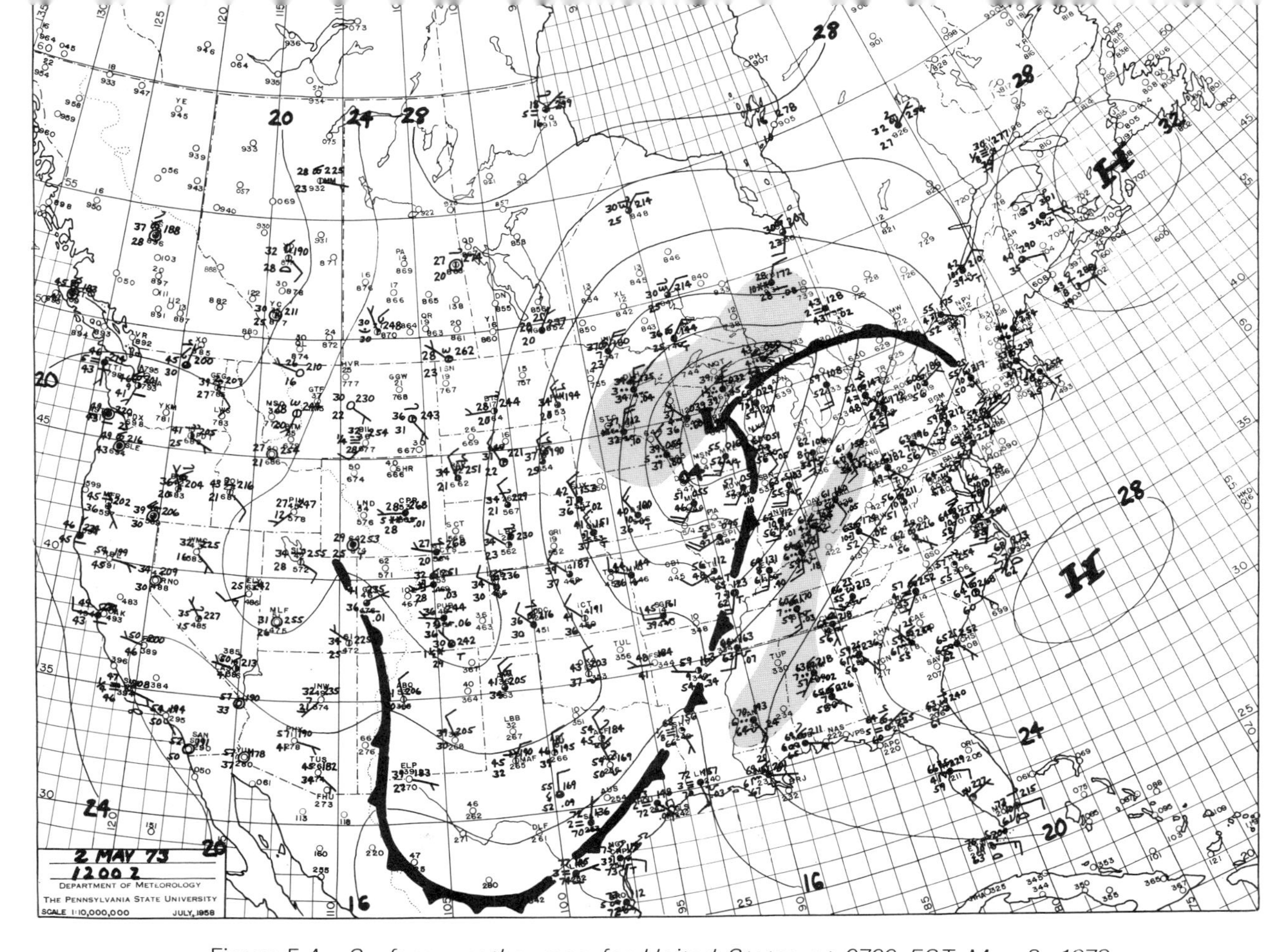

Figure 5.4 *Surface weather map for United States at 0700 EST May 2, 1973.*

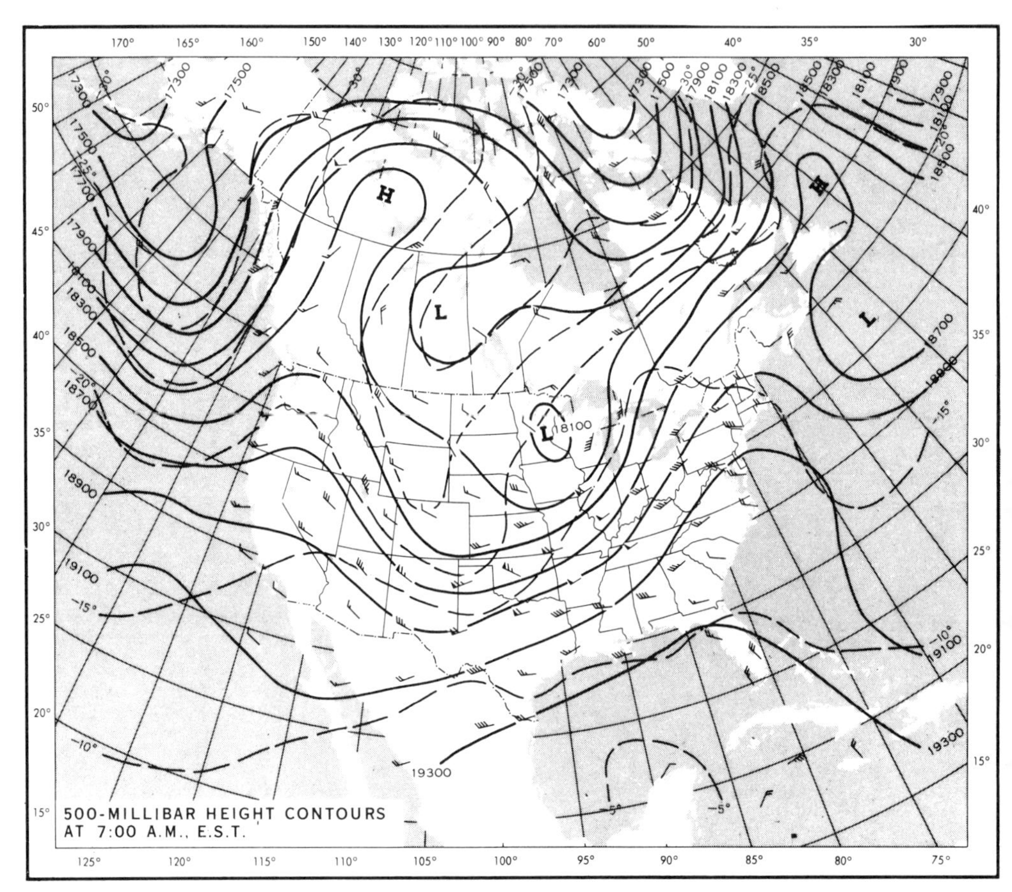

Figure 5.5 *Upper-air (500-mb) map for United States at 0700 EST May 2, 1973.*

winds west of the front over Illinois, Iowa, and Missouri have shifted to west and northwest. If there were another low along the front, near Little Rock, Arkansas, for example, these winds would be different, probably from the northeast, like the winds we had in Colorado and western Kansas on the first day. And if that were the case, the front would move slowly, just as it did on Monday. But in the absence of a third low, the resulting strong west-to-northwest winds across Iowa and Illinois are driving the cold air and its leading edge, the front, eastward quite rapidly. So, we draw the same conclusion from weather maps that we did with our eyes. This front is moving eastward more rapidly today than it was yesterday.

Tonight, the front will overtake us and give us a taste of colder air. It is sticky enough on this warm Indiana morning to make the thought welcome. Today's drive of 400 miles will take us to Pittsburgh, and since we can drive an automobile somewhat faster than the front, we remain in the warm, moist air. As the cyclonically curved isobars on the map show, this moist air is converging near the ground and rising, so we do not see much of the sun. Many low clouds and occasional showers mar the driving.

We watch for signs of the front that evening in Pittsburgh, but the thickening low clouds defeat us. After dark, however, hard showers set in, and the wind shifts from the south around to the west. Now, if we had a barometer, we would observe its lowest reading, followed by a rather sudden upsurge as the cold air rolls in. And we feel the difference in this air after the hard showers end —west winds, cold and drier in the absolute sense even though it is still cloudy. The dew point tumbles from 60° to 50° in just an hour or so.

5.8 The Fourth Day—Cold Advection and Low Clouds over Western Pennsylvania

We take our own morning observation at Pittsburgh at 0700 EST (1200Z) May 3, and what a difference from our previous morning! It was 62°F in Indianapolis at the same latitude yesterday morning, and today in Pittsburgh, it is 52°F. The air feels much less humid, and the wind is more gusty and from the west rather than the south. The low clouds have disappeared altogether; again, we are in the band of "false clearing," behind the cold front. But the clearing affects only the low clouds, and we are able to observe the winds aloft by watching the movement of the high clouds. It is very hard to see the motion of the amorphous layer of altostratus clouds over us, but by looking straight up and watching carefully, we see that they are moving from the southwest. Well, that clinches it! The wind is backing with height, for the surface geostrophic wind must be from west-northwest (about 30 degrees clockwise from the surface wind). The atmosphere over Pittsburgh is cooler, and is getting cooler still with the cold advection. The May sun may shine enough today to drive up the temperature a good deal, but it will very likely be cooler than yesterday, and tonight should be a lot cooler.[3]

Here in Pittsburgh, on the morning of May 3, we bring our road trip to an end, but our observations are not quite finished. We have followed a cyclone family across the United States. Now, let us wrap up the sequence of events with a last synoptic view, that is, an analysis of a number of observations taken at the same time. Figure 5.6 shows the May 3 0700 EST (1200Z) map for the United States for comparison with the other morning surface charts. We see that all of the features that we observed this morning in Pittsburgh are consistent with the map. The cold front has passed east of Pittsburgh, and a westerly flow of cooler and drier air has established itself behind the front over the Ohio Valley. Indianapolis is 17 degrees cooler today than at this time yesterday. On toward the east, the warm, humid air of the warm sector, with 60°F morning temperatures and dew points nearly as high, extends as far northward as Burlington, Vermont, and possibly Montreal, Quebec. Farther north, over New England,

---

[3]The data at Pittsburgh: Yesterday's maximum, May 2, 67°F; May 3, 62°F. Minimum on the first night of our visit, 51°F; the following night, 38°F (with snow flurries!).

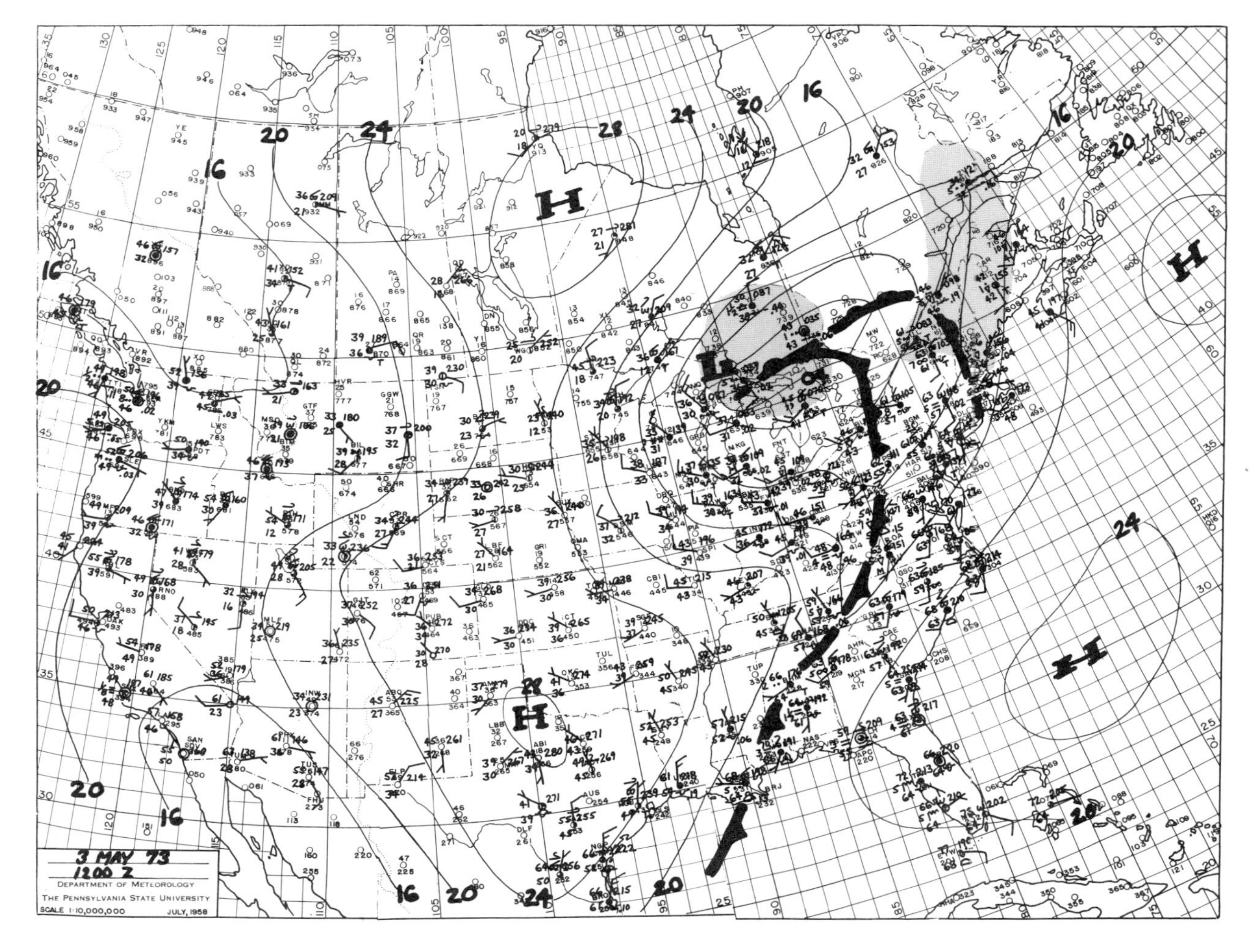

Figure 5.6 *Surface weather map for United States at 0700 EST May 3, 1973.*

the Maritimes, and much of Quebec, cold air at the surface shows that the warm, moist air is overrunning colder air; a surface warm front marks the boundary where the warm air leaves the surface. Portland and Houlton, Maine, Chatham, Ontario, and Yarmouth, Nova Scotia are in the 40s, and Seven Islands, Quebec has rain falling into 34°F air. Sleet is reported at Kapuskasing, Ontario, and it is easy to guess that snow is reaching the ground just north of Seven Islands. This cold side of the storm track is just where we would expect snow or sleet using the Norwegian model. Indeed, we can now identify the great band of clouds that we saw in the composite satellite pictures of Figures 4.1–4.3, which spread northeastward from the central United States all the way to Labrador between May 1 and May 3. Those clouds mark the northeastward movement of these two cyclones under the southwest flow aloft, and, in particular, the overrunning of the cold air in Canada by warm, moist air from the south.

But now, let us look at the satellite pictures at 10 A.M. EST. Figure 5.7 is a photograph taken in the visible wavelengths of the clouds over the United States and Atlantic Ocean. We see a view of a classic middle-latitude cyclone. Notice that the extensive cloudiness over the eastern United States and Canada gives the appearance of counterclockwise (cyclonic) rotation about a point near the Great Lakes; to its rear (west), we can almost sense the air flowing in from the west and northwest. We notice that the skies clear behind the front, which is expected, both because cold advection produces sinking and because the following high-pressure center produces surface divergence and sinking air. What is more, we can detect more extensive bands of clouds along the fronts, where there is overrunning, and where there is strong surface convergence near the cold front. Off to the southeast, the air is nearly clear again, with just a few small cumulus and cirrus clouds and a couple of scattered showers over the Atlantic marring an otherwise clear sky. We can even infer the presence of tiny fair-weather daytime cumulus over Cuba, Florida, and the southeast coast. The individual clouds are too small to see, but their combined effect whitens the land. Notice how few there are over the oceans, which are not heating up as rapidly as the land. And notice that to the rear of the cyclone the air is so dry that there are not even any fair-weather cumulus clouds.

The imaginative observer may even be able to see the wind patterns in the clouds, as the warm air swirls around the western end of the fair-weather Bermuda High.[4] That air is flowing in toward our second cyclone; by tomorrow, it will be feeding the next member of the family over the Gulf of St. Lawrence. But let us look more carefully at the clash of air currents along the cold front. To do this, we look at another picture taken at the same time (10 A.M. EST May 3) from the same satellite, but for this picture, the sensor is tuned to "see" infrared radiation, which is proportional to the temperature of the objects viewed rather than their color. In this picture (Figure 5.8), cold objects are white, and warm ones dark. We notice how warm the surface of the earth is, especially the warm ocean. The Florida peninsula, which is about the hottest land area, shows up distinctly.

But now, we can tell where the deep clouds, with cold tops, are found. Along the cold front is certainly one place. In fact, we can use the infrared picture to locate the front. Comparing the visible and infrared, we can see

[4]The Bermuda High is the semipermanent anticyclone that is centered near Bermuda.

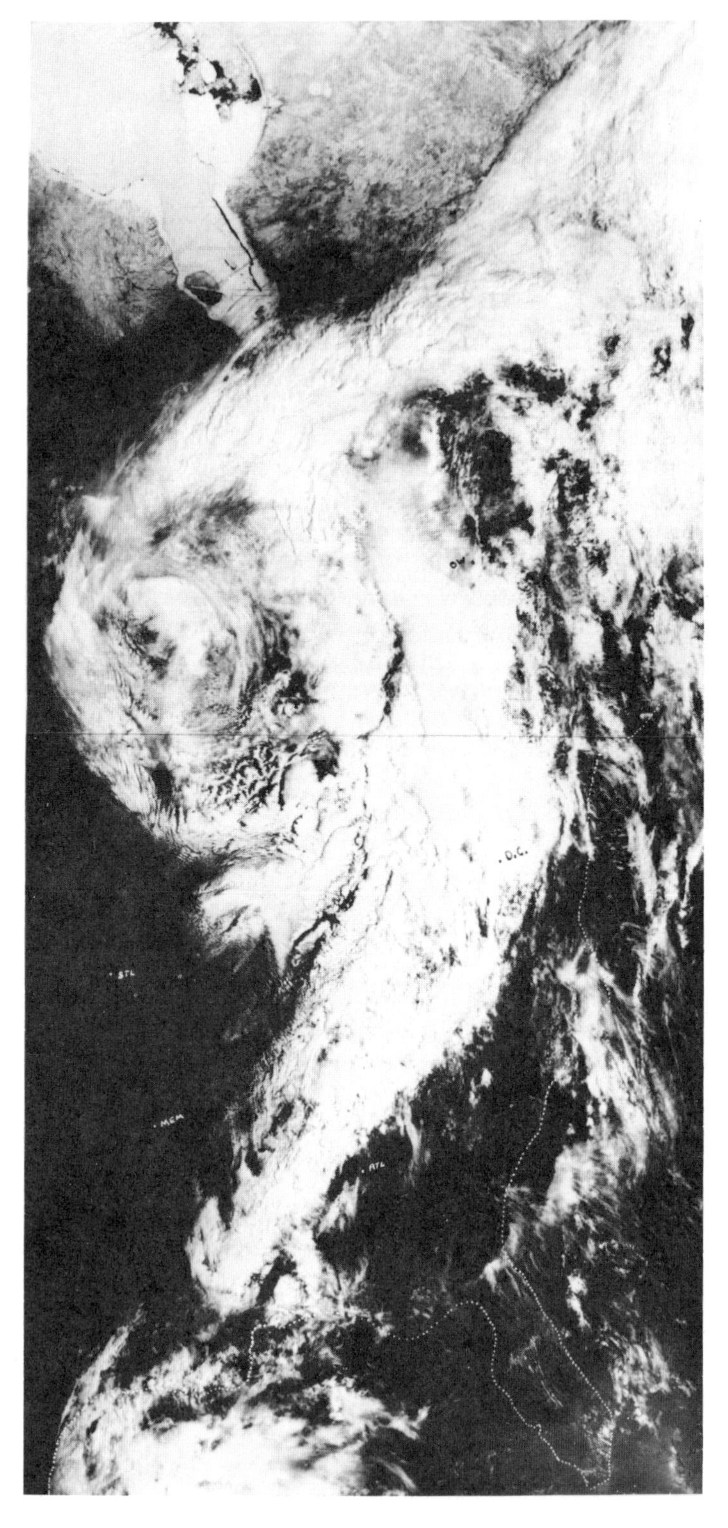

Figure 5.7 *Visible satellite photograph at 1000 EST May 3, 1973.*

Figure 5.8 *Infrared satellite photograph at 1000 EST May 3, 1973.*

that the deep clouds occur along the leading edge of the front. In contrast, behind the front, the clouds are shallow because their tops are warm. (Compare the appearance of this cyclone with the pictures of the cyclone on May 20 in Figures 2.6 and 2.7).

Finally, let us consider the surface weather at Pittsburgh later in the day. Looking outside, we find that low stratocumulus clouds have rolled in since morning, and it looks dark and threatening. We are momentarily baffled; cold air behind a cold front is supposed to sink and produce clearing weather. It is not raining, but the air feels damp and cool, and we wonder what mechanism is producing all these dark, low clouds. Two factors could produce low clouds, and low clouds only. One would be convergence across cyclonically curved surface isobars, the other, air flowing upward over the low Alleghenies. The surface map for 7 A.M. EST (Figure 5.6) shows that the isobars behind the cold front are curved in a cyclonic sense over the Great Lakes and about as far south as Indianapolis, which might explain the cloudiness there. But that is not the situation at Pittsburgh, where the isobars are actually curved anticyclonically. We conclude that these low clouds at Pittsburgh must be very shallow, produced by west winds blowing upward along the western slopes of the Alleghenies. We also conclude that no more than a few sprinkles can fall from these shallow clouds, and the ball game at Three Rivers Stadium will not have to be called off.

We can verify the preceding reasoning on the infrared picture (Figure 5.8). Notice how warm the cloud tops are over Pittsburgh and everywhere to the west and southwest. Warm tops mean low tops, and low tops mean little or no rain.

We call a friend in Roanoke, Virginia. "Has the front reached you yet?" "No, it has been raining off and on all morning. It is very humid, with clouds at many levels, some which are towering like big castles through the low clouds." We could make a forecast for Roanoke, for we know that the front is near. The westerly winds over the Ohio Valley make the relief from the humidity and an end to the showers virtual certainties for Roanoke.

A call to Memphis verifies what we expect. The weather over Tennessee is clear, cool, bright, dry, and beautiful. It is almost too cool (39°F) back near the high center at Kansas City, where the dry, cool air lost much heat last night.

We call a friend in Georgia, near Augusta, where it is a lovely day, with the temperature already 75°F and rising fast. She reports a few cumulus clouds and some high, thin ones. "We know about that; we have seen them on the satellite pictures. Your wind must be from the south; it is probably a bit sticky there. Well, watch the western and the northwestern sky this evening; Georgia will have another northern visitor tonight."

We make a quick call to Augusta, Maine. Just as we thought—at this Augusta, it is another story; rainy, gloomy, foggy, and cold—45°F. The warm front is south of here, and it will be something to watch its approach today as the cyclone runs up the St. Lawrence Valley. Augusta may get into the warm sector, which is over at Burlington, Vermont, and if it does, our friends "down east" are going to witness a dramatic rise in temperature before the cold front arrives to pinch off the warm air. As it happens, the temperature at Boston shoots up close to 80°F, but the cold front comes through Maine before the warm air reaches the surface.

A last call to New York City brings some excitement. The air is warm and humid, and there have already been some showers. The barometers are falling sharply, and while the sky has cleared partially, thunderheads are already appearing in the western sky. Sharply falling morning barometers, warm, humid air, and building cumulus clouds foreshadow a great day for convection. And as we tell of the front rolling toward New York, we know that later today the convection will be invigorated by the frontal convergence. It looks like a good day for a squall line to spring up in the Poconos or the Adirondacks and to sweep across the Hudson River. Over such rough terrain, tornadoes are rare, but they are possible; at least, our friends in New York should be on the lookout for strong winds, lightning, and possible hail today. Tomorrow they can relax, weatherwise.

From sleet at Kapuskasing to hail at New York; from snow in Quebec to steady rains at Roanoke; from high, thin cirrus in the southwestern sky over Newfoundland to the last edge of the shallow boundary-layer stratocumulus just west of Indianapolis; from the cool, dry, clear air over Missouri to the warm, moist, clear air over the Atlantic; from all these facets of the cyclone circulation emerge some sense and order when we consider them all together. And if we know the cyclone model and use our eyes, we can often guess what those reports would be, even if we are hiking in the forest or driving along the highway. The details may vary from cyclone to cyclone and from season to season, but the keen eye is rarely unrewarded.

Plate 7 *Rainbow and dissipating cumulus cloud (photo by Richard Anthes).*

Plate 8 *Waterspout in the Florida Keys (photo by Joseph H. Golden).*

6 Severe weather

The sun sets weeping in the lowly west,
Witnessing storms to come, woe, and unrest.

[Shakespeare, *King Richard the Second,* 2.4.21–22]

6.1 Cyclones, Hurricanes, Thunderstorms, and Tornadoes

The relatively infrequent, severe weather phenomena that most threaten us also hold a fascination that ordinary events do not provide. Few people fail to be stirred, excited, or frightened by the disquieting rumbling of a thunderstorm on a warm summer night, the ominous pounding of the surf as a hurricane approaches, or the terrifying roar of a nearby tornado. These events have rightly been the subjects of a large fraction of meteorological research. Like the news coverage of extraordinary "bad" events, the misbehaviors in meteorology attract most of the attention.

In the previous chapter, we saw how the common, middle-latitude cyclones affect our weather during the year. Many people confuse tornadoes, hurricanes, and cyclones, as evidenced by the following quotation from *The Wizard of Oz,* by L. Frank Baum: "The north and south winds met where the house stood, and made it the exact center of the cyclone." The tornado which swept Dorothy's house from its Kansas foundation and deposited her in the Land of Oz was, in fact, a cyclone; however, the vast majority of cyclones are not tornadoes. Cyclones are merely any circulation around a low-pressure center, regardless of size or intensity. Thus, tornadoes and hurricanes, as well as extratropical cyclones, are all different kinds of cyclones.

The large extratropical cyclones are usually beneficial, providing necessary precipitation, exchanging cold polar air for warm tropical air, and cleansing the air at the same time. Hurricanes and tornadoes are smaller, and frequently more vicious, cousins of the extratropical cyclone, although hurricanes have certain beneficial as well as destructive aspects. Before discussing the origin and structure of these storms, we must first carefully distinguish between these three distinct phenomena.

The *hurricane* is an intense cyclone that forms over tropical oceans. Although smaller in size than the middle-latitude cyclone, it is much larger than either a tornado or thunderstorm, averaging about 800 kilometers (500 miles) in diameter. The most prominent characteristic of the hurricane is a doughnut-shaped ring of strong winds exceeding 75 mi/h surrounding an area of extremely low pressure at the center of the storm. (An idea of the strength of a wind speed of 75 mi/h may be obtained by leaning out the window of a car going 75 mi/h.) It is impossible to stand in such a wind; it is also difficult to breathe. These winds may drive the ocean into waves exceeding 6 meters (20 feet) in height. The hole of the "doughnut," called the eye, is a relatively calm region of clear skies in the middle of the whirlwind (Plate 3). A hurricane may live for three weeks or more as it drifts along in its odyssey over tropical waters.

Tornadoes are similar to hurricanes in that strong winds blow counterclockwise around a center of very low pressure.[1] In fact, the strongest surface winds on earth occur with tornadoes. Although the maximum tornadic velocities that have occurred are somewhat controversial (direct measurements being impossible), there is indirect evidence that winds have exceeded 300 mi/h in extreme cases.

The diameter of the tornado is much smaller than that of the typical hurricane (Figure 6.1), averaging about ¼ kilometer. Thus, it is quite feasible to avoid an approaching tornado by driving, or even running, at right angles to its path of motion.

Size is not the only important difference between tornadoes and hurricanes. While hurricanes originate over warm water in a uniform tropical air mass, tornadoes occur most frequently over land, and are produced when cooler, drier air streaks over warm, moist air, which is less dense. Tornadoes occur most frequently in the spring, when the strongest contrasts between cold and warm air are present. They are also always associated with severe thunderstorms.

[1] A few cases of winds blowing clockwise rather than counterclockwise have been observed, but these are rare.

Figure 6.1 *Tornado (courtesy of Joseph H. Golden, National Severe Storms Laboratory).*

Tornadoes and thunderstorms frequently are found imbedded within the much larger hurricane vortex. Such hydralike storms are not impossible nightmares, but terrifying realities. For example, hurricane Beulah (1967) spawned at least 115 tornadoes as it moved ashore in Texas, killing five persons and causing nearly $2 million damage.

Thunderstorms are so common and familiar to everyone that they scarcely need differentiating from tornadoes and hurricanes. Unlike the winds around the hurricane or tornado, the wind circulation of the thunderstorm is primarily up and down rather than horizontal. Updrafts and downdrafts may exceed 50 mi/h in strong storms as warm, light air rises and cold, dense air sinks. Horizontal winds near the ground in the vicinity of thunderstorms are variable, gusty, and difficult to predict. They do not follow the counterclockwise vortical pattern that typifies cyclones.

By definition, the thunderstorm must produce lightning, which is the cause of thunder. It surprises most people that lightning kills about 200 people a year in the United States, a higher toll than that for hurricanes. The energy from a single lightning stroke can light a 60-watt (60-W) bulb for several months. The total power from lightning in the 2000 thunderstorms in progress at all times over the earth is about 500 billion watts, which is about equal to the total U.S. power generating capacity (1971).

6.2 Hurricanes

In contrast to extratropical cyclones, which derive their energy from horizontal temperature gradients between different air masses, the hurricane winds are generated and maintained by the condensation of water vapor within a uniform, tropical air mass. As we saw earlier, the condensation of one gram of water vapor liberates about 580 calories of heat. The rate of condensation heating in one typical hurricane is 10^{11} kilowatts (kW); in a day, the hurricane produces 24×10^{11} kilowatt-hours (kWh). By comparison, the entire U.S. use of electric energy in 1971 was 15×10^{11} kWh. Anyone who has visited the tropics in summertime can appreciate the enormous amount of latent heat stored in the form of water vapor. A cold can of beer is warmed to air temperature within minutes by the condensation on the can. Day after day, absolute humidities exceed 20 grams of water vapor per kilogram of air, an amount that is twice the annual average in Washington, D. C. Indeed, with the abundance of water vapor available for condensation, it is somewhat remarkable that hurricanes are relatively infrequent, in spite of the numerous thunderstorms and convective showers in the tropics.

On the other hand, with the frequent occurrence of thunderstorms and the associated release of the latent heat of condensation, we might wonder why hurricanes form at all. Let us consider how thunderstorms are organized to form the larger hurricane vortex.

6.2.1 Cooperation makes a hurricane

To understand the genesis of hurricanes, we must consider a cooperation between two weather systems of distinctly different horizontal scales, the hurricane vortex and the much smaller cumulonimbus, or thunderstorm, clouds. We know that if air with the temperature and moisture structure of the tropics is lifted, condensation will liberate enormous amounts of heat. But, theory and observations show that the release of this heat energy will occur in narrow cumulonimbus towers, not in large hurricanes. Why, then, do not all tropical disturbances consist of groups of thunderstorms, with very little organization of horizontal winds on the larger scale? The explanation is that, under the most favorable conditions, the circulations associated with the individual clouds and the hurricane vortex cooperate, with each circulation acting to reinforce the other. If the favorable conditions persist for a sufficiently long time, this mutual cooperation can result in the awesome hurricane.

6.2.2 An example of hurricane genesis

To illustrate the formation of a major hurricane through the cooperation of cumulus clouds and the larger-scale vortex, we consider the warm, humid air mass in the Caribbean in August, 1969, which would eventually spawn the notorious hurricane Camille, one of the most intense hurricanes on record. For weeks, the high summer sun had warmed the ocean and the air and enriched the humidity by evaporation. A weak disturbance in the normally steady easterly trade winds had been drifting westward across the Atlantic for the prior week. Evidence of this disturbance on August 14 is found in the mass of clouds located in the vicinity of latitude 20°N, longitude 80°W in Figure 6.2.

Figure 6.2 *Satellite photograph of early stages of Hurricane Camille in the Caribbean* (*courtesy of William Shenk, NASA*).

The origin of all these disturbances is not fully understood, although some originate over central Africa. Like the wave disturbances in the middle-latitude westerlies, a portion of these disturbances is associated with weak converging air near the surface. As the air flows together, it ascends and cools by expansion. In the moist environment, even a small amount of lifting and cooling is sufficient to trigger intense, but short-lived thunderstorms, which produce copious amounts of rainfall before dissipating. Many such disturbances cross the Atlantic and Caribbean each summer, but most never develop beyond this immature stage.

But the environment of this disturbance was unusually favorable; the water was warmer than normal, and the humidity of the air unusually high. The disturbance itself was moving rather slowly, allowing the numerous cumulonimbus clouds to warm and gradually moisten their environment at higher levels. A day later, the effect of the many individual thunderstorms was a mean warming of the larger-scale environment by 1°C, and the cooperation between the incipient vortex and the embedded clouds began to increase.

The warming of the environment, even by only 1°C, produces a pressure fall at the surface, because warm air weighs less than cooler air. The slowly converging horizontal winds near the surface respond to this slight drop of pressure by accelerating inward, driven by the increased horizontal pressure difference. On the morning of August 14, the growing storm had developed a circulation center, the pressure had fallen to 991 millibars, and winds had increased to 50 mi/h.

But the increased inflow toward the center of falling pressure produces increased lifting of air, so that the thunderstorms become more numerous and intense. The feedback cycle is now established. The inflowing air fuels more intense thunderstorm convection, which gradually warms and moistens the environment. The warmer air in the disturbance weighs less, and so the surface pressure continues to fall. The farther the pressure falls, the greater the inflow and the stronger the convection. The limit to this process would occur when the environment is completely saturated by cumulonimbus clouds. Further condensation heating would not result in additional warming, because the heat released would exactly compensate for the cooling due to the upward expansion of the rising air.

6.2.3 *The disturbance spins faster*

Although we have explained the fall of surface pressure and the inward acceleration that fuels the thunderstorms, we have not yet mentioned why the winds blow counterclockwise around the storm. The production of this *tangential* component to the flow arises from the simple *law of conservation of angular momentum*. This law states: *the product of tangential (rotational) velocity of a parcel of air about an imaginary vertical axis and the distance of the parcel from this axis of rotation is constant,* in the absence of friction or other torques. This law is fundamental to understanding the rotational nature of many atmospheric motions; it is illustrated in Figure 6.3. If a rotating parcel of air moves inward toward the center of rotation, the product of the distance and velocity must remain constant. As the distance decreases, the rotational velocity must increase, so air that originates at a distance of 500

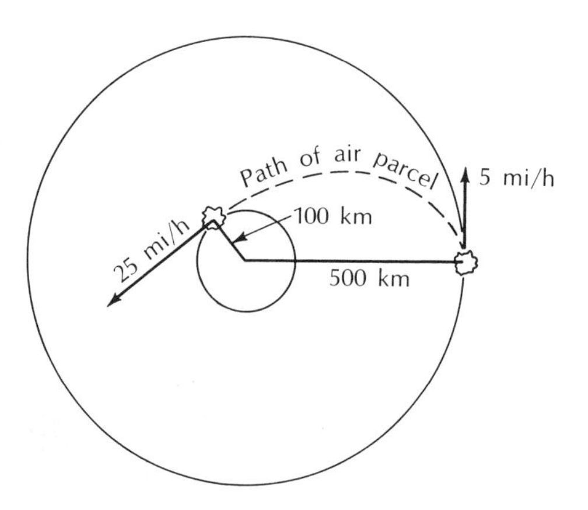

Figure 6.3 *Schematic diagram of conservation of angular momentum.*

kilometers from the center with a tangential velocity of 5 mi/h, upon reaching a point 100 kilometers from the center, would have a velocity of 25 mi/h in the absence of friction. Should the same parcel reach a radius of 10 kilometers, it would be moving at 250 mi/h. Friction reduces these values somewhat. Nevertheless, the basic principle of conservation of angular momentum—that air converging toward the center of a low tends to rotate faster—explains many wind circulations.

6.2.4 *Hurricane-force winds are achieved*

As the young tropical cyclone intensifies, the feedback between the growing vertical circulation and the cumulonimbus clouds increases. As the low-level air moves inward from large distances, the law of conservation of angular momentum demands a more rapid spinning. The intensification process, once started, may proceed rapidly; in this case, hurricane-force winds (exceeding 75 mi/h) were reached on the afternoon of August 15 as Camille approached western Cuba. The pressure in the eye had fallen to 964 millibars by this time.

As Camille passed over the western tip of Cuba during the night of August 15–16, the intensity of the storm weakened slightly. However, upon reentering the Gulf of Mexico, where water temperatures were about 30°C (86°F), the storm rapidly intensified, with the minimum pressure dropping to around 910 millibars during the day of August 16. Figure 6.4 shows a Nimbus satellite photograph of the mature hurricane Camille, and Figure 6.5 shows the surface streamline analysis for 0700 EST August 16. (*Streamlines* are lines which are parallel to the direction of the wind at every point, and therefore indicate the flow of air at a given time.) In Figure 6.5, there are no surface observations near the center of the storm where the winds are greatest. However, the observations around the storm show clearly the general counterclockwise circulation

Figure 6.4 *Nimbus satellite photograph of mature Hurricane Camille, 1210 EST August 16, 1969.*

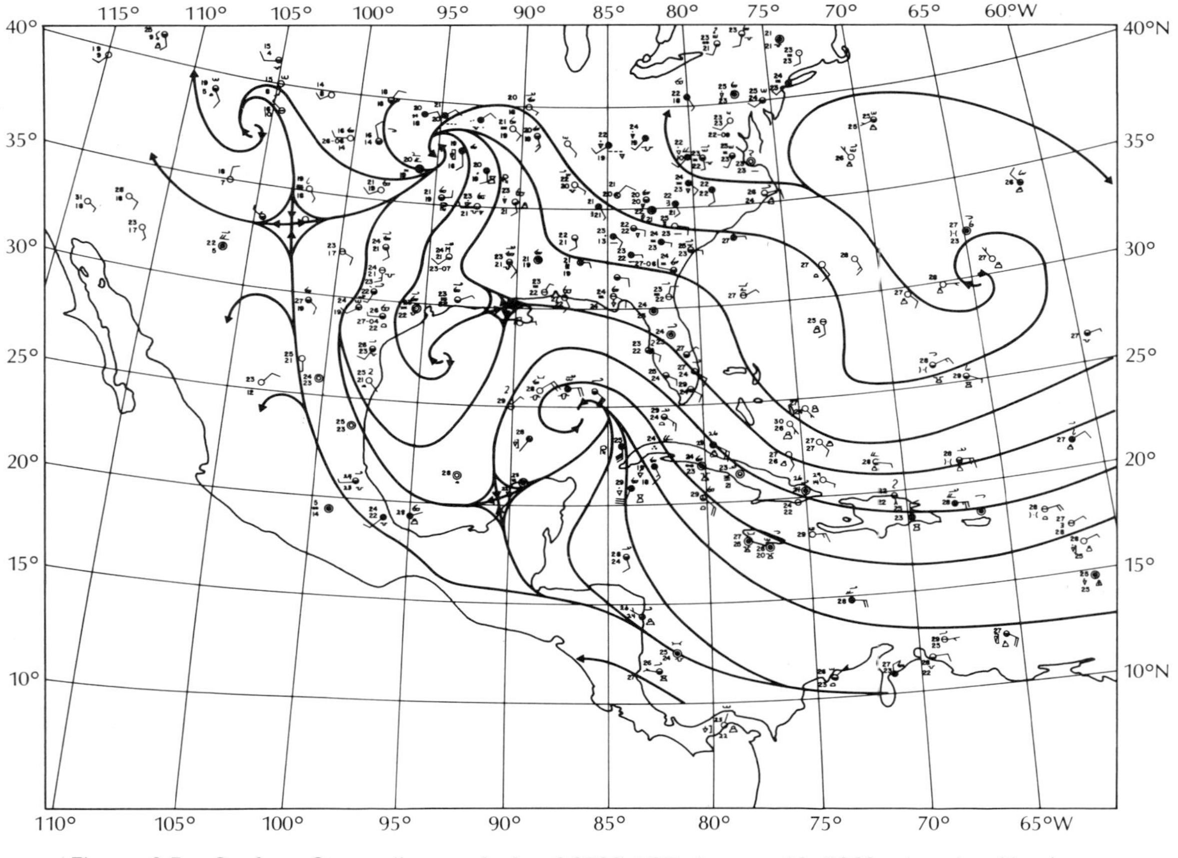

Figure 6.5 *Surface Streamline analysis of 0700 EST August 16, 1969, showing Hurricane Camille in the Gulf of Mexico (courtesy of William Shenk, NASA).*

around the storm center, which is located at about 24°N and 86°W. In addition, there is an appreciable component of inflow, especially in the southeast quadrant. The fastest winds occur where the inflowing air penetrates closest to the center, thus explaining the "doughnut" of strong winds surrounding the storm center.

But, the inflowing air cannot reach all the way to the center. As the radius becomes very small (less than 30 kilometers or so), the conservation of angular momentum demands higher speeds (and therefore kinetic energy, which is proportional to the square of the wind speed) than can be produced by the available amount of condensation heating. In other words, there is a limit to the total energy, which prohibits velocities higher than about 200 mi/h. Therefore, instead of penetrating all the way to the center, the inward-moving air turns upward and outward, as indicated in the vertical cross section of Figure 6.6. In this spinning ring of rapid upward motion, condensation and rainfall are intense; a ship unlucky enough to remain under this practically continuous ring of thunderstorms would collect about 100 inches of rain per day.

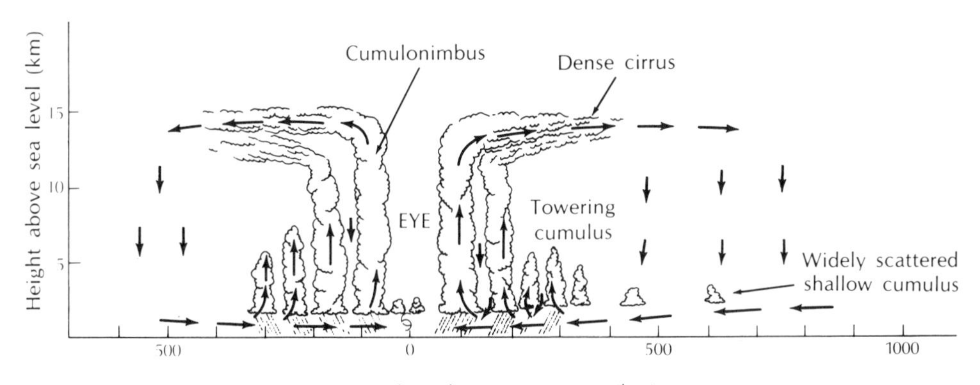

Figure 6.6 *Vertical cross section through a hurricane.* A vertical slice through the center of a hurricane shows typical cloud distribution and direction of flow.

Finally, in the cold upper atmosphere, the now-dry air turns outward and moves away from the storm center. As the distance from the center increases, the law of conservation of angular momentum operates in reverse, and the cyclonic rotational speed subsides, eventually becoming anticyclonic at a distance of 300 kilometers or so from the center. At greater distances, the air merges with the large-scale circulation of the tropics.

6.2.5 *Death of the hurricane*

If the hurricane remains in a favorable environment, i.e., over warm waters and away from hostile cold, dry air masses, the storm may continue indefinitely. A particularly long-lived storm, Hurricane Ginger (1971), traveled halfway across the Atlantic twice over a period of a month (Figure 6.7). In fact, hurricane paths may be extremely erratic, as the sample tracks in

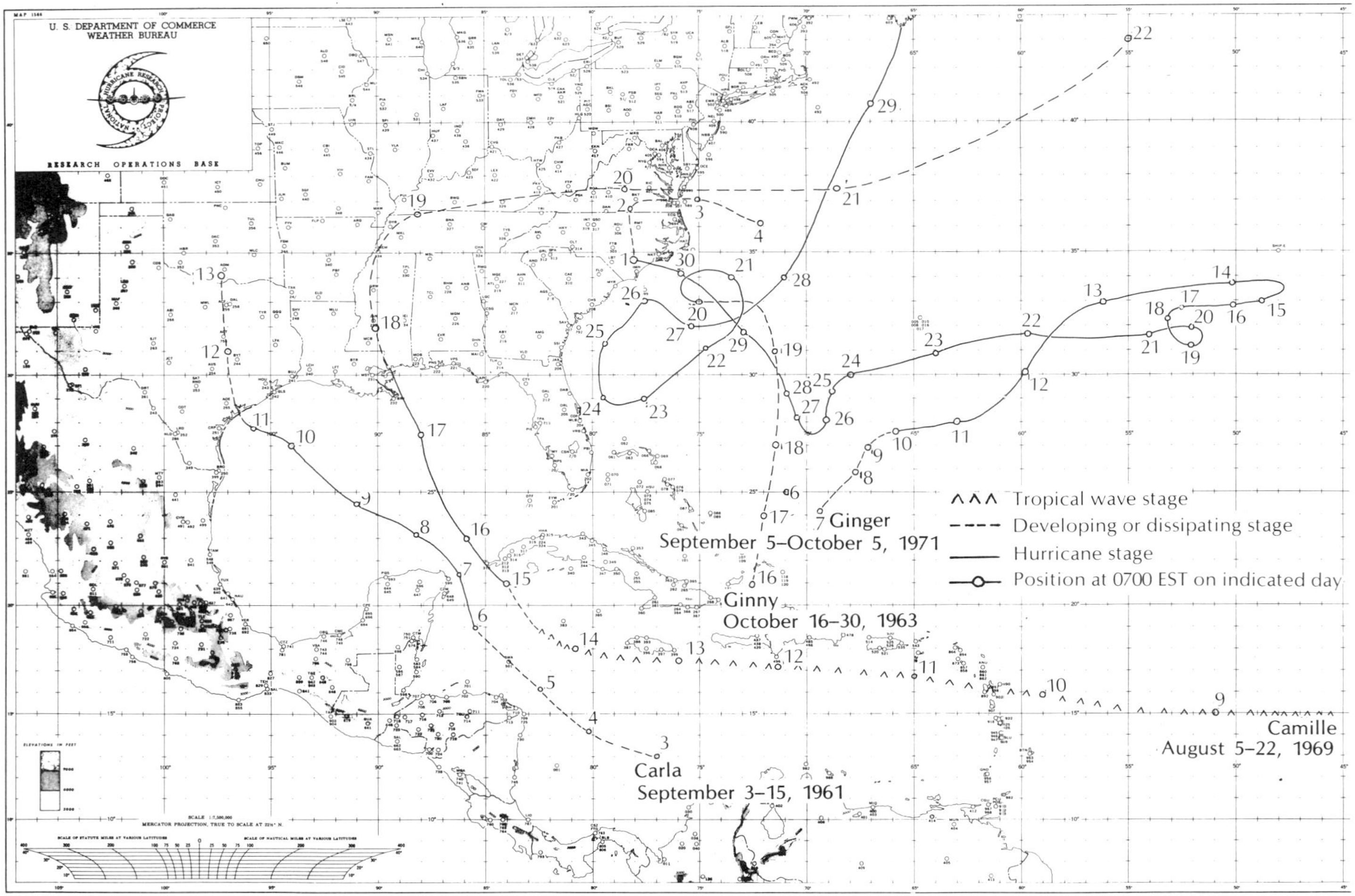

Figure 6.7 *Examples of hurricane tracks.*

Figure 6.7 indicate. We can easily appreciate the forecaster's problem when dealing with such behavior.

Eventually, however, most hurricanes drift northward and are captured by the middle-latitude westerlies. When this occurs, the storm may quickly die, or it may acquire some characteristics of the middle-latitude cyclone as contrasting air masses are brought into the circulation. Hurricane Hazel (1954) is an example of the latter type. After moving inland on the Virginia coast, it moved nearly due north into Canada, increasing in diameter and producing copious amounts of rainfall. The storm retained some of its tropical character, however, as residents of Toronto were treated to the rare spectacle of the "hurricane eye" as it passed overhead.

Hurricane Camille, whose birth and growth we have followed, met the fate of many hurricanes that move north. After reaching a maximum intensity of 200 mi/h on August 17, it moved ashore in Mississippi, killing 135 people, causing losses to over 63,000 families, and leaving a billion dollars damage in Mississippi alone. As with the majority of hurricanes, most of the damage was caused not by wind, but by water in the form of extreme tides, called the *storm surge*. In Pass Christian, Mississippi, the storm surge was 24.6 feet, higher than any previous tide on record.

After moving ashore, the sudden cutting off of evaporation from the sea starves the hurricane of its vital water supply, and a rapid reduction of the maximum wind occurs. The increased friction associated with the roughness of the land also contributes to the wind reduction. Killer winds in Camille dropped from around 180 mi/h to 60 mi/h only six hours after landfall. Moving northward into Mississippi and Tennessee, Camille's rainfall proved beneficial, breaking a severe drought.

Even though the maximum sustained hurricane winds are reduced by 50 percent or more in just a few hours, the rainfall and flooding risk may continue for days as the remnants of the storm continue to move overland. For example, after drifting slowly northward over Tennessee, Camille swung eastward into Kentucky, West Virginia, and Virginia. As the moist air was lifted by flowing over the Appalachians, rainfall increased to record amounts. Unofficial reports in Virginia claimed 25 inches of rain in the mountainous areas, enough to cause massive mud slides and flooding. The flooding in Virginia's mountains claimed 112 lives, nearly as many as were lost when the hurricane moved ashore on the Gulf Coast three days earlier. Finally, on August 20, Camille moved back off the Atlantic Coast, and its old circulation was absorbed in a frontal zone separating polar air to the north and tropical air to the south (Figure 6.8).

6.3 A Computer Model of the Hurricane

Having explained in considerable detail the life history and structure of one intense storm, it is instructive to consider how a hurricane might be modeled using a high-speed computer. Besides the hope of increasing our understanding of the hurricane, there are several practical motivations behind such an effort. The first practical use of a realistic computer model would be the prediction of the motion and

Figure 6.8 *Nimbus-3 satellite photograph showing Hurricane Camille dying in polar front off Virginia coast, August 20, 1969 (courtesy of William Shenk, NASA).*

intensity of real storms. The advantages in terms of warning and preparation from an accurate prediction are obvious. It is also true that an improvement in hurricane forecasting would obviate many unnecessary and expensive preparations. For example, because of the uncertainty in the hurricane's landfall position, a large number of poeple either evacuate or make preparations which later turn out to have been unnecessary because of a slight deviation of the storm's path. The second practical motivation for modeling the hurricane on a computer is to produce a realistic model that may be used as a guinea pig for testing hurricane-modification hypotheses. It is relatively inexpensive, and far safer, to test modification theories on a computer model than on real storms.

6.3.1 *A numerical model of the atmosphere*

Although we will discuss how a computer model of a hurricane might work, the concept of modeling the atmosphere on computers is quite general. Our daily forecasts are based to a large extent on a computer model of the atmosphere that is run on a daily basis at the National Meteorological Center, Camp Springs, Maryland. This model, as well as many others, is very similar in principle to the hurricane model discussed in this section.

What is the basic idea behind computer models of the atmosphere? The computer certainly does not produce a physical model of the hurricane, with miniature clouds and scaled-down winds. Instead, the computer allows us to solve complicated mathematical equations that describe the physical laws of the atmosphere. For example, the force of gravity is a physical effect that can be described completely by one equation, and a mathematical model of a stone dropped from a cliff is relatively easy to construct. In a similar way, other effects, such as the expansion of air as it is lifted, or the warming of air as it is heated, can also be expressed by mathematical equations.

6.3.2 *Equations of the atmosphere*

The mathematical forms of the equations that describe the motion, temperature, pressure, and humidity of the atmosphere are treated in many texts [for example, Horace R. Byers, *General Meteorology* (New York: McGraw-Hill Book Co., 1974)]. Although necessary for *quantitative* calculations, they are not necessary for a *qualitative* understanding of the physical effects they describe. Here, we schematically write each equation and briefly explain the meaning of each term.

The equation that predicts the change of wind over short time intervals is the *equation of motion,* which may be written:

Change of velocity / at a point in space = advection of nearby air of different velocity + acceleration due to pressure gradient forces (toward low pressure)

+ Coriolis force (due to earth's rotation, deflects moving air current to right) + friction (slows down air) **(6.1)**

Advection represents the movement of air with higher or lower velocities to the point at which the change is being computed. The pressure gradient force represents the acceleration of the air caused by horizontal differences in atmospheric pressure. A nonmeteorological example of the pressure gradient force is the acceleration of toothpaste when the pressure is increased at the end of the tube. Like the motion of the toothpaste, air is accelerated from high to low pressure whenever pressure differences exist. If there were no other forces, air would flow directly from high to low pressure.

As discussed in Section 4.5.1, the Coriolis force represents the effect of the earth's rotation on moving air, and, in the Northern Hemisphere, appears as an acceleration to the right (facing downstream) of the flow. It is primarily the Coriolis force that prevents air from flowing directly from high to low pressure. Instead, a balance between these two forces results in a wind that blows parallel to the isobars, with low pressure on the left and high pressure on the right of the moving air current.

The equation that predicts the rate of change of temperature with time is the *thermodynamic equation,* which may be expressed as:

$$\frac{\text{Change of temperature}}{\text{at a point in space}} = \begin{array}{l}\text{advection of nearby}\\ \text{air of different}\\ \text{temperature}\end{array} + \begin{array}{l}\text{cooling or warming}\\ \text{caused by expansion}\\ \text{or compression of air}\end{array}$$

$$+ \text{ change due to } \begin{array}{l}\text{(a) radiation}\\ \text{(b) condensation}\\ \quad\text{or evaporation}\\ \text{(c) conduction}\end{array} \qquad \textbf{(6.2)}$$

The processes that can cause the temperature to change at a point are easier to understand than the processes that cause the velocity to change. Advection of temperature is the transport of warmer or colder air to the point. The rapid fall of temperature as a cold front passes is primarily caused by advection. The cooling of air by expansion and the warming by compression are also easily described in terms of familiar examples. As compressed gas rushes out of an aerosol can, it expands rapidly and cools by 10 degrees or more. On the other hand, the temperature and pressure of air in a tire rise as more air is pumped in.

Finally, the addition or removal of heat energy (expressed in calories) may change the temperature at a point. The important heating and cooling processes in the atmosphere are radiation, condensation, and conduction. Short-wave radiation absorbed from the sun warms the air, while long-wave radiation emitted by the atmosphere usually cools the air. Radiative temperature changes above the ground are normally small, about 1–2°C per day. Condensation of water vapor or evaporation of liquid water can release or absorb enormous amounts of heat, as discussed earlier. Changes in temperature of 100°C per day would result in heavy rain situations if the compensating cooling effects of advection and expansion were not present. Of course, such extreme changes are never observed, so that these compensating effects must nearly offset the release of latent heat.

Two other equations describe the conservation of mass and water (vapor + liquid + solid). For *mass,* we have:

$$\frac{\text{Change of mass}}{\text{in a volume}} = \begin{array}{l}\text{net amount flowing}\\ \text{into (or out of)}\\ \text{column by}\\ \text{horizontal motions}\end{array} + \begin{array}{l}\text{net amount flowing}\\ \text{into (or out of)}\\ \text{column by vertical}\\ \text{motions}\end{array} \qquad \textbf{(6.3)}$$

Because the change of mass in an atmospheric column is usually very small, the conservation of mass equation provides a simple relationship between vertical and horizontal velocities, as discussed in Section 4.5.5 (see Figure 4.14). The equation describing the *water vapor budget* is:

$$\frac{\text{Change of water vapor}}{\text{in a volume}} = \begin{array}{l}\text{amount of vapor}\\ \text{flowing into}\\ \text{volume}\end{array} - \begin{array}{l}\text{amount of vapor}\\ \text{flowing out of}\\ \text{volume}\end{array}$$

$$+ \text{ evaporation} - \text{condensation} - \text{sublimation} \qquad \textbf{(6.4)}$$

The last two equations needed to describe completely the state of the atmosphere are the *hydrostatic equation,* which simply relates the pressure at any level to the total mass of air above that level:

$$\text{Pressure at a level} = \text{total weight per unit area of atmosphere above that level} \quad (6.5)$$

and the *equation of state,* which relates pressure, temperature, and density at a point:

$$\text{Pressure} = \text{density} \times \text{temperature} \times \text{constant} \quad (6.6)$$

We thus have six equations and six unknowns (vertical velocity, horizontal velocity, pressure, temperature, density, and water vapor). We are now ready to solve these equations for the six variables, which will then provide a detailed description of the atmosphere. Note that such principles as the conservation of angular momentum do not appear explicitly, but are implied within the preceding equations.

6.3.3 *Data for the model*

Because the solutions to the predictive equations for wind, temperature, pressure, and water vapor depend on the three-dimensional structure of the atmosphere at a given moment, we must first make analyses of these variables at many points on a three-dimensional weather map. Figure 6.9 shows a schematic three-dimensional streamline analysis of a tropical cyclone. Such an analysis is based upon measurements from weather balloons, aircraft, satellites, and surface observing systems.

Starting with a three-dimensional analysis of wind, temperature, pressure, and moisture, the prediction equations are solved at each point on the map for a short time later (for example, 10 minutes), thereby creating a weather map of the future. Then, this future analysis is used as a starting point for a new forecast 10 minutes farther in the future. The process is repeated many times, until the desired forecast time is reached. The complexity of the equations, the large number of points needed to resolve properly the detailed structure of the atmosphere, and the large number of successive short forecasts needed to arrive at a 24-hour forecast necessitate extremely fast computers. For example, a 24-hour forecast of the weather over the Northern Hemisphere made by the operational model at the National Meteorological Center requires about 300 million calculations!

6.3.4 *The computer forecasts a hurricane*

The earlier part of this chapter described the formation of hurricane Camille. To illustrate a numerical model of an atmospheric phenomenon, let us follow the development of a hypothetical hurricane from a weak disturbance as predicted by a computer model. Figure 6.10 illustrates the

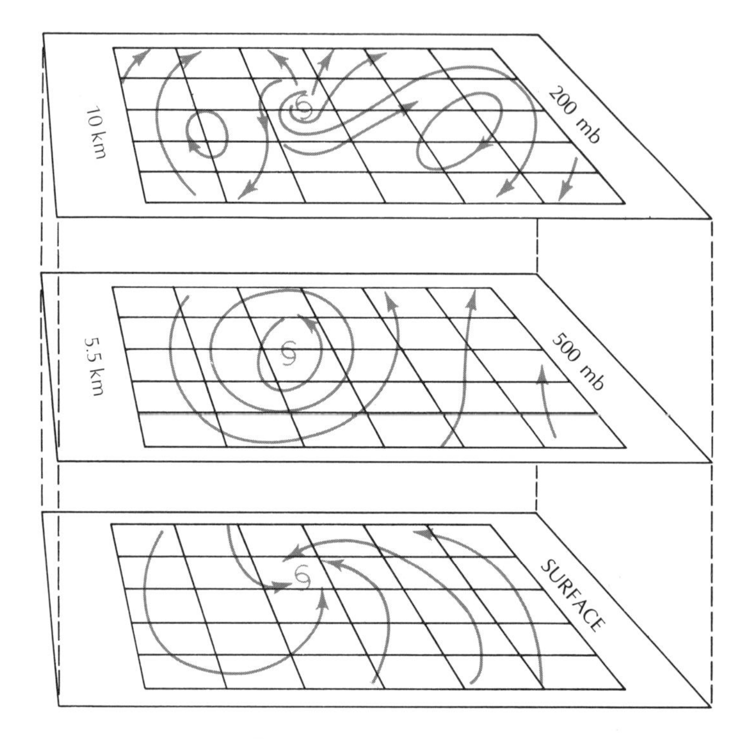

Figure 6.9 *Three-dimensional airflow in a mature hurricane.* Schematic three-dimensional streamline analysis of a hurricane. Note the low-level inflow toward the center of the storm (represented by the symbol ◎) and the upper-level outflow away from the center.

structure of a circulation associated with a weak low-pressure center. Now, let us follow the computer through several forecast cycles.

At the initial instant of time, the equation of motion, which accounts for advection, accelerations due to horizontal pressure differences, the rotation of the earth, and friction, predicts a small net acceleration of the wind inward toward the center of low pressure. This acceleration is the result of a slight predominance of the pressure gradient force over the other forces. One result of the increased inflow toward the storm center is an increase of the rotational speed, a consequence of the conservation of angular momentum discussed earlier. Thus, the equation of motion predicts a slight increase in both the inward and tangential parts of the wind. The horizontally converging air must eventually rise, so another consequence of the increased horizontal inflow is a slight increase in upward motion, which affects the thermodynamic equation in two ways. The first effect is an increased cooling due to the expansion of the air as it is forced to rise. However, because the cooler air is able to hold less water vapor, condensation occurs quickly. The latent heat of condensation then acts

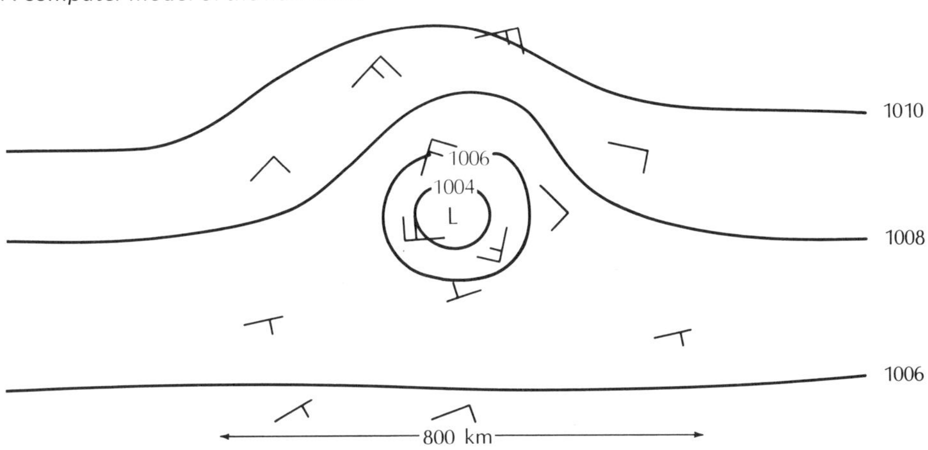

Figure 6.10 *Circulation around weak low-pressure center in tropics.* Each full barb represents 10 knots. Isobars are labeled in millibars.

to warm the ascending air. Although the net result of these two opposing effects is not immediately obvious, the latter predominates under the right conditions, so that the mean temperature of the inner region rises.

The water vapor equation predicts a loss of moisture by condensation, but at the same time, predicts an increase from inward horizontal advection and from evaporation from the sea, which enables the convection and condensation to continue.

As the inner region of the developing storm warms slightly and expands upward, an outflow of mass away from the center at high levels is produced. The upper-level outflow exceeds slightly the low-level inflow, and the surface pressure drops a small amount.

All of these changes predicted in the 10-minute forecast made by machine are small in magnitude; for example, the wind might increase by 0.1 mi/h, the mean temperature might increase by 0.05°C, and the surface pressure might drop by 0.1 millibar. But, then, we have a new, slightly stronger storm, and the new values of wind, temperature, and moisture may be used to generate another 10-minute forecast.

A repetition of the sequence described in the preceding paragraph over many 10-minute intervals gradually describes the intensification of the model storm: A slight predominance of the inward-directed pressure gradient force continues to accelerate the air inward. The increased inflow also produces stronger vertical velocities and increased condensation. Water vapor fuel for the convection is supplied at faster rates as the wind speed increases. The thermodynamic equation predicts warmer temperatures as a result of the condensation, and the warmer temperatures lead to still further decreases in surface pressure.

The intensification process does not continue indefinitely, however, because the balance of forces gradually changes. For example, as the winds increase, so does the effect of friction, which tends to decrease the wind speed. The increased upward motion, besides increasing the convection, also increases the cooling due to expansion. Both the frictional and expansion cooling effects tend to weaken the circulation. At some point, these two weakening effects may cancel, or even slightly exceed, the forces of intensification. At this point, the storm reaches a steady state in which the generation and the dissipation effects are balanced. Such a mature state is shown in Figure 6.11,

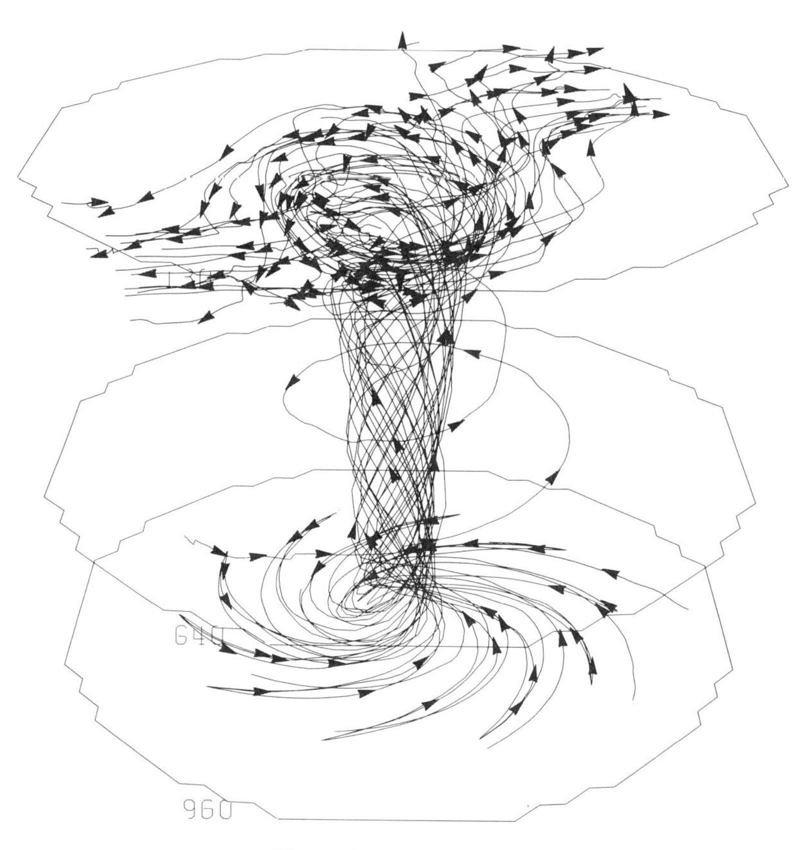

Figure 6.11 *Three-dimensional trajectories in model hurricane.*

which illustrates the three-dimensional trajectories of air parcels in a model hurricane. A comparison with Figures 6.6 and 6.9 shows that the storm model accurately reproduces many aspects of the mature hurricane. In particular, the model vortex consists of strong cyclonic low-level inflow, upward motion near the center, and outflow of air in the upper atmosphere. A view of the rainfall rates predicted by the model (Figure 6.12) shows intense rainfall near the center, and even some spiral bands which resemble the spiral bands of rainfall that are visible in radar and satellite pictures of real storms (see Plate 3, Figure 6.4 and 6.13).

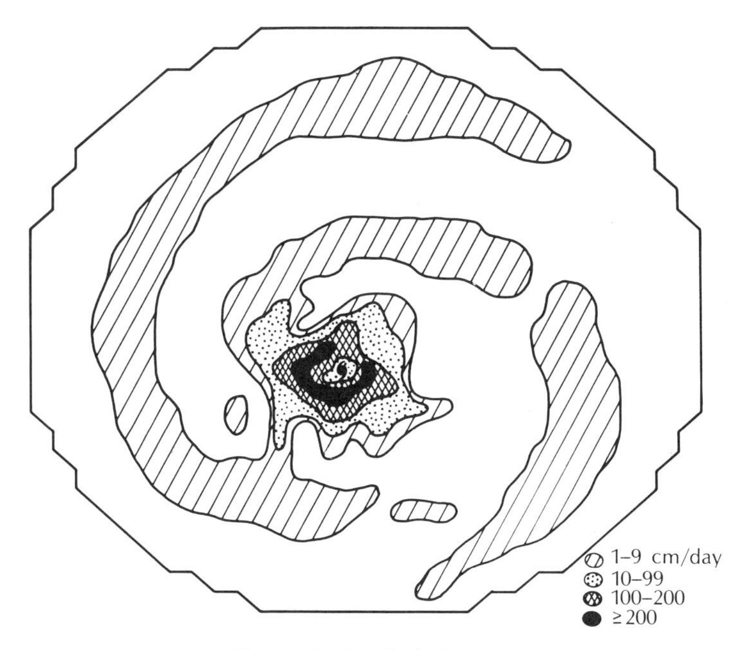

Figure 6.12 *Spiral bands of rainfall computed from model hurricane.*

Several physical effects may contribute to the model hurricane's downfall. Movement of the storm over land or colder water may reduce evaporation from the sea, so that the storm starves from a lack of water vapor, or, cool, dry air from middle latitudes may be drawn into the circulation, destroying the core of warm, light air. In spite of its enormous amounts of energy, the hurricane is quite fragile, and may weaken quickly under unfavorable conditions. All of these effects may be represented in a mathematical model of the hurricane.

6.3.5 *A word of caution*

Because the preceding example of a computer model appears quite realistic, it is easy to be overimpressed by the capability of computer models in general to predict the weather. As everyone knows, however, weather forecasts are still far from perfect, even with the greatly increased use of computers, because computer models are only as good as the physical models upon which they are based. Because the number of points in a model is limited by the finite speed of computers, it is not possible to resolve simultaneously all scales of motion. Thus, the hurricane model cannot predict both small cumulus clouds and the larger-scale hurricane vortex. However, the net effect of the cumulus clouds must be accounted for in some simple and approximate way. Such approximation produces errors. The uncertainties in

Figure 6.13 *Radar picture of hurricane rainbands in Hurricane Betsy, 1965 (courtesy of Peter Black, National Hurricane Research Laboratory).*

initial data fields also lead to errors in the forecast. Thus, although many models have demonstrated a skill in predicting behavior of the atmosphere, they are not completely dependable.

6.4 Thunderstorms

It is a warm, humid, but clear July morning in Miami, Florida. Soon after sunrise, small puffy cumulus clouds appear over the south Florida peninsula. By noon, purple-black thunderheads drench the Everglades with cooling rainfall. It is a sunny, pleasant morning in the Rockies. By early afternoon, the mountain peaks are shrouded in clouds, and thunderstorms begin to drift eastward over the plains, persisting long after sunset. In the Midwest, after four days of a sultry heat wave, rumbles of thunder herald the approach of a squall line from the west. Shortly after the line of thunderstorms moves through the area, a cold front with brisk northwesterly winds brings in drier air and a welcome drop in temperature.

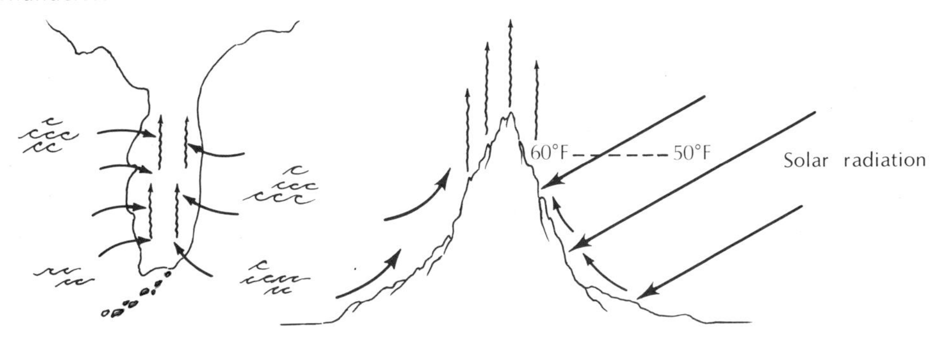

(a) Daytime heating over Florida produces lifting.

(b) Air over mountain peaks warms faster than air at same level over plains.

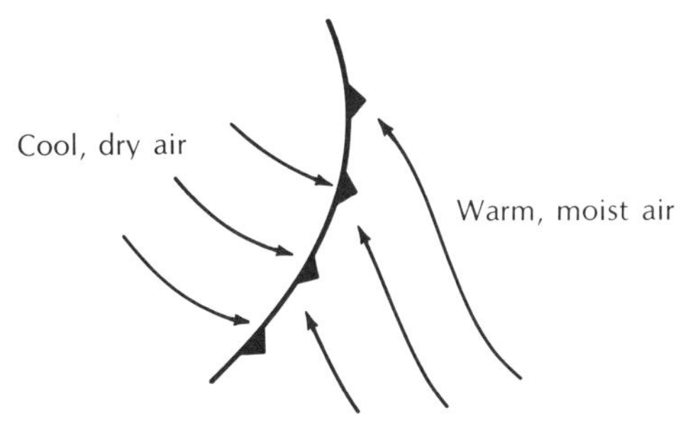

(c) Convergence of low level air in vicinity of cold front.

Figure 6.14 *Three lifting mechanisms that can produce thunderstorms.*

6.4.1 *Lifting produces thunderstorms*

The preceding three examples of thunderstorms (illustrated schematically in Figure 6.14) have one thing in common—a rapid lifting of moist, warm, low-level air to the high troposphere. The lifting in the first two examples was caused by differential (uneven) heating on a small scale. Just as the differential heating on a global scale produces the general circulation, horizontal differences in heating on the small scale produce local circulations. As the sun warms the Florida land faster than the surrounding water, a mean lifting of the air over the land occurs. Within this lifted air, smaller-scale buoyant parcels of air are created that are warmer than their environment, and therefore rise (Figure 6.14a). In the Rockies, the sun heats the air in contact with the mountain peaks more than air at the same level over the plains, again producing buoyancy and lifting (Figure 6.14b). In the Midwest example, the lifting is associated with horizontal convergence of low-level moist air in the vicinity of a cold front (Figure 6.14c). Under such conditions, thunderstorms tend to align themselves in squall lines (squall lines will be discussed later in this chapter). It is important to note that here the processes which produce the lifting occur on a larger scale than the thunderstorm cells themselves. As discussed in the hurricane section, large-scale lifting of a warm, moist air mass produces small-scale convection; i.e., local "hot spots" develop on the ground and generate bubbles of warmer, lighter air, which rise through their denser environment and produce the actual thunderstorm cells.

6.4.2 *Survival of the fittest*

Most hot bubbles of air never survive to become visible clouds or to develop into a mature thunderstorm. As with newborn insects, nature produces far more infant thermals than ever reach the adult stage. Normally, the rate of expansion cooling in the rising parcel (10°C/km) exceeds the rate of decrease of environmental temperature, so that parcels soon become cooler and heavier than the air around them and return downward, as illustrated by parcel *A* in Figure 6.15. However, a few large and unusually warm (or moist) bubbles may rise high enough to reach the condensation level, the level at which expansion cooling has lowered the temperature of the cloud to its dew point. This situation is illustrated by parcel *B* in Figure 6.15. With the onset of condensation, the chances for the growing cloud's continued rise are increased greatly, because condensation adds heat to the parcel and helps negate the

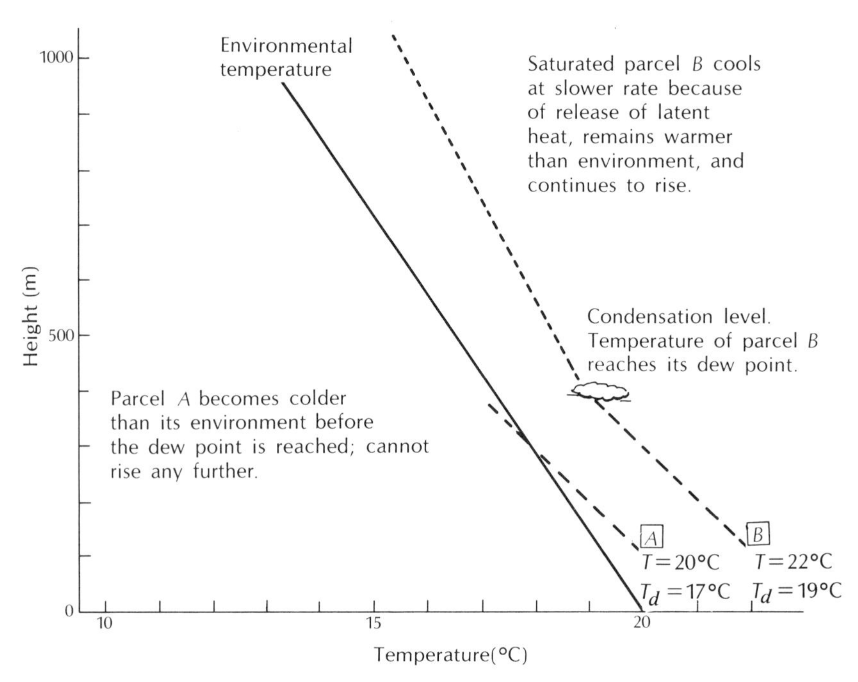

Figure 6.15 *Temperature variation with height of a parcel of air reaching condensation level.*

expansion-cooling effect. The net result of these two processes is a cooling rate in the parcel (neglecting *entrainment,* which is the mixing of the cloud with its environment) of about 6°C/km instead of 10°C/km. It is therefore more likely, although by no means certain, that the new cloud will remain slightly warmer than its environment and continue to rise.

Figure 6.16 *Four stages of thunderstorm development.*

A typical sequence of events leading to isolated thunderstorm cells on a hot summer day is illustrated in Figure 6.16. The isolated cumulus stage, as shown in the Earth Resources Technology Satellite (ERTS) photograph of cumulus clouds over Florida (Plate 4), consists of many small, white cumulus clouds with low tops. As we know from watching the hundreds of fleeting cumulus clouds form and die on a summer day, very few thunderstorms are actually produced from the many little clouds. The fledgling cloud faces a hostile environment of dry, cool air. Mixing of this air through the sides and top of the cloud evaporates water droplets and cools the parcel. Thus, most cumulus clouds decay a short time after they become visible. However, in their death by evaporation, the clouds enrich the humidity of the air, so that future bubbles find a slightly more favorable environment and are able to last a little longer and grow a little taller. Time-lapse photographs show a series of rising bubbles or towers of cumulus clouds, each one growing bigger than its predecessor, until several clouds merge to form a cloud that is large enough to grow into the mature thunderstorm stage. In this larger cloud, the updrafts near the center are protected from the cooling effects of entrainment by the outer portion of the cloud.

Occasionally, when the updraft of air in the growing thunderstorm is very rapid and the air above the growing cumulus cloud top is moist, a very thin veil of clouds forms above the visible top of the cumulus cloud (Figure 6.17). This thin web of clouds, which is frequently comprised of ice crystals, is called the *pileus,* or *cap cloud,* and is caused by the rapid expansion and cooling above the rising cumulonimbus.

The mature thunderstorm is an impressive sight—a dark mass of clouds towering 5–10 miles over the surface and capped by an anvil of brilliant white cirrus clouds. This familiar anvil (Plate 5) is created when the rising air currents reach the tropopause and encounter the stable stratosphere. The stra-

Figure 6.17 *Pileus, or cap clouds, forming over three growing cumulus clouds* (*photo by Ronald Holle*).

tosphere is so stable that the rising air parcels quickly become cooler than their environment and lose their buoyancy. The updrafts then spread out horizontally, fanning the ice crystals into the characteristic anvil pattern. About 15 minutes to a half hour after the mature stage, the isolated thunderstorm cell begins to die, and the large number of falling raindrops drag the air downward, producing downdrafts instead of updrafts. The rain also cools the surface underneath the storm, cutting off its supply of warm, buoyant air. As the cool downdraft hits the ground and spreads out horizontally, it forms the *gust front,* a miniature cold front separating the cool downdraft air from the warmer environmental air. This front can propagate several miles away from the thunderstorm that generates it, producing cool, gusty winds as it passes. An example of a gust front is shown in Plate 6.

Rainbows are most common during the dissipation stage of the thunderstorm (Plate 7), when the sun is able to shine through the thinning clouds onto the still-falling raindrops. As discussed in Chapter 9, the rainbow is formed when parallel light rays from the sun are bent (refracted) upon entering the raindrop, reflect off the inner rear surface of the drop, and are bent again as they emerge from the forward side of the drop.

6.4.3 *Lightning and thunder*

Lightning and thunder occur during the mature stage of the thunderstorm (Figure 6.18). Lightning, which is simply a giant electric spark between regions of oppositely charged particles, is produced by the separation of charges within the growing thunderstorm. Electrons (negative charges) are

Figure 6.18 *Lightning at night (photo by Robert McAlister).*

stripped off water droplets and ice crystals and accumulate near the base of the cloud. Because the earth's surface below the cloud is induced to be positively charged by the negatively charged cloud base, a potential gradient of about 1000 volts per meter is created between the cloud and the ground. When this potential becomes large enough, an electrical discharge between the cloud and the ground occurs. In this spark, electrons flow from cloud to ground and reduce the voltage.

High-speed photography has provided us with a detailed picture of the rapid sequence of events in a typical lightning stroke. An initial, nearly invisible, discharge of electrons starts downward from the cloud toward the earth. As this negatively charged *leader* approaches the ground, upward-moving positive charges rush to meet it, thereby establishing a conducting path of charged particles. A much larger "return stroke" then surges upward from the ground to the cloud. All of the preceding events happen in about 50 millionths of a second, and so to our eyes appear as a single flash. It is interesting that most people visualize the visible lightning stroke as propagating from the cloud to the ground, when actually the reverse is true—at least for the return stroke, which is the brightest flash.

The rapid heating of the air in the channel of the lightning stroke produces a violent expansion of the air. This expansion initiates a sound wave, which propagates outward at a speed of about 600 mi/h. Because the speed of light is much greater (671 million mi/h), we see the lightning stroke almost the instant it occurs, but the slower-moving sound wave takes about five seconds to travel a mile. Thus, by counting the number of seconds between the sighting of lightning and the arrival of the thunder, a good estimate of the distance from the lightning may be obtained.

Because of its threat to life and property, meteorologists are seeking methods of suppressing lightning, although it is not known whether such a suppression would modify other aspects of the weather. One promising technique consists of "seeding" thunderstorm clouds with aluminum-coated chaff fibers. These fibers should conduct electricity between oppositely charged regions of the thunderstorm and prevent the buildup of the enormous electrical potential that precedes the lightning discharge. Instead of a sudden lightning stroke, the more or less continuous weak discharge might produce a glow similar to St. Elmo's fire, a glowing halo seen around aircraft wings, ship masts, and power lines during electrical storms.

6.4.4 *Hail*

One of the most damaging byproducts of severe thunderstorms is *hail,* stones of ice which may, under extreme conditions, reach the ground the size of baseballs. Although hail causes few injuries, economic losses in the United States are about $300 million annually. Hailstones are formed when a small embryonic ice pellet remains for a sufficient time in a region of the thunderstorm in which supercooled water (liquid water below freezing temperatures) is present. Under such a favorable condition, the hailstone grows as the water droplets strike the ice pellet, freezing on contact. To grow to a size large enough to survive the final fall through the warm air to the ground, the hailstone must remain in the presence of abundant supercooled water for at least several minutes. Also, in order to hold the growing stone above the freezing level, very strong updrafts are required, because the typical hailstone's terminal velocity (velocity a hailstone would fall in still air) is 10–40 mi/h. The updrafts in most thunderstorm cells are somewhat lower than these high velocities, which is one reason why hailstones that reach the ground are relatively rare. But in the most severe thunderstorms, updraft velocities of sufficient magnitude to keep the stones suspended above the freezing level for long periods of time are possible. Then, the hailstones grow until they finally become too heavy or until they are tossed into a part of the cloud where the updraft is weak or absent entirely. When the latter happens, the hailstones fall rapidly to the ground.

Occasionally, a hailstone may get carried up and down past the freezing level several times in successive up- and downdrafts before finally reaching the ground. The partial melting that occurs in the warm air, and differences in the cloud droplet size and concentration above the freezing level, produce a stone with laminations of ice of varying textures and appearances.

Because of the enormous crop damage caused by hailstones, considerable research has been directed toward reducing their size. Results have been promising enough that the Soviet Union has established an operational hail suppression program in the Caucasus Mountains. The Soviet program consists of firing cloud-seeding agents, such as silver iodide crystals, into growing cumulonimbus clouds. These seeding crystals encourage the formation of many small ice crystals rather than a few large ones. The hailstones that eventually form, although more numerous, are too small to survive the fall to the ground through the warm air, and therefore arrive as beneficial raindrops rather than destructive hailstones.

In spite of the success in hail suppression claimed by the Soviet Union, controversy over the possibility of significantly reducing hail continues. Experiments in the United States with cloud seeding have produced insignificant results. As with any weather modification program, the main difficulty is that there is no way of knowing for sure what would have happened without the seeding. To circumvent this obstacle, statistical techniques must be used; that is, the behavior of a large number of unseeded (control) thunderstorms must be compared with a large number of seeded storms.

An additional problem with weather modification programs is political rather than meteorological. For example, cloud seeding may reduce hail (a beneficial result for almost everybody), but may also increase rain, perhaps producing local floods. Such a controversy arose in the Black Hills of South Dakota in June of 1972, when a freak thunderstorm produced over a foot of rain. Unfortunately, this cloud had been seeded in an experiment. While most of the evidence suggests that very unusual local wind-flow patterns, and not the seeding, caused the floods, some people blamed the flood on the seeding. Thus, even if thunderstorm modification by cloud seeding became a proven fact, important political problems would remain concerning the control and application of this weather modification tool.

6.4.5 *Squall lines and severe thunderstorms*

The types of thunderstorms considered so far in this chapter have been more or less isolated, created by local differential heating within a horizontally homogeneous air mass. In many parts of the United States, however, thunderstorms frequently occur in a highly organized pattern, with individual cells strung out in a line extending hundreds of miles. These *squall lines* are generated and maintained by especially favorable large-scale weather patterns, and are often forerunners of cold fronts. Unlike the isolated storms, the squall-line thunderstorms move rapidly eastward and may persist for several hours. The most severe thunderstorms, including those that produce tornadoes, are usually embedded in a squall line.

The basic large-scale weather patterns that produce squall lines are fairly well known. Because some of the most severe squall lines in the world occur on the Great Plains, we will discuss the circulation pattern that is favorable to squall line formation in this region.

The typical weather situation conducive to squall line and severe thunderstorm formation occurs often in the spring months over Texas and Oklahoma. With a high-pressure center (anticyclone) located over the southeastern Atlantic states, and a cold front with its associated trough of low pressure approaching from the west, extremely warm, moist air is carried northward at low levels into the Great Plains by the clockwise circulation around the anticyclone. At the top of this layer of warm, humid air (around 2 kilometers), an inversion is usually present, as shown in Figure 6.19. (An *inversion* is a layer of the atmosphere in which the temperature rises instead of falls with height, which is an "inversion" of the normal situation.) At higher levels above the inversion, westerly flow brings cool, dry air into the region and over the Gulf air.

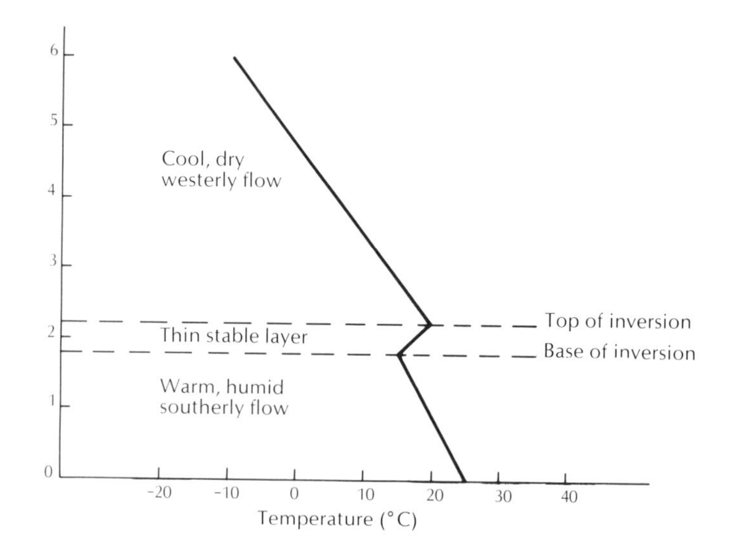

Figure 6.19 *Typical vertical temperature structure prior to severe thunderstorms and tornadoes.*

The temperature in this upper-level flow decreases rapidly with height, which means that the air is unstable for saturated air parcels. However, the thin layer between the base and the top of the inversion (see Figure 6.19) is very stable—the temperature increases with height. This stable layer acts as a "lid" to the convective parcels in the lower layer, preventing them from rising into the unstable layer above.

As long as the inversion is maintained, or the convective parcels of air rising from the ground fail to reach saturation, only relatively shallow convection is produced. However, as the ground becomes hotter under the influence of strong solar heating, or as the inversion weakens, some of the warm bubbles of humid air may penetrate the inversion, reach saturation, and extend into the unstable air aloft, thereby triggering explosive thunderstorms.

In the typical weather pattern, the lifting of the low-level air may be produced by at least two conditions. First, as discussed in Section 4.5.5, the dynamical laws of atmospheric motion require a large-scale upward motion ahead of the moving troughs of low pressure. This requirement explains why bad weather generally occurs in the vicinity of fronts, which always lie in troughs of low pressure. Second, radiative heating of the ground by the sun, which is quite strong in spring, heats the air next to the ground and produces rising thermals of hot air, which may eventually become buoyant enough to break through the inversion. These two conditions favor the development of squall lines, which form parallel to and ahead of the advancing cold front. These lines then move eastward at 10–40 mi/h, frequently maintaining their identity for several hours. The movement of the cells and the strong vertical wind shear (increase of wind with height) allow the raindrops to fall out of one side of the thunderstorm instead of uniformly throughout the entire cell. Thus, the updraft may continue in another part of the cloud, keeping the storm alive.

Within the squall line, individual thunderstorm cells grow, nurtured by the supply of water vapor and warm air accelerating toward the squall line. Under the most extreme cases, a "super" rotating thunderstorm cell may form. These rotating cells are likely to spawn tornadoes, which will be discussed in the next section.

Although idealized histories of squall line development are useful conceptual models, it is perhaps more interesting to consider a real occurrence. For this purpose, we consider the squall line of June 8–9, 1966, which crossed most of Oklahoma. A special observational network, consisting of 11 rawinsonde stations (see Appendix 1 for a discussion of rawinsondes) in southwestern Oklahoma plus many surface stations manned by volunteers provided a reasonably complete picture of the life cycle of the intense squall line.

Figure 6.20 shows the large-scale weather map at 12 noon CST on June 8, 1966. An advancing cold front separates cooler, drier air to the west and northwest from hot, moist air to the southeast. Notice the wind shift from southeasterly winds flowing around the high ahead of the front to westerly winds behind the front. The converging of these surface winds provides a large-scale lifting that was favorable for thunderstorm formation.

Figure 6.21 shows the mesoscale surface weather map 6½ hours later, at 6:30 P.M. CST. This mesoscale map was constructed using the special

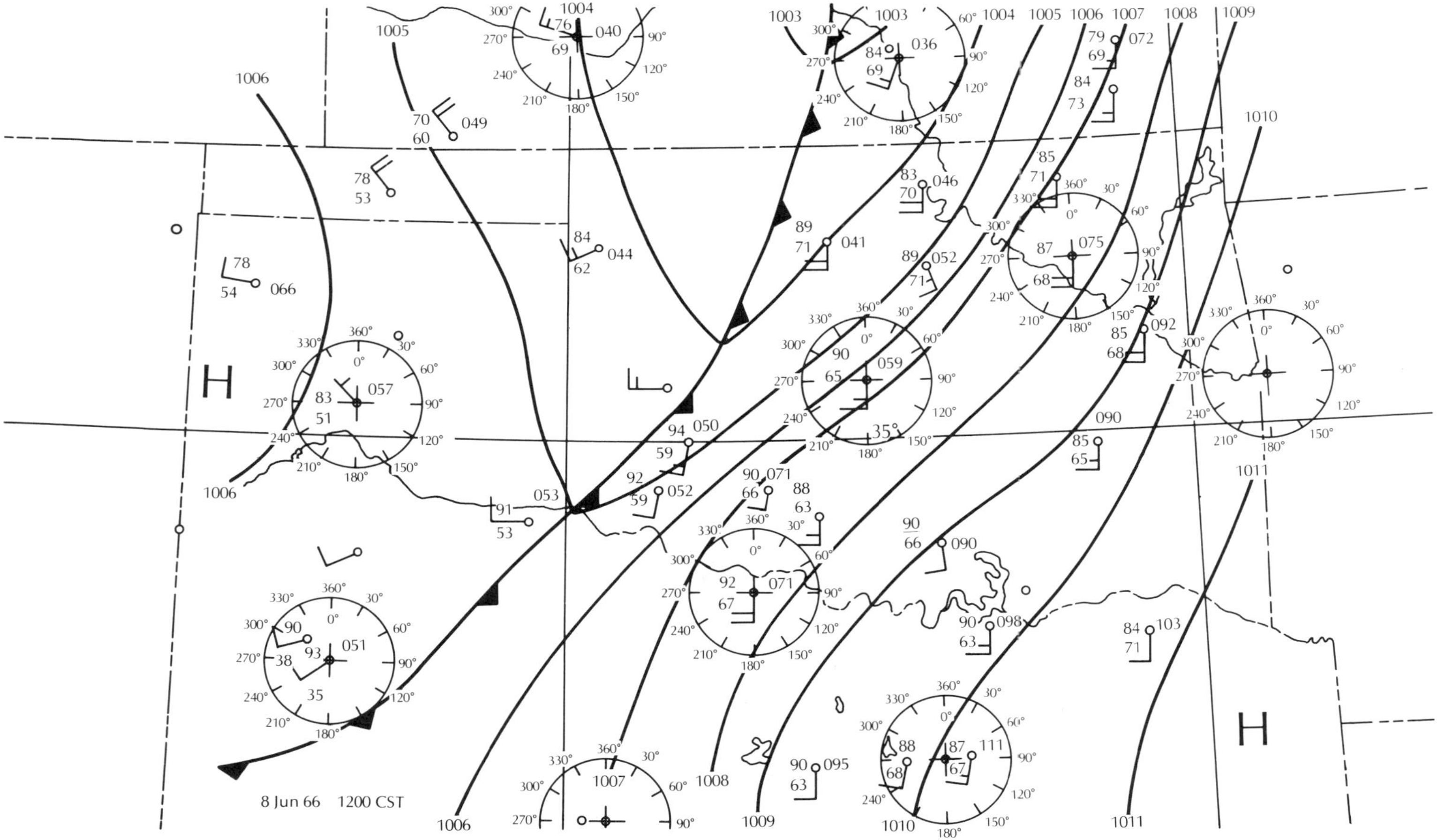

Figure 6.20 *Surface weather map (large-scale, pre-squall line) for 1200 CST June 8, 1966.*

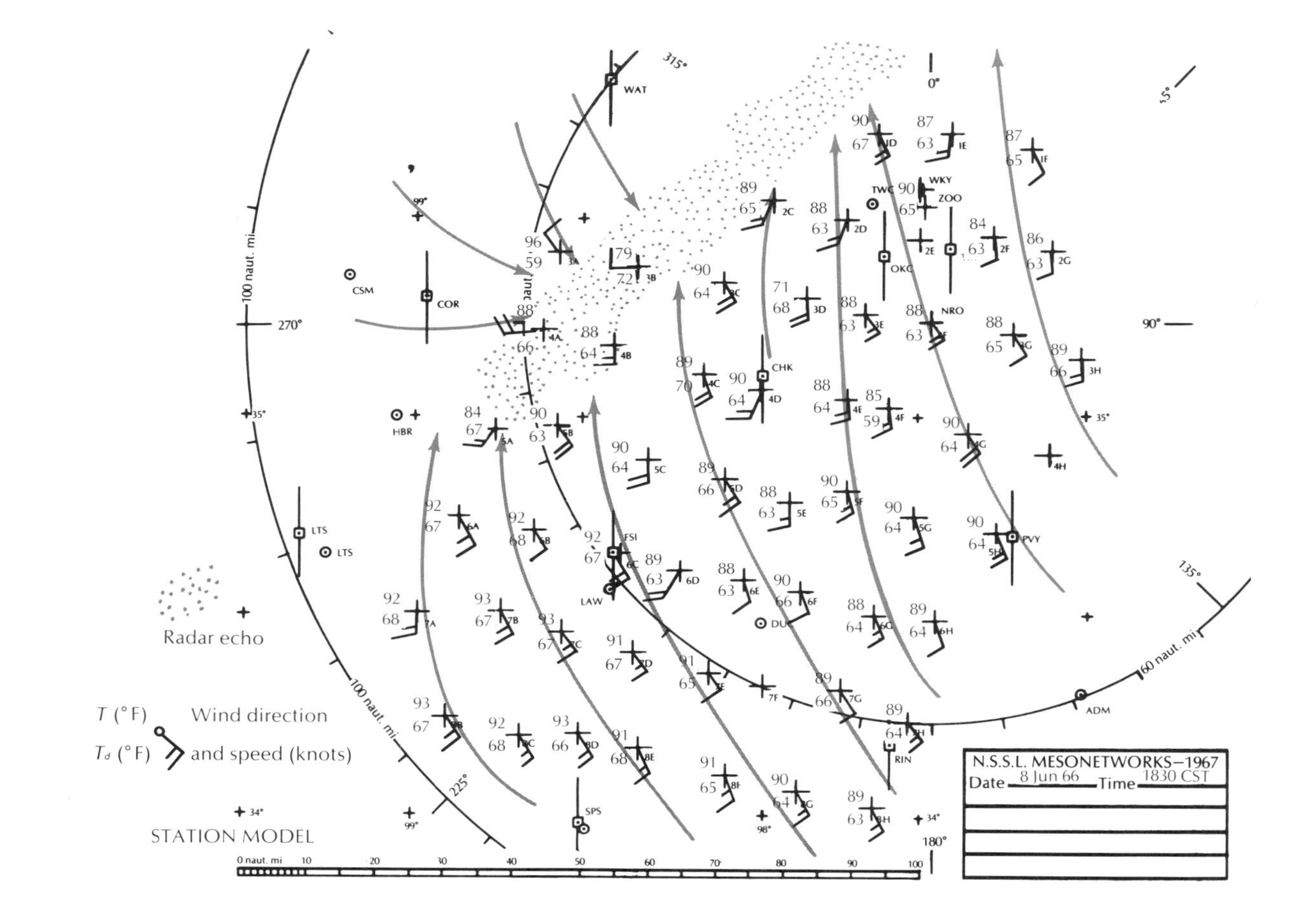

Figure 6.21 *Mesoscale surface analysis showing inflow into squall line and radar echoes (1830 CST June 8, 1966).*

data network and shows details in the flow that would escape the coarse network of the regularly reporting stations. A line of thunderstorms, about 30 miles wide and 200 miles long, has formed ahead of the cold front. The position of the precipitation cells, as revealed by radar, are superimposed on Figure 6.21. The smooth, larger-scale wind flow of the earlier map has been perturbed by local circulations associated with the squall line. In general, air is flowing into the squall line at low levels and away from the line at higher levels.

At the ground, the passage of the squall line is marked by an abrupt change in wind direction from southeast to northwest and a rapid drop in temperature. The surface pressure also varies strongly across the squall line, rising to a maximum directly under the squall line, then falling slightly as the squall passes. Behind the line of thunderstorms, the air becomes drier, the winds become northwesterly, and the pressure begins a slow rise as the next high-pressure system drifts eastward.

As the thunderstorms within the squall line reach maturity, the combined downdrafts reach the ground, and cool, moist air spreads out horizontally, forming the gust front. Figure 6.22 shows the beginnings of the gust front associated with the Oklahoma squall line at 8 P.M. CST. Notice that brisk northerly winds at the surface are now outrunning the line of thunderstorm cells, marked by the precipitation echoes. An hour and a half later, at 9:30 P.M. CST, the gust front is well developed (Figure 6.23) and is located about 25 miles ahead of the line of precipitation echoes. Plate 6 shows a photograph of the early stage of a gust front.

6.5 Tornadoes

Without a doubt, the most terrifying weather phenomenon is the tornado, a twisting vortex of winds that can exceed 300 mi/h. These winds rotate about a center of pressure so low that buildings explode outward if the tornado passes overhead. This region of low pressure causes the air to expand and cool below its dew point, producing the ominous twisting funnel that identifies the tornado (Figure 6.1). No man-made structure is capable of withstanding a direct hit by a strong tornado; railroad cars have been lifted from their tracks and carried hundreds of yards away, and whole brick houses have been demolished, the debris swept cleanly upward into the sky. Figure 6.24 shows an aerial view of the nearly complete devastation in Crystal Lake, Illinois wreaked by the Palm Sunday tornadoes of April 11, 1965. In rather bizarre instances, small tornadoes cause freakish events such as the stripping of feathers off live chickens without killing them, or the lifting of roofs and walls from houses without disturbing their contents.

Tornadoes are extremely variable, and because of their small size (usually less than a mile in diameter) and short lifetime (a few minutes), are impossible to forecast precisely. However, the large-scale weather conditions that are favorable to severe thunderstorms and tornado formation are well known, as discussed in the preceding section. When these conditions are expected, the National Weather Service issues watches for the areas affected. Fortunately, even under conditions favorable for tornado formation, any one geographical point is unlikely to be hit because of the tornado's small diameter.

Plate 9 (above) *Pollution-induced fog near Tyrone, Pennsylvania (photo by Alistair B. Fraser).*

Plate 10a (below) *Cumulus cloud over power station (photo by Richard Anthes).*

Plate 10b *Cumulus cloud over power station (photo by Frank Schiermeier).*

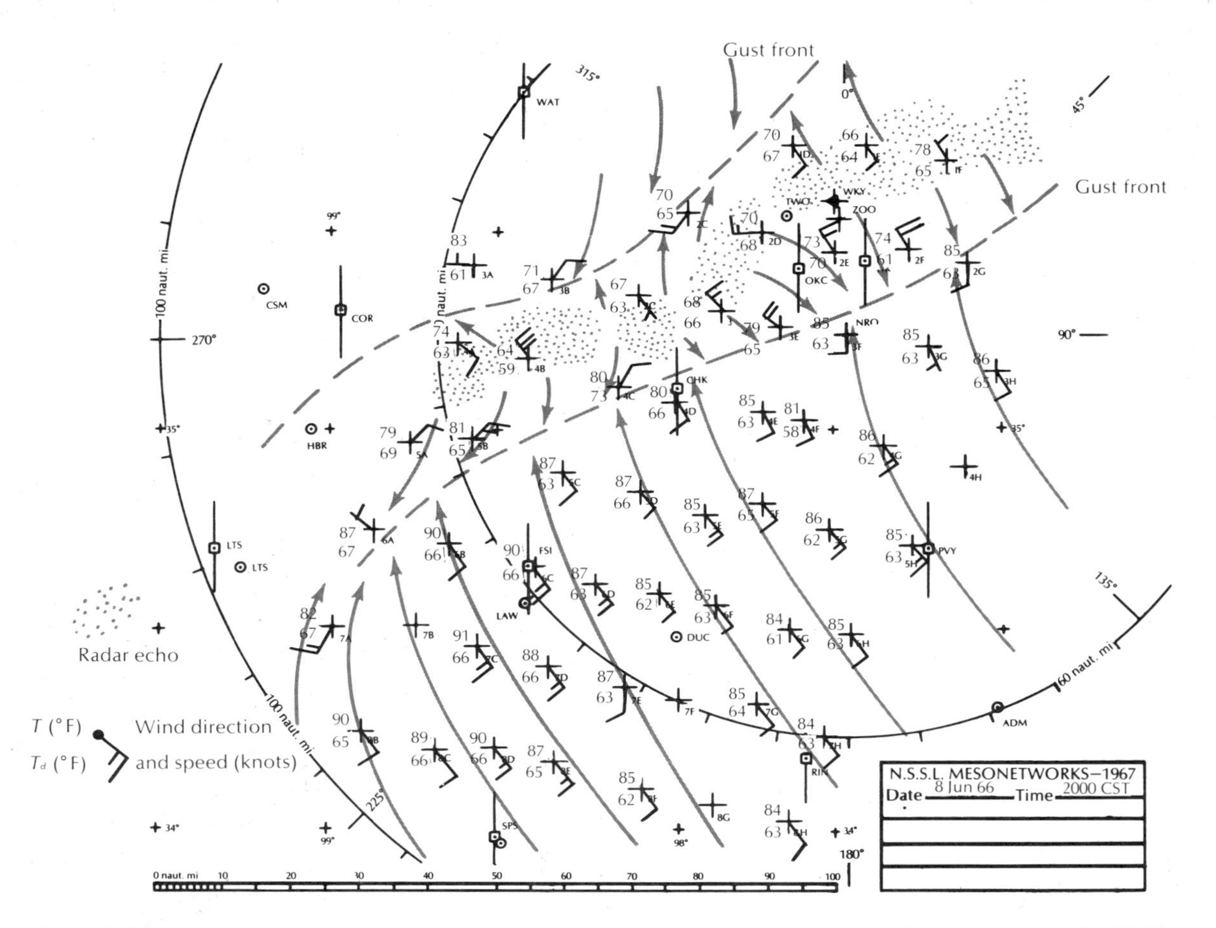

Figure 6.22 *Mesoscale surface analysis showing early stage of gust front (2000 CST June 8, 1966).*

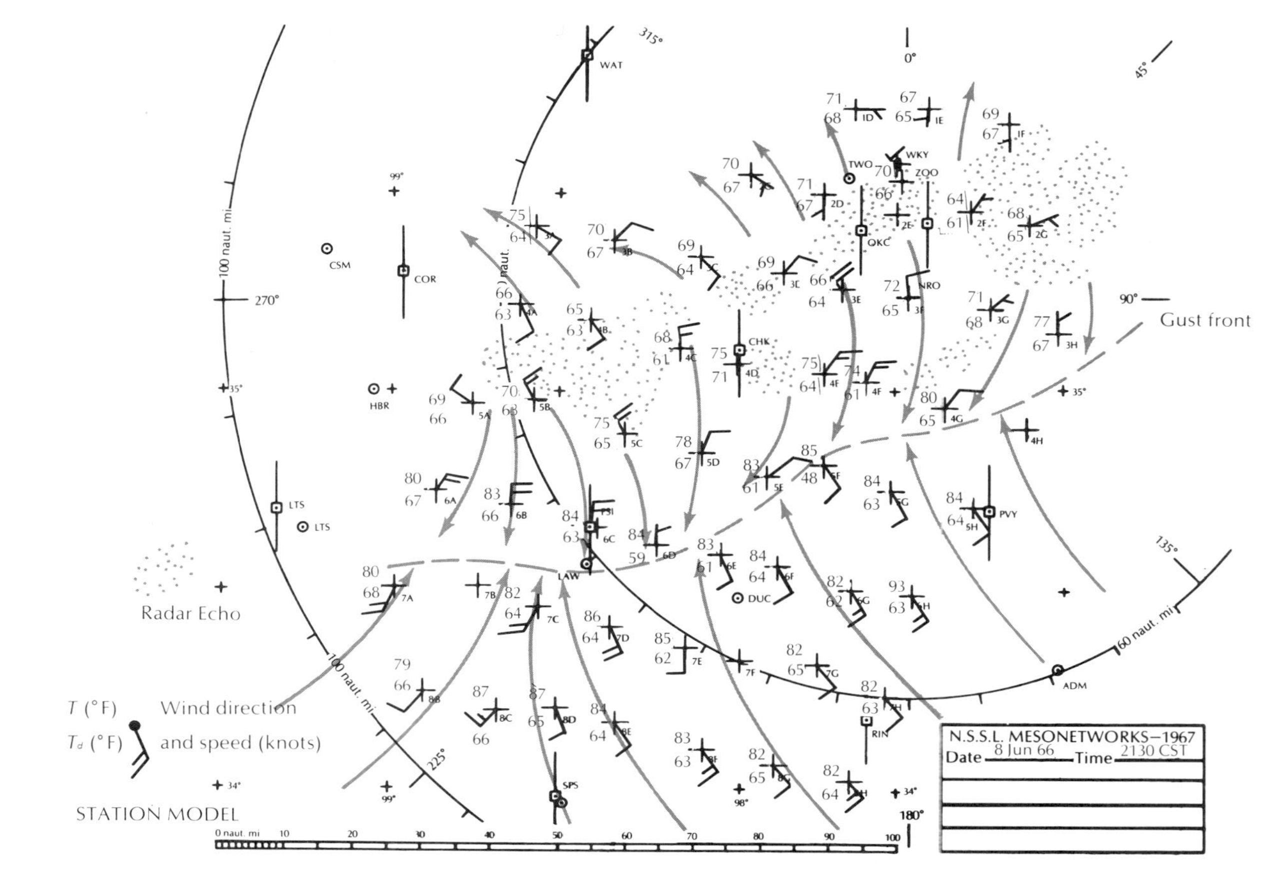

Figure 6.23 *Mesoscale surface analysis showing well-developed gust front (2130 CST June 8, 1966).*

Figure 6.24 *Aerial view of damage in Crystal Lake, Illinois by Palm Sunday tornadoes (courtesy of Professor T. Fujita).*

Although the details of tornado formation are not fully understood, the conservation of angular momentum plays a fundamental role, as it does in hurricane formation. A violent updraft caused by a severe thunderstorm and a concomitant fall of surface pressure would initiate a rapid convergence of low-level air toward the center of the updraft. Because the air in the vicinity of severe thunderstorms already possesses considerable wind speeds (e.g., 40 mi/h), it is not hard to see how tornadic velocities could be rapidly generated. For example, a 40-mi/h tangential wind at a radius of 5 miles from a low-pressure center could achieve 200 mi/h if it were accelerated inward to a radius of 1 mile while conserving its angular momentum. Indeed, many tornadoes appear to be spawned by a parent cyclone, a rotating severe thunderstorm several miles in diameter.

Evidence of the extraordinarily rapid updraft in tornado-producing thunderstorms can be seen in satellite photographs. Figures 6.25 and 6.26 show two views of a line of thunderstorms occurring over the southern United States at 12 noon CST May 27, 1973. Figure 6.25 shows the appearance of the clouds in the visible wavelengths. This is the view we would see with our eyes or photograph with ordinary cameras. In Figure 6.25, the Gulf of Mexico appears dark against the lighter continent. Three main lines of clouds appear to be streaking northeastward from points along the Gulf Coast states. From this picture, we might think that rainfall is occurring more or less uniformly along these bands.

Figure 6.25 *Visible satellite photograph of line of thunderstorms along Gulf Coast (courtesy of NASA).*

However, in Figure 6.26, which is an infrared view of the same area, we see a considerably different and more revealing picture. In contrast to the visible picture, in the infrared view the cool Gulf of Mexico appears lighter than the warm continent. The infrared view of the clouds along the Gulf Coast shows considerably more detail than the visible. Because low and high clouds have nearly the same albedo (reflectivity), their appearance in the visible picture is nearly uniform. But because the temperature of low and high clouds varies greatly (40°C), the infrared picture shows a great difference in the clouds. In particular, four nearly circular cumulonimbus clouds stand out in the line that parallels the coast. In contrast, the other two lines are mainly comprised of lower and warmer clouds, and hence appear darker. Thus, the infrared view gives us a much better idea where the tall thunderstorms (and greatest likelihood of severe weather) are occurring.

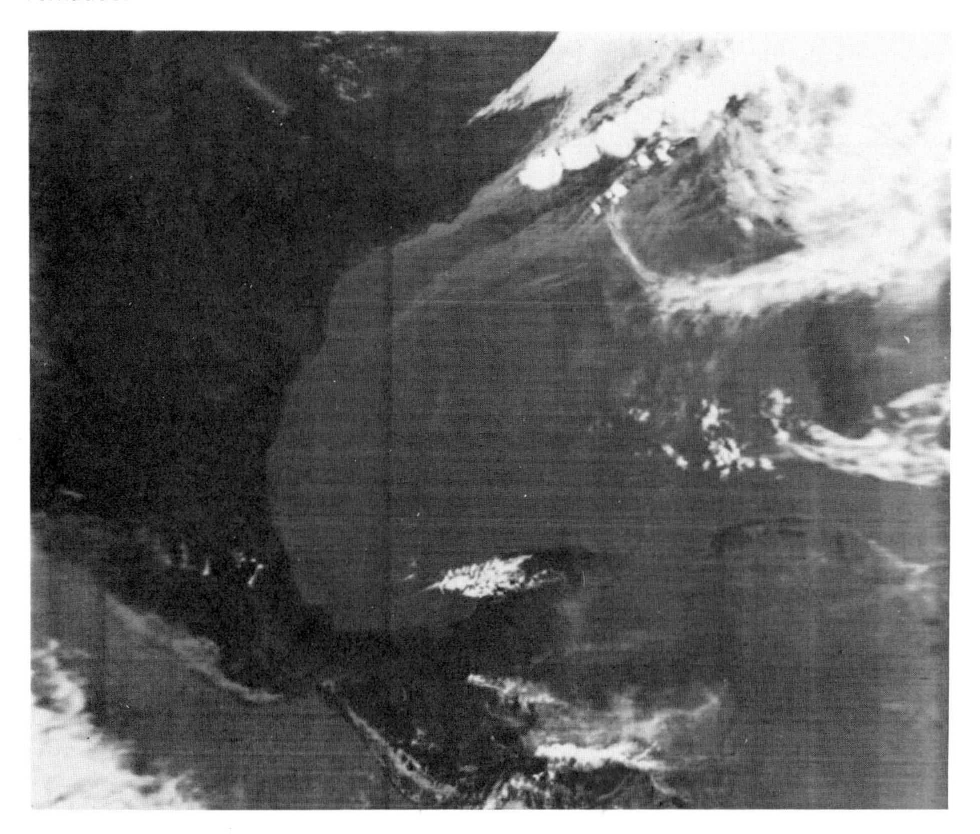

Figure 6.26 *Infrared satellite photograph of line of thunderstorms depicted in Figure 6.25 (courtesy of NASA).*

In this example, the thunderstorm cells over Louisiana produced several tornadoes. The strong vertical currents associated with severe thunderstorms carry vast amounts of water vapor into the upper tropopause, where it spreads out horizontally and moves with the winds at this level. From the satellite's elevation, the thunderstorm resembles a smoking chimney with cloud droplets and ice crystals streaking downstream. Such a picture is a good indication of an extremely rapid updraft and suggests the likelihood of violent weather below.

The electrical properties of tornadoes are interesting, even though they are probably results rather than causes of the tornado funnel. In some cases, lightning flickers eerily inside the funnel itself. Tornadoes also emit electromagnetic radiation which may be picked up on ordinary television receivers, as discussed in Section 5.4.

The possibility of tornado modification has attracted considerable interest, but any theories are speculative at this time. Because of the large rate of energy production in a severe thunderstorm (roughly equal to the total power-generating capacity of the United States in 1970), any modification will have to be directed at triggering natural changes in the circulations of these storms by application of relatively small amounts of energy. Perhaps cloud seeding could beneficially alter the tornado-producing clouds, or increase the growth of nearby clouds, thus providing competition for the finite amount of available water vapor. On the other hand, if a possibility for decreasing the frequency or intensity of tornadoes exists, a possibility for an increase must also be admitted. Some of the answers may eventually come from numerical modeling of tornadoes and their parent thunderstorms, but the answers are not likely to come easily or soon.

6.6 Waterspouts

Waterspouts are not produced in cold weather. [Aristotle]

A sailboat skims along the surface of the emerald waters off the Florida Keys on a bright June morning. The blue skies are interrupted by a few towering cumulus clouds which speckle the water with shadows, but the cloud tops are not very high, and no rain is falling. The winds blow steadily in from the southeast at 7 mi/h.

Suddenly, the base of one of the towering cumulus clouds lowers, and a distinct whirling is visible. The water underneath the clouds is disturbed by some unseen force. The whirling cloud material builds downward from the cloud base, forming an unmistakable funnel. Now, the water underneath the lengthening funnel whips around in a frenzy; spray is thrown upward and outward. The funnel finally reaches the water, merging freshwater cloud droplets with saltwater spray. The cloud and its appendage drift toward the sailboat, which frantically jibes to escape the ominous whirlwind. A few minutes later, the funnel passes the boat barely 100 meters to port. At the sailboat's position, the wind and water are scarcely disturbed and the sun is shining; a near miss—as good as a mile.

Although waterspouts are not usually nearly as intense as their big brothers, the tornadoes, they can nevertheless produce wind speeds of over 100 mi/h. The average waterspout, however, is much smaller than the tornado, perhaps only 50 meters in diameter, and has maximum wind speeds on the order of 50 mi/h. Boats have occasionally passed through the weaker waterspouts with little damage, but the larger spouts can completely destroy small craft.

Waterspouts are of interest because of their dynamic similarity to tornadoes. Both are caused by a rapid updraft and associated inflow of low-level air, which, under the principle of conservation of angular momentum, spins faster and faster until it reaches a point close to the center of low pressure. In moderate-to-intense waterspouts, the pressure at the vortex center is low enough to cause air to cool by expansion below its dew point, so that a visible funnel appears, much like the tornado (Plate 8). Thus, the waterspout funnel is not *ot* wa sucked up from the sea, but merely cloud droplets.

Tornado research in the past has been slowed by the very understandable fact that no one wants to take direct observations of well-developed tornadoes, at least not from very close range. Because the dynamics of tornadoes and waterspouts are similar, however, there is hope that by understanding the far less dangerous waterspouts, which are amenable to direct observation, insight into tornadoes can be gained.

There are some significant differences between tornadoes and waterspouts, however. A striking difference, obvious to those who have witnessed waterspouts, is that waterspouts frequently occur in basically fair weather, at least compared to the severe weather invariably associated with tornadoes. Frequently, the cumulus clouds from which waterspouts descend are not even raining and may extend to only 8000 feet or so, well below the freezing level. In comparison, severe tornado-producing thunderstorms extend to 50,000 feet and higher, well into freezing temperatures aloft. Waterspouts are also much more frequent than tornadoes, especially over the warm, shallow waters surrounding the Florida Keys. For example, Key West alone observes nearly 100 waterspouts per month during the summer.

Most Florida waterspouts are short-lived, the average duration being about 15 minutes. Their formation is favored by high surface temperatures and humidities (low densities) and, therefore, unstable conditions in the lowest 2000 feet. In this respect, they are like dust devils, which form over desert regions when the ground is heated to a very high temperature.

Waterspouts are common over coastal waters from Florida to the Gulf Coast and occur most often in the summer, when the water temperatures are high. Waterspouts also occur in other parts of the world, including the Indian Ocean, the equatorial Atlantic, and the Mediterranean Sea.

Occasionally, a tornado which forms in the usual manner over land moves over water and becomes a waterspout. In this case, the waterspout can be very severe, causing the same degree of damage to structures in and on the water as tornadoes do over land.

7 Air pollution meteorology

Many of the creations of people add undesirable particles or gases to the atmosphere. Some of the most important pollutants include coal dust, incompletely burned rubbish, photochemical "smog," sulfur dioxide, and carbon monoxide. Different cities have different pollution problems; for example, carbon monoxide is the major pollutant in Washington, D.C., rubbish particles and sulfur dioxide plague New York City, and Los Angeles is infamous for its photochemical smog.

In the United States, about half of the total pollution produced in 1974 came from the automobile. Car exhausts spit unburned hydrocarbons, carbon monoxide, and oxides of nitrogen into the atmosphere. In sunny cities such as Los Angeles or Phoenix, Arizona, the oxides of nitrogen combine with the hydrocarbons (left over from the gasoline), first to form ozone, and then complicated organic particles. These particles look like a smoky fog, and therefore are called smog, although they are really neither smoke nor fog. Smog irritates human and animal eyes and kills vegetation. Carbon monoxide from cars can also be an important health hazard, especially on busy streets.

Stationary sources such as power plants or home space heaters that burn coal are primarily responsible for the sulfur dioxide and coal dust in the atmosphere. Coal dust is the easier to control. It used to blacken the skies in

London and St. Louis, Missouri; in fact, the famous "London fogs" were caused almost entirely by coal dust from home coal-burning fireplaces. When these were banned, the fogs disappeared. In the following sections, we discuss some of the meteorological aspects of the air pollution problem.

7.1 Air Pollution and the Weather

There are really two completely different kinds of air pollution problems related to meteorology; one kind deals with the effects of weather on pollution, the other with the effects of pollution on weather and climate. Weather influences pollution in a number of ways: the winds transport pollution from one place to another, and the dilution of pollution depends on various meteorological factors. Therefore, a knowledge of atmospheric conditions is useful in the planning, execution, and evaluation of air pollution control. On the other hand, weather can be changed by pollution: increased pollution decreases visibility, causes raindrops to become drops of acid, and may increase precipitation, although this latter possibility is somewhat controversial.

The effects of pollution on climate are even more controversial: increased amounts of carbon dioxide from combustion processes increase the absorption of outgoing radiation from the ground, thereby reducing the overall loss of heat to space and producing a gradual warming trend. The opposite effect is possible when particles emitted by industry into the atmosphere reduce the intensity of sunlight reaching the surface, thereby causing lower temperatures. Much is uncertain about the magnitude of these effects, but they must be studied in detail so that measures can be taken in time to prevent either disastrous warming, which would produce widespread melting of polar ice and consequent flooding, or the equally undesirable extension of ice sheets associated with a reduction in mean temperature.

7.2 Effects of the Atmosphere on Pollution

The atmospheric temperature, moisture, and wind structure determine the fate of pollutants from the time they leave the source to the time they reach receptors, such as plants, animals, or people. Thus, a meteorologist is asked where pollution from a given source is going, what the concentration will be at the receptor, and whether pollution levels will be much higher tomorrow than today. Based on the meteorologist's advice, industry may be asked to reduce operations under adverse conditions, or switch to a cleaner fuel.

For convenience, we will discuss the effects of atmospheric variables on three parts of the pollution cycle: the initial rise of the pollutant from the source, the transport downwind from the source, and the dilution or dispersion of the contaminant.

The initial rise from a smokestack is important because the higher the material rises initially, the smaller will be the concentrations near the ground later. The actual amount of rise depends on five variables, three of which are nonmeteorological and may be partially controlled by the polluter. These three are the *temperature of the effluent,* the *initial exit velocity,* and the *cross sectional area of the source,* e.g., the area of the discharging stack. Then there are two meteorological variables, *wind speed* and the variation of temperature with height, called the *lapse rate,* over which the polluter has little control. On a windy day, smoke plumes do not rise very high, as can easily be verified visually. The effect of temperature variation with height is more subtle. Generally, when the air is more unstable, with warm, light air near the surface and colder air aloft, the smoke rises higher. In the morning, when there is frequently a low-level inversion, with cold, dense air underlying warm air, smoke spreads out horizontally soon after leaving the source. Sometimes, under the stable conditions of early morning when relative humidities are high, smoke and water vapor emitted by industry may produce dense fogs which drift slowly downwind. Such a case of pollution-induced fog in the Bald Eagle Valley of Pennsylvania is shown in Plate 9.

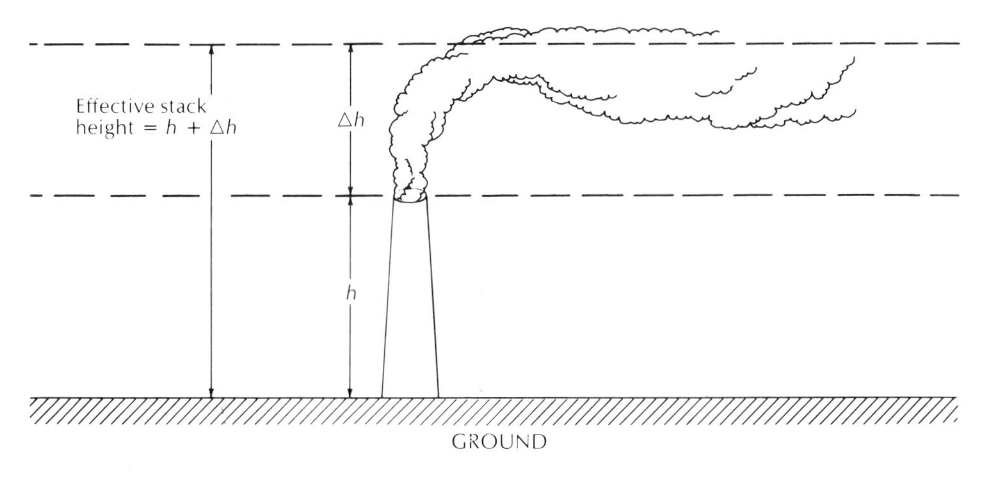

Figure 7.1 *Effective stack height.*

In order to compute the characteristics of a plume, it is convenient to define an "effective" stack height (Figure 7.1), which is the sum of the actual stack height, h, plus the additional rise, Δh, caused by the exit velocity and buoyancy of the warm plume. Although Δh obviously depends on the five variables previously mentioned, there is still no generally accepted formula for this complicated relationship. Because it is advantageous to make the effective stack height as great as possible in order to reduce concentrations near the ground, modern plants are often equipped with high stacks, and the effluent is emitted at high temperatures. Of course, when there is a strong wind, the effect of the high temperature on raising the effective height is diminished; however, conditions of fast winds usually do not lead to serious pollution problems, as we shall see.

7.2.1 *Horizontal transport of pollutants*

The question of horizontal transport of pollution appears at first to be quite simple: the pollutants travel with the wind, so all we need in order to forecast the trajectory of the pollutants are wind speed and direction. Unfortunately, the wind behaves in a very complicated manner over time and space, especially near the ground. In the first place, wind speed and direction are not steady, but vary from minute to minute and hour to hour. There are theoretical reasons why it is convenient to consider hour-average concentrations of pollutants. To estimate such concentrations, we should use hour-average winds. Consistent with this requirement, winds are reported at many weather stations every hour. One difficulty with the hourly wind reports, however, is that these winds are not really hour-average winds, but, instead, represent about one-minute averages. This difference may cause serious errors in computing the transport of pollutants.

Within an hour, wind directions fluctuate over perhaps 30 degrees. The effect of these fluctuations is to spray the pollutants in different directions, very much as water is spread from a waving garden hose. Even though the center of the pollutant "cloud" will move with the hour-average wind, the effect of the fluctuations is to dilute the concentrations and spread the material over a wider area. Thus, wind fluctuations lead to dispersion of pollutant material.

To make use of the fact that the center of the pollutant plume moves with the hour-mean wind, we need to know the wind at the source of the pollution and all along the plume's trajectory. Unfortunately, wind observations are almost never available exactly where they are needed. Hourly winds are usually measured at airports, not near the centers of industrial activity or human habitation. In addition, it is rarely legitimate to assume that the surface wind directions and speeds are uniform across an area as large as a city. Most cities are near large water bodies, or have hills nearby, and some, like Los Angeles, are close to high mountains. All these topographic features produce local circulations on a scale (mesoscale) much smaller than what we can see on ordinary weather maps.

In some cities where air pollution has been especially serious, mesoscale wind circulations have been studied in detail. Such studies utilize constant-level balloons, numerous rawinsonde stations, or helicopters. For example, we know the local circulations in considerable detail for New York, Chicago, the San Francisco Bay area, and Los Angeles. From simple models of typical flow patterns over these cities, if we have measurements of winds at a few locations, we can make a good estimate of winds at other locations in the same cities. However, detailed wind surveys are expensive. An alternate technique is to construct mathematical models of the flow patterns in a city, given various characteristics of the large-scale flow and the detailed topography of the city. This type of modeling is quite difficult, and efforts in this direction have only just started.

One problem with estimating the wind field is that the most serious air pollution problems arise with very light winds. Under these conditions, the mesoscale circulation features tend to dominate the larger-scale circulations, and the transport patterns are quite erratic. We do not know how to deal with such strong local variations of concentrations which can arise under these circumstances. All we can really say is that light winds permit pockets of very strong pollution to develop.

Not only does the wind vary horizontally from place to place and time to time, but it also varies with height. Over flat terrain, the wind speed increases very rapidly in the lowest few meters, but then changes quite slowly with increasing height. Also, the direction is nearly constant with elevations up to 100 meters (300 feet) or so, unless the air is quite stable. But more significant changes in direction occur higher up.

Typically, the wind is measured at heights about 10 meters above the ground. These observations are quite good indicators of the general wind in the lowest 100 meters. Above a level of about 100 meters, a slow rotation of the wind direction begins. This rotation is clockwise (veering) in the northern hemisphere; that is, a south wind near the ground becomes more westerly aloft. Also, the speed increases slowly with height, a fact which should be considered for accurate dispersion calculations. These general rules apply when the air is unstable. When the air is stable and the winds are light, the vertical distribution of wind is erratic, contributing to the difficulty of quantitative computations. In summary, measuring or calculating the transport winds for a given city is usually a very difficult problem.

7.2.2 *Dilution of pollutants by mixing*

After the transport wind has been determined, it is necessary to find out how fast the pollutant is diluted. If the dilution, or dispersion, is rapid, pollutant concentrations will generally be low, and no serious problems arise. But, on certain days, pollutants are diluted very slowly, and harmful pollutants may reach dangerous levels.

The rate of dispersion is primarily governed by the vertical variation of temperature and by the wind speed. To a lesser degree, it depends on the roughness of the terrain. Also, the topography, such as hills and mountains, can limit pollutant dilution.

The effect of the wind speed on dispersion can be seen in Figure 7.2. Suppose that a stack emits a burst of smoke every second. If there is a wind of 10 m/s (about 20 mi/h), the distance between smoke puffs will be 10 meters. If the wind speed is 5 m/s, the distance between puffs will be 5 meters. In general, the faster the wind, the greater the distance between puffs, and the smaller the concentration. Thus, pollution problems are never found with strong winds; instead, so-called air pollution episodes occur only when the wind is weak.

There is another, but less important, effect of strong winds. Wind blowing over any terrain, but particularly over rough terrain, will produce vertical eddies. Such eddies (turbulence) mix polluted air with clear air, so that the clean air becomes less clean, and the dirty air less dirty. In fact, turbulence acts like Robin Hood—it takes pollution from regions that have too much and gives to regions where there is little.

Small-scale horizontal variations, or eddies, in the mean wind direction also act to disperse the pollution across the mean wind, as indicated by the increasing size of the puffs downwind of the source in Figure 7.2. These eddies

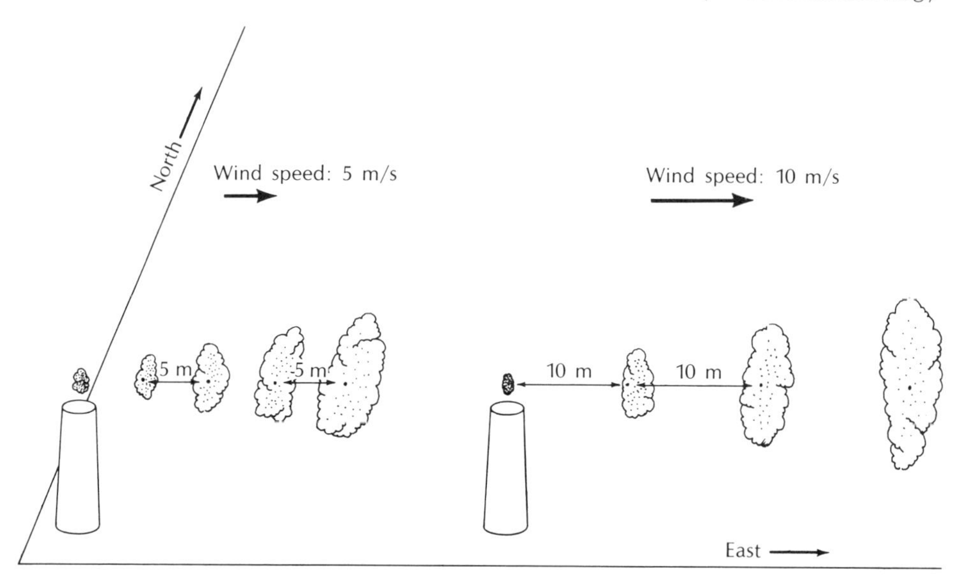

Figure 7.2 *Effect of wind speed and horizontal eddies on dispersion of pollutants.*

produce a horizontal fanning of the pollution downwind, which reduces the concentration per unit area in the plume.

7.2.3 *The mixed layer*

More important than wind for producing eddies, however, is thermally produced convection, which depends strongly on the surface heating and the vertical temperature distribution. Whenever air is heated from below, as on a sunny day, convection currents, which are another form of turbulence, are generated. Convection currents mix air with different characteristics, so that polluted air becomes cleaner. In terms of the temperature distribution, mixing by eddies occurs when temperature decreases rapidly with height, that is, when air is *unstable*.

Usually, on sunny days, the heating from the ground is so strong that the well-mixed portion of the atmosphere extends to fairly high levels, to one kilometer or more. Above that mixed layer, an inversion usually exists. Inversions are quite stable and tend to prevent mixing with air above. Hence, the distance from the ground to the bottom of the first inversion is called the *mixing depth*. This depth plays an important role in air pollution prediction. The mixing depth limits the extent to which pollution can spread vertically: the smaller the mixing depth, the greater the concentration.

Figure 7.3 shows a map of the average annual morning mixing depths (in hundreds of meters) over the United States. The greatest mixing depths occur near the coasts, where the warm waters (compared to the land) maintain a relatively unstable layer of air near the ground.

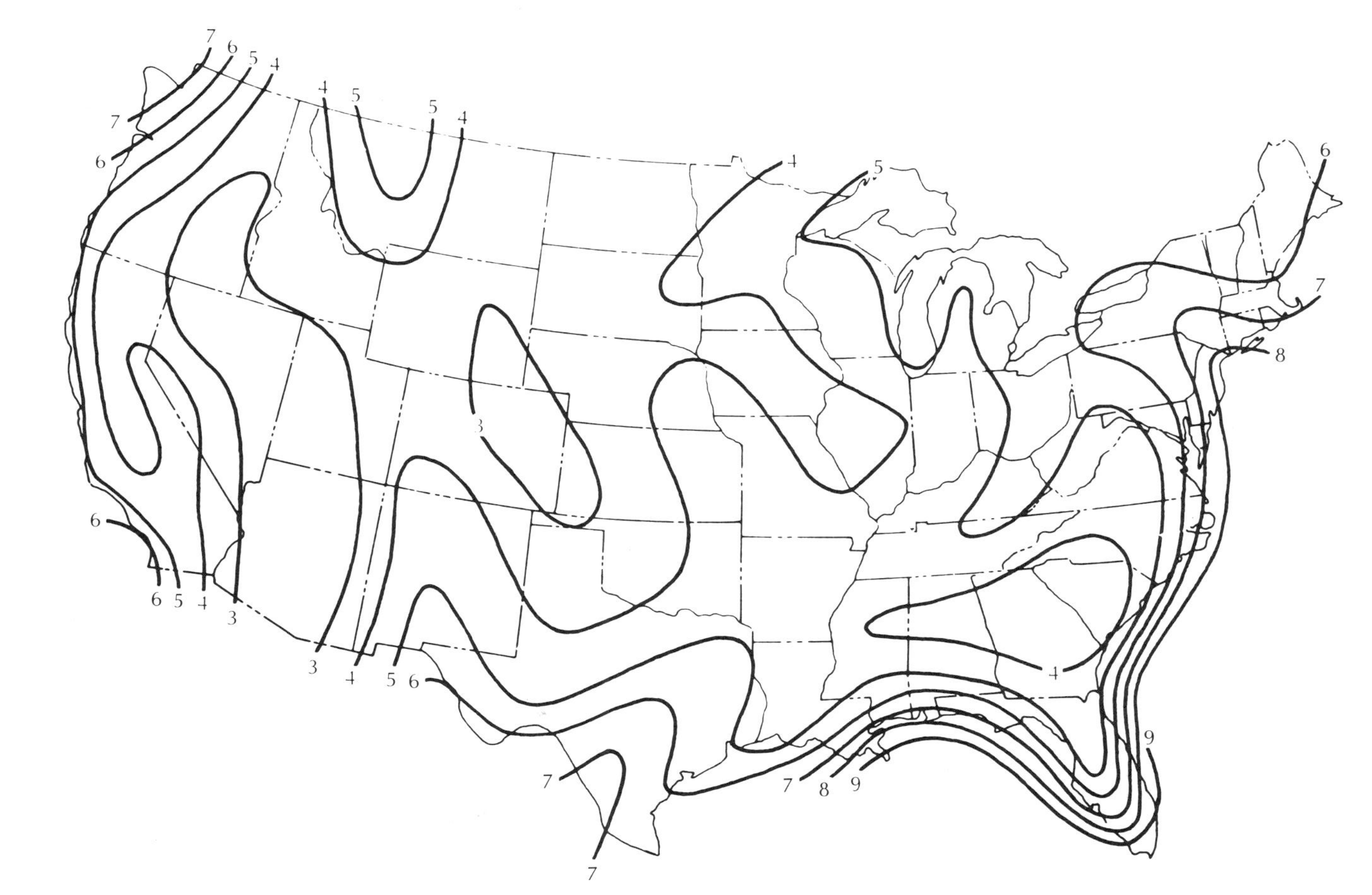

Figure 7.3 *Mean annual morning mixing depths (in hundreds of meters) (source: George Holzworth, "Mixing Heights, Winds Speeds, and Pollution throughout the Contiguous United States," Environmental Protection Agency, Research Triangle Park, North Carolina, January, 1972).*

Again, it is quite clear that the daytime mixing depth increases with increased heating of the ground, and will be, for example, greater in summer than in winter. For example, Figure 7.4 shows the annual average mixing depths in the afternoon, which may be compared with the morning values in Figure 7.3. In many inland places, the afternoon mixing depths are four times greater than the morning values, reflecting the solar heating at the ground. Variations near the coasts, where the diurnal variations in water temperature are small, are much less. But the daytime mixing depth depends on other things besides the heat input at the surface. In particular, it depends on the stability of the atmosphere before the heating begins. For example, if the air aloft is slowly sinking (subsiding), it warms as it is compressed by the higher pressure. This warming aloft produces stability, and even strong heating from below under such conditions cannot produce a large mixing depth. General subsidence and stability occur most often in anticyclones, especially near the high-pressure centers. This is where daytime mixing depths are smallest and winds are lightest.

The effects of both the wind speed and the depth of the mixed layer are combined in the *ventilation factor,* which is the product of the mixing depth and the average wind between the ground and the top of the mixed layer. The concentration varies inversely as the ventilation factor; for example, if the mixing depth and the wind speed are both doubled, the pollution concentration is reduced by a factor of four.

Figure 7.5 shows an analysis of the ventilation factors over the United States on the day when the ventilation was the least over the five-year period from 1960 to 1964. On this day, the atmosphere's ability to dilute pollution was a minimum. In this figure, the lowest numbers represent the greatest potential for the development of high concentrations of pollution. In general, the western United States had the lowest ventilation factors, and, therefore, the greatest potential for air pollution episodes. Of course, it takes more than low ventilation factors to produce severe air pollution; it also requires sources of pollution. Since many of the sparsely settled regions of the West have few sources of pollutants, the air may be quite clean there even with low ventilation factors.

As we have seen, both mixing depth and wind speed tend to be low in anticyclones; hence, high-pressure areas are often associated with large concentrations of pollutants. Normally, anticyclones move fairly rapidly across a weather map, so that conditions at any one point will not get out of hand. But occasionally, most likely in the summer or fall, highs may remain in the same area for a week or more, giving rise to a *stagnation* situation. (Winter anticyclones seldom stagnate, but when they do, their especially low mixing depths have produced some of the most devastating pollution episodes.) Under such conditions, air pollution meteorologists must alert the public and industry in order to reduce the concentration of harmful effluents to a minimum.

In some parts of the world, highs may stagnate for much longer periods. For example, in summer, west coasts of continents are influenced by anticyclones for months at a time. Not only that, but cold ocean currents increase the stability of the air and reduce mixing depths. This is why especially strict air pollution regulations are needed for west coast cities such as Los Angeles.

Another factor influencing the pollution climate is topography; mountains can restrict horizontal spreading. There are mountains both to the east

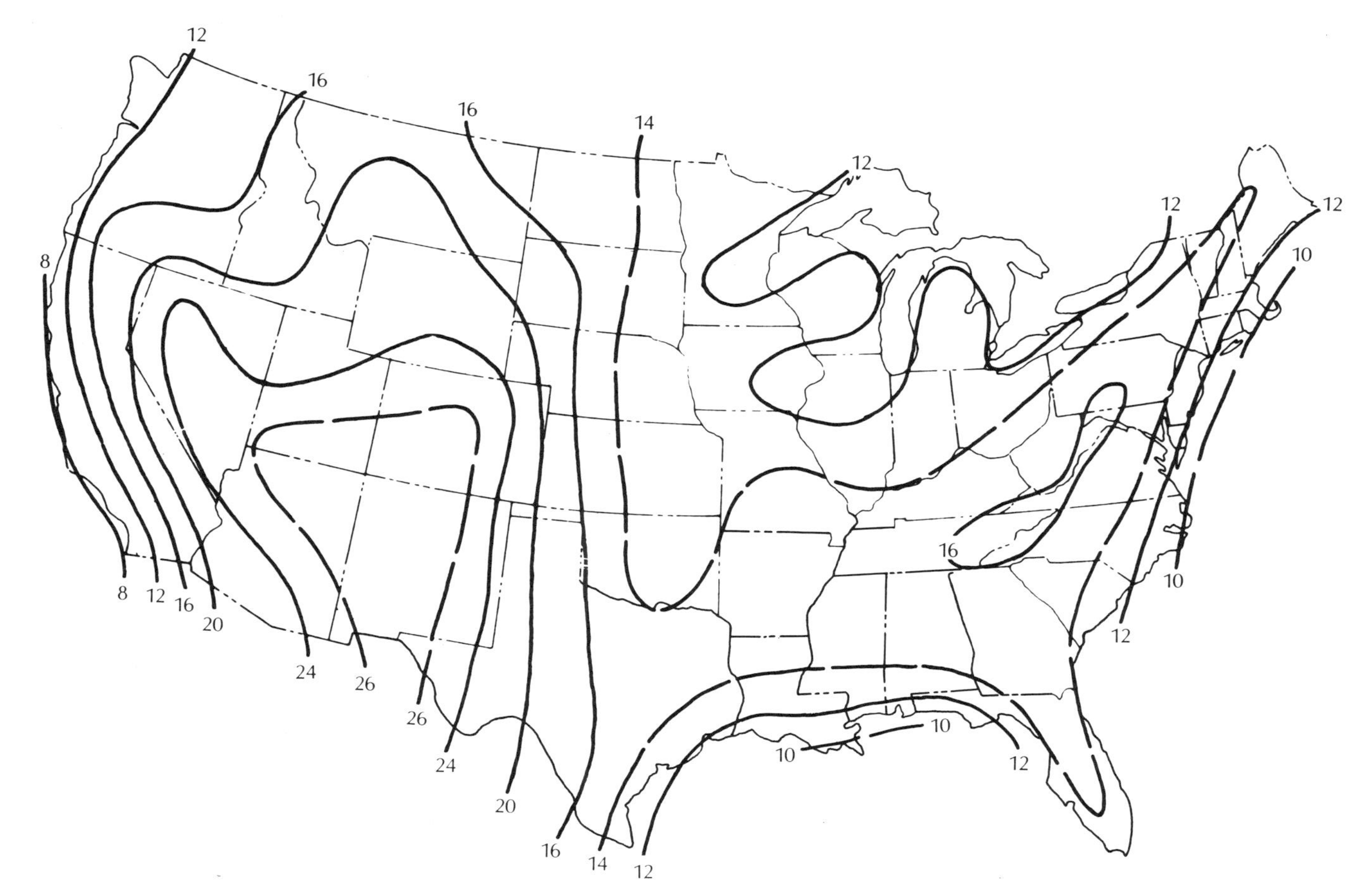

Figure 7.4 *Mean afternoon mixing depths (in hundreds of meters) (source: George Holzworth, "Mixing Heights, Wind Speeds, and Pollution throughout the Contiguous United States," Environmental Protection Agency, Research Triangle Park, North Carolina, January, 1972).*

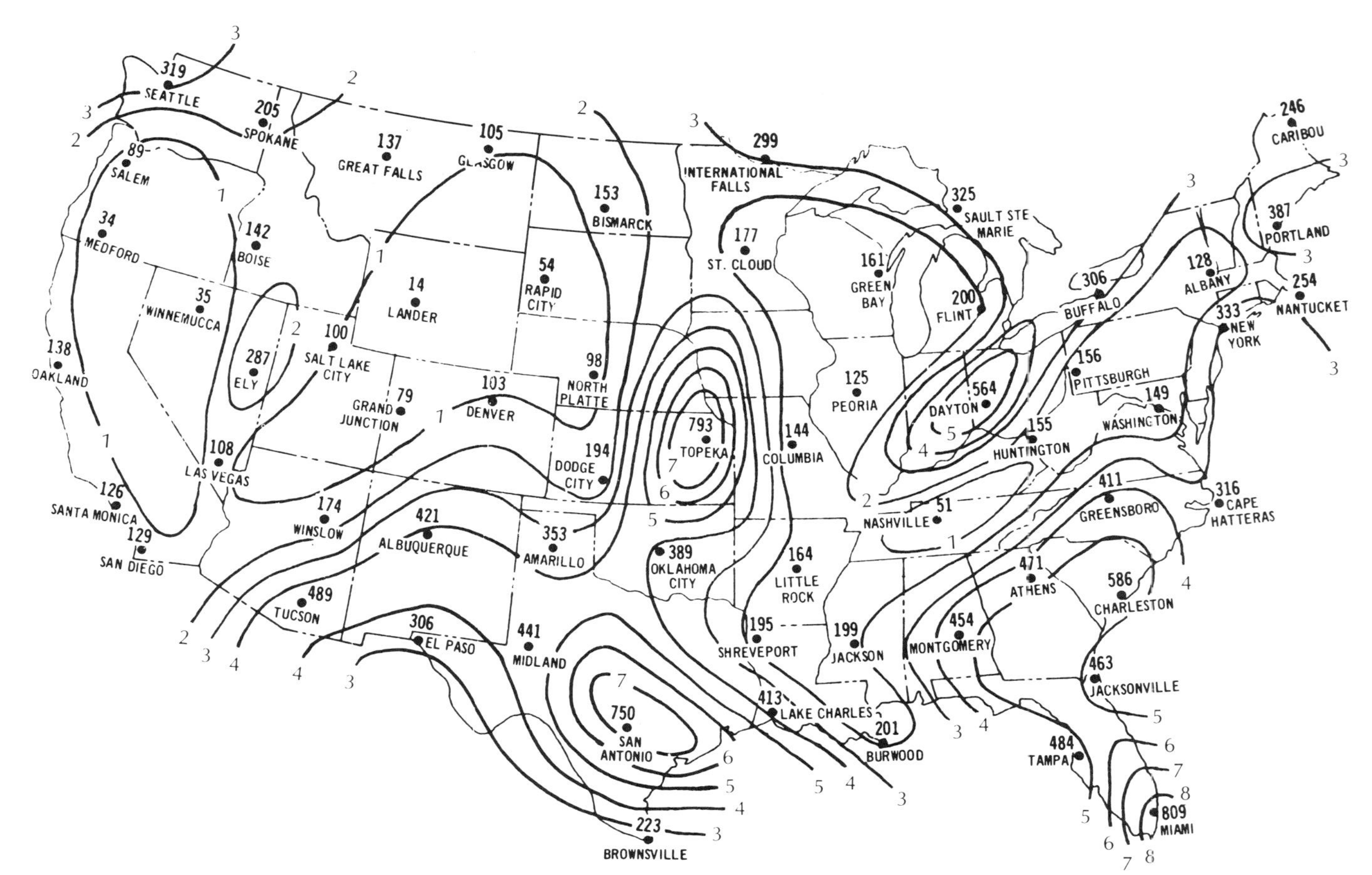

Figure 7.5 *Ventilation factors (product of mixing depth and wind speed in mixed layer) for one-day period when ventilation was a minimum during the period 1960–1964 (source: George Holzworth, ''Meteorological Episodes of Slowest Dilution in the Contiguous United States,'' Environmental Protection Agency, Research Triangle Park, North Carolina, February, 1974).*

and north of Los Angeles, adding to the problems there. Topography plays an especially important role in the case of deep, narrow mountain valleys, which are frequently sites of pollution disasters. Such disasters occurred in the Meuse Valley of Belgium in 1930, when 63 people died, and in the Monongahela Valley of western Pennsylvania (Donora) in 1948, when 22 succumbed. In both places low-level inversions persisted for several days, trapping industrial pollutants in a stagnant layer of air trapped between the valley walls.

So far, we have considered mixing depths and air pollution potential only for the daytime. At night, wind speeds tend to be less than during the daytime, and, as we will see, the same is true for mixing depths, so that concentrations tend to be larger at night than during the daytime. Of course, industry likes to dispose of its unpopular pollution at night when it cannot be seen; but, unfortunately, that turns out to be the worst time.

Concentrations are not always higher at the ground level at night, however. If the air is very stable, with the coldest air at the ground and the temperature increasing upward, the pollutants emitted from high stacks may accumulate in a thin layer a short distance above the ground during the night (Figure 7.6). After sunrise, when the inversion is destroyed by solar heating,

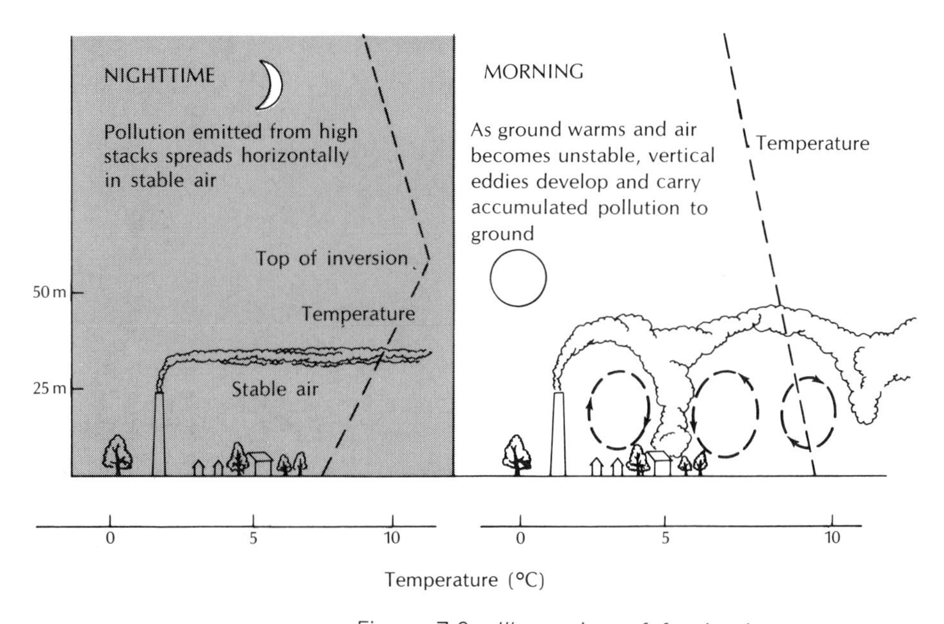

Figure 7.6 *Illustration of fumigation process.*

vertical convective eddies may mix the accumulated pollutants downward to the surface, a process known as *fumigation.*

7.2.4 *Mixing depths over cities*

At night, a mixed layer does form near the ground in cities. However, the mechanism is different from that during the daytime. Before air reaches a city, it is usually quite stable, particularly on clear, dry nights with little wind. Under such conditions, there is almost no vertical mixing, and a plume from an elevated source may not reach the ground for tens of miles. Under these conditions, the temperature usually increases vertically to a height of one kilometer or more; we have a *surface inversion* (see Figure 7.7). We noted in Chapter 3 that cities, because of their high heat capacities and artificial heat sources, are heat islands. Therefore, nighttime surface temperatures are several degrees warmer in the city than in the surrounding countryside.

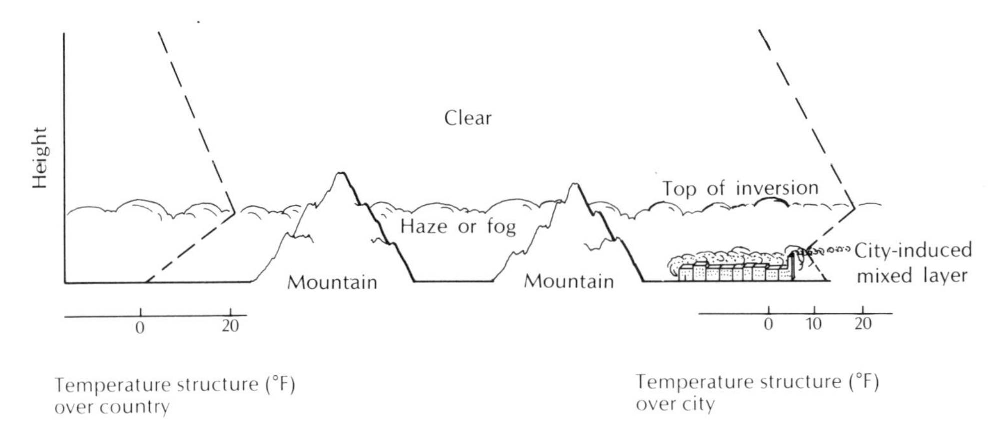

Figure 7.7 *Nocturnal lapse rates over city and surrounding countryside.*

The effect of the urban heating is to establish an unstable layer in the otherwise stable air close to the ground as illustrated in Figure 7.7. Thus, a new mixing depth over the city is established from the surface to the inversion. An example of the urban mixing depth for Cincinnati, Ohio at sunrise is shown in Figure 7.8. A general inversion covers the entire region at an elevation of 366 meters (1200 feet), where the temperature is a warm 77°F (25°C). In the rural areas surrounding downtown Cincinnati, the temperatures are as low as 65°F. In the city, however, the heat island produces a shallow mixed layer in which the temperature decreases from 72°F to 68°F from the surface to a height of about 220 feet above the ground. This mixed layer of warm air is carried downwind from the city and overspreads the cooler air over the rural surface. Such city-induced mixing depths are different in every city, generally increasing with the size of the city. For example, typical mixing depths are 300 meters in New York City, 150 meters in Columbus, Ohio, and 100 meters in Johnstown, Pennsylvania. Although the heat-island effect is subject to intensive research, there is as yet no simple method to estimate accurately nighttime mixing depths, which depend on meteorological as well as city characteristics.

Because nighttime mixing depths in cities are generally smaller than daytime mixing depths, pollution from low-level sources (e.g., apartment house incinerators) is worse at night. However, very high stacks often extend to levels above the mixing depth and will not contribute to the pollution of the city night air at the surface (Figure 7.7).

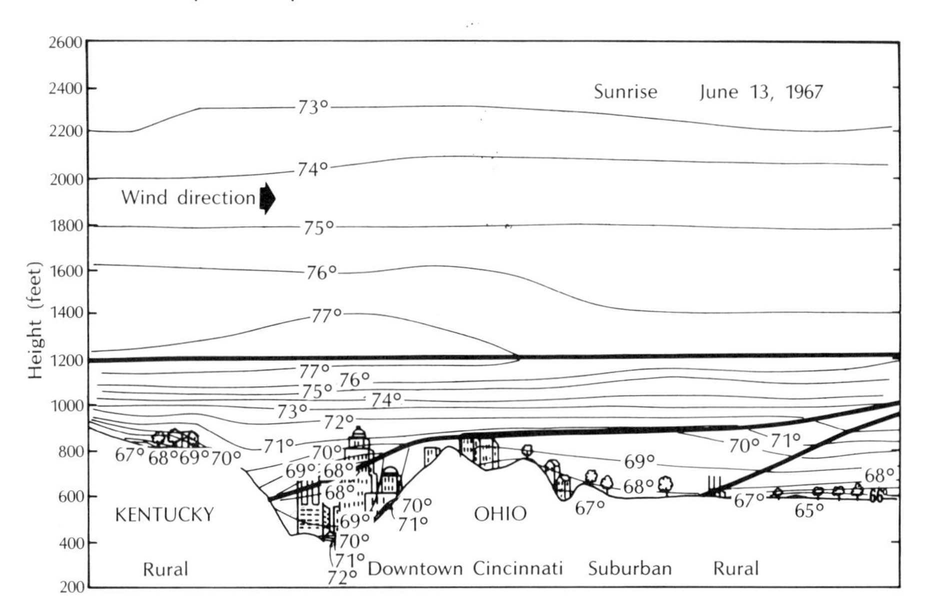

Figure 7.8 *Vertical cross section showing temperatures and mixing depth over Cincinnati, Ohio at sunrise (source: John F. Clarke, "Nocturnal Urban Boundary Layer over Cincinnati, Ohio," Monthly Weather Review, August, 1969).*

Reasonably good techniques exist for estimating concentrations of pollution from individual continuous sources to a distance of 10 kilometers or so away from the source, given the source strength, wind, lapse rate, and mixing depth. In the case of cities, there are numerous sources. A great deal of work has been done on modeling the pollution in individual cities by estimating the total pollution emitted over each square mile. However, separate calculations have to be made for different weather situations and for different pollutants. The purpose of these city models is to determine how much of the pollution in a given area is caused by each source, so that the effect of curtailing certain sources can be established. Thus, for example, it has been shown that in central New York City almost 95 percent of the sulphur dioxide (SO_2) came from apartment house heating nearby (before recent control measures), and only a very small fraction from power plants, even though the power plants emitted about the same amount of SO_2 as the apartments. The reason was that the power plants had high stacks, and, therefore, produced little effect nearby, although they had a relatively large effect on the suburbs. It is clear that in this case the use of cleaner fuels in the power plants would have had little effect on the quality of the air in the inner city. It is also noteworthy that even in such examples where the meteorological question of the cause and dispersion of the pollutant is answered, the political question of what to do about it remains, and, in many cases, turns out to be more difficult to solve than the original scientific problem.

7.2.5 *Evaluation of background pollution*

An important problem in air pollution control is an evaluation of *background pollution,* which is the normal amount of pollution present at a given place. With certain wind directions, air quality may be greatly influenced by distant pollution sources, creating both meteorological and political problems. As an example, much of the pollution at Windsor, Ontario comes from Detroit, and fully half of the pollution at Düsseldorf, West Germany comes from the Ruhr Valley.

It has become possible to extend some of the techniques designed for estimation of short-range diffusion to distances of 100 kilometers or so. New difficulties arise in such estimates, but, on the other hand, some simplifications are possible. One difficulty is that the large distances involved mean that the pollutants must travel many hours, over which time meteorological variables such as wind and mixing depth undergo important changes. In addition, the pollutants change chemically, interact with the ground, and may be affected by rain and clouds. At present, we know much better how the pollution gets into the atmosphere than how it eventually disappears, because there still are many unknowns in the chemical behavior of the various pollutants.

On the other hand, it is possible to simplify the problem by representing whole cities as point sources some distance upstream of the actual cities, and to arrive at good concentration estimates for considerable distances downstream.

Under unusual weather conditions, pollutants can be traced over more than 1500 kilometers; e.g., particles in Oklahoma have been traced to sources in the northwestern Indiana industrial area; and the acid rains in Scandinavia (which will be discussed later) have been ascribed to sources in West Germany. Such estimates are naturally quite controversial.

7.3 Effects of Pollution on the Atmosphere

It is quite likely that pollution modifies the weather now, and that these effects will grow in the future. However, the details of the future influence of pollution on weather are quite debatable and speculative.

The most obvious effects occur on the local scale. For example, the visibility is sharply reduced on "smoggy" days which occur occasionally in some cities, and frequently in others, such as Los Angeles. Along with visibility, direct sunlight reaching the ground is reduced. In a recent study in St. Louis, Missouri, it was found that direct sunlight was reduced in the city by as much as 40 percent on "dirty" days. However, it is not clear whether even such dramatic reductions of sunlight have significant effects on temperature, because much of the light will reach the ground anyway due to scattering by the polluting aerosols (particles).

7.3.1 *Effects of pollution on clouds and rainfall*

There are many reports of air pollution changing regional weather. Perhaps the most important modification is known as the *Laporte effect.* Laporte, Indiana is located east of major steel mills and other industries south of Chicago. Rainfall at Laporte has been observed to be systematically higher than at other stations nearby. If this effect does not occur by chance, the increase could be caused by the excess particles, heat, or moisture produced by the industries upwind of Laporte. Unfortunately, some observations appear unreliable, and the matter is not yet settled, even after lengthy and controversial debate between concerned meteorologists.

An increase of precipitation similar to that observed in Laporte has been reported downwind from some pulp mills in western Washington. Analysis of the raindrops there suggests that some particles introduced into the clouds by the plumes are large enough to capture water drops when falling, and thus attain sufficient size to reach the ground as precipitation.

In western Pennsylvania, large power plants seem to have increased the precipitation downstream compared to what it had been prior to the construction of the plants; unfortunately, the natural variability of precipitation is so large that this result was not "statistically significant." Two dramatic examples of the effect of the heat emitted by such large power plants are shown in Plate 10, which depicts growing cumulus clouds situated directly over the smokestacks of the Keystone power plant, located near Indiana, Pennsylvania. On these two days, the atmosphere was very nearly unstable and therefore sensitive to the extra amount of heat and moisture produced by the power plant. This extra heat triggered the release of the natural instability of the atmosphere, and the result was the cumulus clouds.

Another spectacular example of the effect of pollution on local weather is illustrated in Plate 11, which shows two areas of growing cumulus clouds generated by the smoke and heat from grass fires in the Everglades of Florida. Intense wildfires can also produce strong enough updrafts and convergence of low-level air to spawn whirlwinds of fire—miniature tornadoes that resemble dust devils.

As the previous examples prove, pollution can, at least under just the right conditions, significantly modify the local weather. On a larger scale, we have a number of indications that clouds and precipitation can be increased by pollution emitted tens of miles upwind, but none of these observations are conclusive.

A damaging effect of pollution on a regional scale, which has important political implications, is the occurrence of *acid rain*. Acid rain, which is simply dilute liquid sulfuric acid falling from the sky, has damaged forests, particularly in Norway and Sweden. Construction of air trajectories and estimates of concentration of sulphur dioxide (SO_2) have indicated that as much as 80 percent of the sulphate in the rain (which causes the acidity) falling in Scandinavia is produced by the emissions of sulphur dioxide in England and West Germany. The acidity of the rain is particularly large only on days when winds blow effluents from one of these two countries over Scandinavia. Similarly, pH values as low as 3.1 (considerable acidity) have been observed downwind from the high-stack power plants in western Pennsylvania. This result suggests that, although sufficiently high stacks can be built to prevent large concentrations of SO_2 near the ground, the plume can still pollute clouds and rain downwind.

A different form of regional pollution by industry is the increase in number of liquid drops in the atmosphere introduced by huge cooling towers. The hot water in these towers also evaporates, adding large amounts of water vapor to the air. In cold air with high relative humidity, the additional water added to the air can produce widespread fogging, and even highway icing to distances of over 100 miles.

7.3.2 *Can pollution change the climate?*

Of all the possible effects of pollution on weather, those on climate are the most nebulous, but are nevertheless the most important. A mean cooling of the atmosphere by only a few degrees Celsius, according to some estimates, could cause the polar ice to spread over the whole world. A warming by a similar amount could melt the polar ice caps and inundate large low-lying areas of the world, including cities such as New York, Leningrad, and London.

Although some of the effects on a given meteorological variable (e.g., temperature) by individual polluting agents can be estimated fairly well, the ultimate effect on climate is very difficult to estimate, because once one variable changes, others will, too. For example, if the temperature increases, the absolute amount of water vapor is likely to increase; wind circulations will change, and so will the amount and types of clouds. All these changes may feed back to affect the temperature, either magnifying or diminishing the original change. So far, no mathematical models have been able to predict the original effect and all the possible feedbacks. Another basic difficulty is that the causes of natural climatic changes are not known, so that we do not know whether a certain observed change is man-made or natural. Of course, eventually the man-made effects may be so large that they override natural climatic changes.

Of all the pollutants likely to change climate, carbon dioxide (CO_2) has been studied most thoroughly. We know the amount of CO_2 put out by industrial and other combustion, and we have quite a good idea what fraction of this actually stays in the atmosphere (about 50 percent). Most of the remainder presumably is absorbed by the oceans, although a small part goes into vegetation, creating thicker jungles. Figure 7.9 shows that the average annual concentration of CO_2 increased from about 310 to 320 parts per million between 1958 and 1970. The figure also shows an extrapolation for the concentration in the year 2000, when the expected value is 380 parts per million, or almost 20 percent higher than the present concentration. Such an extrapolation is based on certain assumptions, which may be hazardous. One such assumption, for example, is that the ocean will continue to absorb a large fraction of the CO_2 in the atmosphere. Also, this prediction ignores the worldwide energy shortage, which may increase the shift from oil and natural gas to atomic fuel and decrease the emission of CO_2. On the other hand, a shift to coal would make no difference.

Additional CO_2 in the atmosphere does not affect incoming short-wave solar radiation, but absorbs more long-wave radiation emitted by the earth and reradiates additional infrared radiation to the ground, thus raising the temperature of the air near the surface. It also increases radiation emitted into space, and therefore cools the upper atmosphere. The warming of the lower layers presumably means that evaporation increases at the surface, thus further increasing the infrared absorption and reradiation of atmospheric radiation. Therefore, the temperature would probably be further increased by this feedback effect.

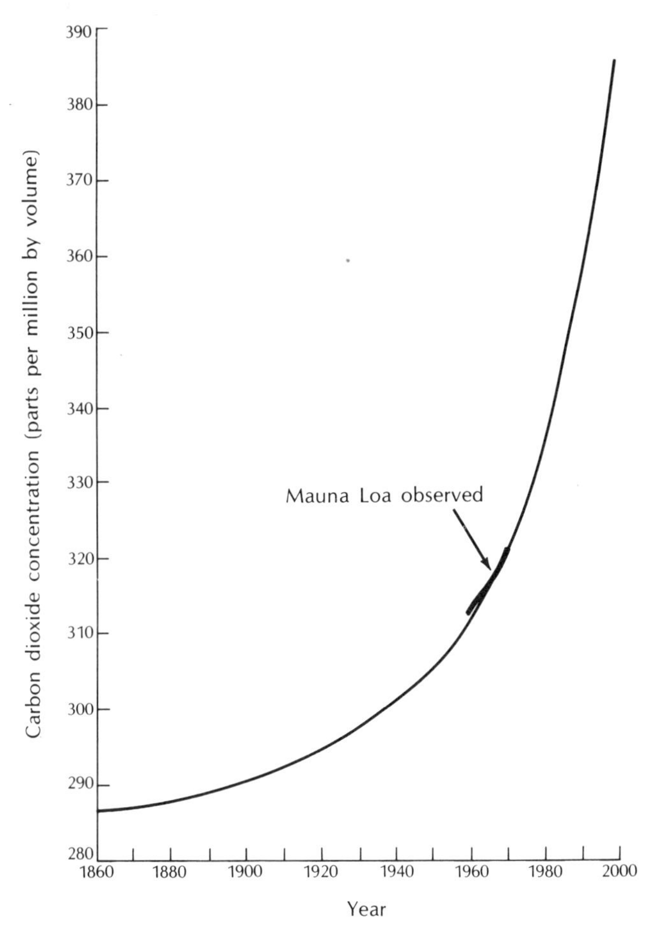

Figure 7.9 *Carbon dioxide in the atmosphere (courtesy of L. Machta).*

Taking these factors into account, but not allowing for large-scale atmospheric circulation changes, recent studies have shown that a doubling of CO_2 in the atmosphere would raise temperatures near the ground by about 2°C. Thus, the anticipated increase of atmospheric CO_2 by the year 2000 should raise the surface temperature by about ½°C. Such an increase is not believed to be serious, but computations suggest that climate modifications by CO_2 probably will become important in the twenty-first century. Therefore, it is extremely important to monitor worldwide concentrations of CO_2 and to plan measures to counteract the exponentially increasing growth of carbon dioxide concentrations.

Of the other gases, excess water vapor and sulfur dioxide may eventually affect climate, not so much directly, but after sulfate particles have formed from them. Also, there is some evidence that cirrus cloudiness has increased proportionately to increased aircraft traffic. Preliminary computations suggest that such increases have no large effect on surface temperature; however, some measurements have shown that the solar radiation may be reduced significantly. If confirmed, this reduction could be damaging to agriculture.

Sulphur dioxide is converted to sulphate particles, which comprise an important part of the total particle load in the stratosphere. Even in the troposphere, sulphates are the most plentiful man-made particles. Other particles in this category include coal dust and other substances emitted as particles. Also, some particles (e.g., smog) form chemically from oxides of nitrogen and hydrocarbons. The estimated concentrations of all such particles are quite uncertain; however, it seems clear that tropospheric particles from natural sources (volcanic particles, sea salt, dust) are much more plentiful than man-made particles. At present, only about 20 percent of all the particles are man-made. However, some crude estimates suggest that by the year 2000, perhaps half the tropospheric particles will be introduced by human activity.

Increased particle concentrations can affect two aspects of climate —precipitation and radiative transfer. Of these, the second is believed to be the more important. Thus, it seems likely that the reflectivity (albedo) of the atmosphere is changed by the addition of particles. But even the sign of the change is not known. For example, if the particles are bright, the albedo is likely to increase, reflecting more sunlight back into space and cooling the air. However, if dark particles are emitted over an otherwise bright surface (an extreme case being snow), the albedo will decrease, and more radiation will be available to heat the atmosphere.

In any case, the magnitude of the effect of particles on the radiation budget is extremely uncertain. It depends on the optical properties of the particles, which in turn depend on particle composition and size distribution. Furthermore, particles radiate in the infrared, and therefore can alter the outgoing long-wave radiation. It is clear, therefore, that our knowledge of the effects of particles in the atmosphere is badly inadequate. We must know more about scattering, absorption, and emission of natural and man-made particles and must predict what kinds of particles are likely to be added to the atmosphere. At the same time, we must monitor changes of particle concentrations, reflected sunlight, and emitted radiation.

So far, we have been concerned entirely with the troposphere. Stratospheric pollution is also important for two reasons. First, the stratosphere has lower densities than the troposphere, so that a given amount of pollution has relatively larger effects in the stratosphere. Second, the residence time in the stratosphere is long because of the weak vertical mixing. For example, at 20 kilometers, the residence time of fine particles is about a year. One year after introduction, about one-third of the initial number still remains.

Nature occasionally pollutes the stratosphere with particles from volcanic eruptions. At about 20 kilometers, there exists a rather permanent layer of particles which is attributed to continuing volcanic eruptions. The two most severe eruptions in the nineteenth century may have been associated with measurable cooling at ground level; the 1963 eruption of Mt. Agung in Indonesia was followed by a stratospheric warming of about 6°C. Measurements of the quantity of particles introduced and computations based on their optical properties have suggested that a 5°C rise is quite reasonable. However, no definite change in surface weather has been linked to the eruption of Mt. Agung.

Because volcanic particles in the stratosphere may have had important effects on the stratosphere, and in extreme cases, even on the troposphere, the question has been asked whether man-made pollution adds perceptibly to such effects. At present, human activity does not contribute much to the concentration of contaminants in the stratosphere. European supersonic aircraft and military high-altitude planes have had no detectable effect while flying mostly at levels below 20 kilometers, where the residence time is shorter. Atomic testing in the 1950s introduced particles into the stratosphere but no definite effect on weather has been demonstrated. And, although a possible decrease of stratospheric ozone following certain atomic tests has been suggested on the basis of theoretical chemical arguments, it has not been conclusively confirmed by observations.

7.4 Supersonic Transports and Possible Modification of Climate

Most of the concern about stratospheric pollution has centered about the introduction of the proposed American supersonic transports (SSTs). These are relatively large, and will, if built, fly at an altitude of about 20 kilometers. It is estimated that about 500 SSTs must fly to make using them profitable. Thus, a number of calculations have been made of the possible effects of various pollutants introduced by this number of planes, each flying seven hours per day, over routes most likely to be popular.

The principal emissions of concern are SO_2, H_2O, oxides of nitrogen, and soot. Carbon dioxide is also emitted, but the SSTs will be relatively unimportant sources compared to the private and industrial sources which were discussed earlier. It is estimated that enough water vapor will be emitted to increase the water vapor content of the stratosphere (which is just a few parts per million) by 10 percent. Three potential effects of this increased water vapor would be:

(1) Increased cloudiness in the very cold regions of the winter pole (about –80°C), where there now exist some thin, iridescent clouds, called *mother-of-pearl* clouds;
(2) Change in the infrared radiation to space and toward the ground;
(3) Interaction of the water vapor with stratospheric ozone.

Of these three items, quantitative estimates have shown that the effect of the additional long-wave radiation would be negligible. Almost nothing is known about the first item. We know that condensation trails are almost never seen behind stratospheric planes, because the relative humidities are so low in most of the stratosphere. But, then, no stratospheric aircraft has been observed in regions around the cold polar night, which is an area where mother-of-pearl clouds sometimes exist. Presumably, not many SSTs will fly through these regions, so that the problem of additional cloudiness and radiative changes associated with them may be unimportant. Also, arctic clouds are rare in the stratosphere, very thin, and are therefore unlikely to affect climate significantly.

Any interference with ozone is potentially dangerous, for ozone protects us from ultraviolet light. If ozone is removed, ultraviolet light is increased, thus increasing the incidence of skin cancer. It has been estimated that a 10 percent decrease of ozone will produce a 20 percent increase in skin cancer.

Water vapor will, by itself, attack ozone. However, it is now clear that this effect is not serious. On the contrary, addition of water vapor is, in a roundabout way, helpful to people. This paradox results because oxides of nitrogen (NO_x) in the exhaust of SSTs are much more damaging to the ozone. These substances remain intact while they break up the ozone; therefore, a very small amount of NO_x can destroy a great deal of ozone. Here is where water vapor is helpful; it converts some of the NO_x into nitric acid, which is removed rapidly from the stratosphere. Still, a large fleet of SSTs with conventional jet engines will release enough NO_x to make the expected increase of cancer intolerable. Further, the loss of ozone can have climatic consequences, which cannot be estimated accurately now, but which may be considerable. For these reasons, it is likely that large SSTs will not be allowed to fly unless their engines are redesigned so that they burn at a lower temperature and thus emit a negligible amount of NO_x.

A further problem with SSTs is that jet fuel contains small amounts of sulphur, which causes the emission of sulphur dioxide into the stratosphere. As we have seen, sulphur dioxide may form particles, which reflect sunlight. Again, we cannot rule out the possibility that climate will be adversely affected in parts of the world. So, as "insurance," it is proposed to incur the modest cost of removing almost all the sulphur from the fuel to be used for SSTs.

Oxides of nitrogen are not the only substances that interfere with stratospheric ozone; chlorine is potentially worse. Certain chemicals that are used widely as refrigerants and as propellants in spray cans contain chlorine. When these chemicals slowly seep into the stratosphere, the chlorine is released by solar radiation and starts to destroy ozone. It is believed that chlorine has already destroyed about 1 percent of the ozone (1975). Continued use of these chemicals is bound to exacerbate the problem.

In summary, human activity, particularly industry, can affect climate in various ways. In most cases, we do not know the magnitude of the changes, and in some cases, not even the sign. But, because this matter is so important, we must attack these questions by using mathematical models and measurements of the chemical and physical characteristics of the atmosphere, and by carefully monitoring the worldwide increase of a number of contaminants.

8 Weather and water

In regions of plentiful annual precipitation, we frequently underestimate the importance of water, and yet water, next to air, is our most vital need. Without water, there would still be weather, but it would bear little resemblance to the weather as it is now. Cyclones and anticyclones would still exist, spawned by the north-south temperature difference. But without the tempering effects of the oceans, these arid storms would probably be nightmarish whirlwinds of dust, first scorching the earth with torrid heat from the tropical deserts, then chilling the hostile landscape with glacial cold. This chapter considers some of the many aspects of water, from the formation of precipitation to the use of fresh water by man.

8.1 Water—Too Much and Too Little

When meteorologists predict a heavy rainfall, their job is not over; in fact, in some cases, it has just begun. When a drenching cyclonic storm occurs in a river basin, hydrometeorologists try to predict how much of the precipitation will reach the river, and when. Their efforts are directed toward providing necessary flood information to reservoir operators, industrial and domestic interests on the river's floodplain, generators of electric power, and agriculturalists. In times of drought, the same hydrometeorologists are responsible for forecasting river low-flows for water pollution regulation, power generation, and municipal water supplies. In order to meet the needs of the water-dependent community, these river forecasters must have an intimate knowledge of the complicated and sometimes tortuous route that precipitation must follow to reach the stream channel.

In some parts of the world, such as the western United States, an even more critical water problem is prevalent. There is not enough water to satisfy everyone's desires, and a rationing system has been established in the form of a water law which gives first priority to those who hold the oldest water rights. After the oldest users are supplied, the more recent claimants have their chance, if any water remains. In years with drier than average weather conditions (about every other year), there is not enough streamflow to supply everyone. In humid regions of the United States, people can hardly believe that such water rationing exists, because there is an adequate amount of water for nearly everyone. The only time that shortages may occur in humid areas is when drought conditions persist for an extended period, such as those which occurred in the northeastern United States in the years 1962–1966. Such shortages have fortunately been temporary, however, and normally, residents in humid areas have water to spare.

Plans for transporting water from humid areas to persistently dry areas have been suggested; for example, the North American Water and Power Alliance (NAWAPA) proposed to transport water from Alaska and Canada to the Canadian prairies, the arid southwestern United States, and Mexico. However, means of conserving existing water in dry regions seem more practical. To do this, water managers (and river forecasters) must have detailed information about how much precipitation reaches the stream channel. Such movement of water above, on, and beneath the surface of the earth is called the *hydrologic cycle*. To better understand this cycle, it may be useful to examine more closely the properties of water and its distribution over the earth.

8.2 Unusual Properties of Water

Water, although common, is very unusual because of its many strange physical properties. One of these properties is that it is the only substance that exists in all three states—vapor, liquid, and solid—at temperatures normally found at the earth's surface. The fact that the vapor phase of water commonly exists is of direct importance to the meteorologist. Of course, condensation of water vapor to liquid is an important weather process because it is necessary before precipitation can develop. This change of phase is doubly significant because the amount of latent heat released during condensation is quite large. The latent heat of condensation of water is about 600 calories per gram (which is greater than that of any other commonly occurring substance) and is sufficient to raise the temperature of six grams of water from the melting point to the boiling point. As a result, when significant amounts of water vapor condense, the latent heat released becomes an important source of energy for the maintenance of atmospheric processes. Water vapor is also an extremely good absorber of selected wavelengths of radiation emitted by the earth, and, as a result, influences the energy balance of the atmosphere.

The percent by volume concentration of water vapor in the atmosphere varies both temporally and spatially, from near zero at certain times in desert regions to as high as 4 percent at other times in tropical jungles.

In addition to using up atmospheric heat through evaporation, water bodies can absorb a great deal of heat without becoming much warmer themselves because of the high heat capacity of water. (In fact, the only liquid having a higher heat capacity than water is ammonia.) This characteristic means that water bodies such as large lakes, seas, and oceans have a marked moderating effect on climate, as was discussed in Chapter 3.

8.3 The Many Faces of Precipitation

In previous chapters, we noted again and again how warm air is able to contain more water vapor than an equal mass of cold air (Figure 8.1) and how cooling the air below its

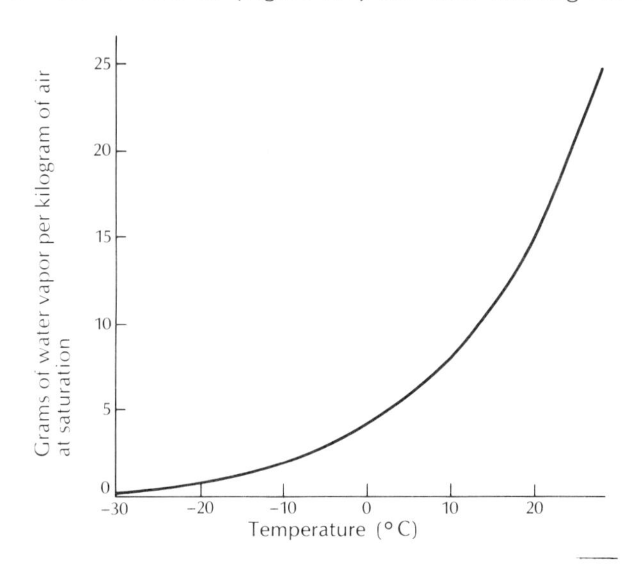

Figure 8.1 *Maximum possible amount of water vapor a kilogram of air can hold at different temperatures.* Note that air at 20°C can hold approximately twice as much water vapor as air at 10°C.

dew point first produces clouds, and then rain or snow. But, we have only touched on how the various types of precipitation—rain, sleet, graupel (soft hail), hail, snowflakes, and snow grains actually form, grow, and fall to the earth. It is important to understand the *microphysical* processes (physical processes which occur on a very small scale) by which these varied forms of liquid and solid water are produced in order to appreciate and evaluate some of the weather modification schemes likely to be proposed repeatedly in the future (e.g., rainmaking, fog dispersal, hail prevention, and even hurricane modification).

To describe the formation of water drops and ice crystals, we set aside the large-scale motions of the atmosphere and center our attention on the tiny world of the cloud, where relevant distances are measured in micrometers rather than miles. The typical sizes of the components of the precipitation process are illustrated in Figure 8.2. We first discuss the condensation process when temperatures are above freezing, and then introduce the complicated, but extremely important, effects of subfreezing temperatures.

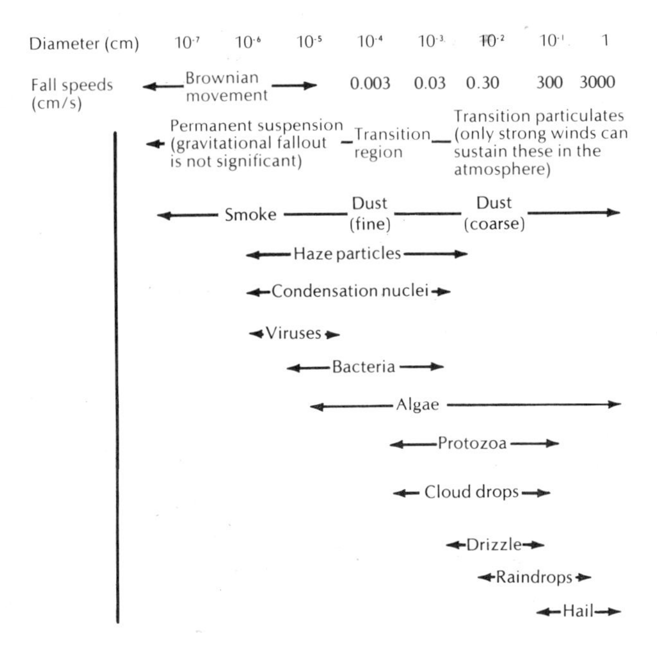

Figure 8.2 *Typical sizes and terminal velocities of particles and forms of precipitation in the atmosphere (from Patterns and Perspectives in Environmental Science, National Science Foundation, 1973).*

8.3.1 Condensation in warm clouds

The beginnings of all visible water droplets in nature start with microscopic (typical radius 10^{-5} centimeter) particles called *condensation nuclei,* which are simply any of the abundant particles of dust that are al-

Plate 11 (above) *Cumulus clouds over Everglades grass fire (photo by Ronald Holle).*

Plate 12 (below) *ERTS-1 view of the flooded Mississippi River near St. Louis.*

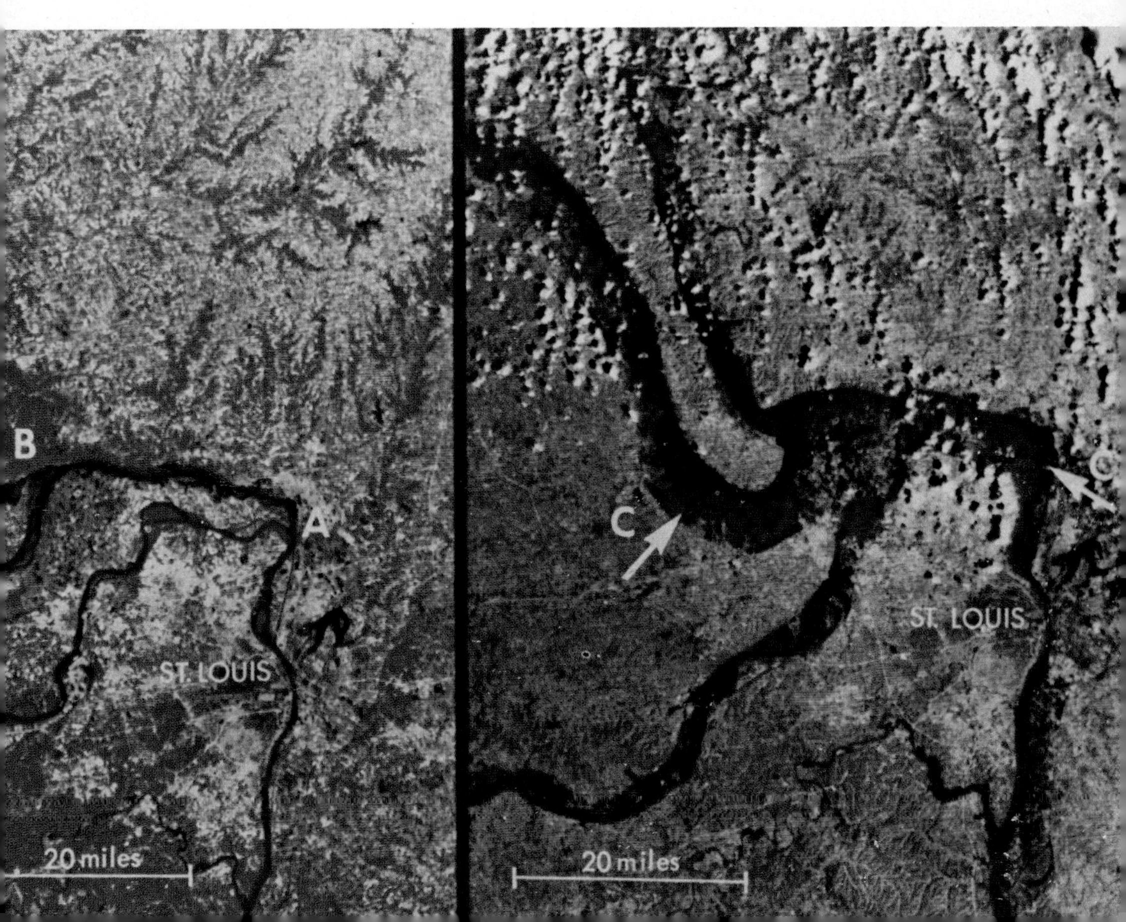

Plate 13 *Circumhorizontal arc as seen over Mt. Olympus, Washington State (photo by Steven Hodge. © Alistair B. Fraser).*

ways present, even in the purest arctic air. Although condensation may occur on any speck of dirt if the humidity becomes high enough (it sometimes must exceed 100 percent if *hydrophobic*—"having no affinity for water"—particles are present), certain particles are much preferred for condensation. These particles, which have a special affinity for converting water vapor to liquid drops even when the relative humidity (*RH*) is less than 100 percent, are called *hygroscopic* particles. Ordinary table salt is hygroscopic, turning wet with moisture on humid summer days, in spite of advertising claims to the contrary. Garden fertilizer, which includes nitrates of ammonia, is also very hygroscopic.

The perceptive reader will sense in the preceding discussion that there is nothing particularly magic about the 100 percent relative humidity figure when condensation of water vapor into water drops is considered. It is entirely possible to have cloud droplets forming vigorously in air with 98 percent relative humidity, just as it is possible, though not as likely in the atmosphere (which always seems to have enough hygroscopic "dirt" in it), to have perfectly clear air with a relative humidity of 101 percent. To understand these paradoxes, we note that relative humidity is defined as the ratio of the actual water vapor pressure (amount of water vapor in the air) to the saturation vapor pressure (equilibrium amount of water vapor that would occur in a closed volume of air over a flat plane of pure water).[1] The subtle point is that the equilibrium amount of vapor at saturation is defined with respect to a flat plane of pure water, but newborn cloud droplets are neither flat nor pure. The *effect of curvature* alone is to require higher relative humidities—perhaps 101 percent—before condensation will occur. On the other hand, impurities in the form of hygroscopic chemicals reduce the ambient relative humidity required for the onset of condensation, down to as low as 95 percent for large sea-salt particles. This latter effect, called the *solute effect,* is an important basis for warm cloud seeding, in which hygroscopic particles such as ordinary table salt are scattered into a cloud, increasing the rate of condensation.

In any case, at relative humidities near 100 percent, water vapor condenses around the hygroscopic particles, and the growth of the cloud droplets begins. Sea-salt particles, abundant near and over oceans where the evaporation of the salty water in breaking waves leaves behind giant salt nuclei (as large as 10^{-3} centimeter), are very efficient for the initial condensation process. These particles are responsible for the bluish haze over the oceans, even miles out to sea. Other hygroscopic particles that are efficient condensers of water vapor are produced by combustion, and are therefore abundant over cities.

[1] Saturation is an equilibrium state in which the pressure of the water vapor in contact with the liquid equals the surface tension on the liquid. The surface tension for curved surfaces is greater than for flat surfaces of water, and, therefore, curved droplets require a higher vapor pressure (and relative humidity) in the surrounding air for saturation to exist.

But the formation of cloud droplets does not water the cornfield, for the terminal velocity of a cloud drop with radius 10^{-3} centimeter is about 1 centimeter per second; at this rate of fall, it would take 84 hours for the water to reach the ground from the middle cloud level, which is about three kilometers. Furthermore, raindrops must usually survive a fall through an unsaturated layer between the base of the clouds and the ground. A droplet of radius 10^{-3} centimeter would survive for a distance of only a few meters through a layer of 90 percent humidity near the surface. Therefore, some process must increase the size of the billions of cloud droplets to a size which will make them heavy enough to fall to the ground in a reasonable amount of time (before they evaporate or are carried away by the horizontal winds).

It might seem plausible that the droplets would continue to grow by the condensation process; however, the growth rate by continued condensation on existing water droplets is too slow, because, as a drop grows in size, the hygroscopic solution responsible for the condensation in the first place becomes more diluted, and, therefore, less efficient in condensing water vapor. Indeed, most clouds never produce rain, as larger-scale changes in the synoptic flows (such as subsidence) dry the cloud before sufficient growth can occur through the simple condensation process. For the occasional active clouds which do exhibit rapid growth and fallout of precipitation, *collision* and *coalescence* come to the rescue. Figure 8.3a illustrates these two processes in warm clouds.

Collision is simply the bumping together of two cloud droplets. Collision is favored between drops of different sizes and, therefore, different terminal velocities. The large drops overtake the slower-moving small drops.

In many cases, the drops will bounce off each other, but if they merge after the collision, they are said to coalesce. Therefore, coalescence is the combination of two drops into one larger drop as a result of collision. Coalescence is favored if the droplets have opposite charges or if the charges on each drop are separated, since opposite charges attract each other. Thus, atmospheric electricity plays a role in the growth of cloud droplets.

While collision and coalescence are sufficient to produce copious rainfall in the tropics, where relative humidities are high and hygroscopic nuclei are abundant, another process, involving the growth of ice crystals rather than raindrops, is efficient in middle latitudes. Even in summer, much of the rain that falls over the United States begins as snowflakes at elevations of 3–5 kilometers, melting before reaching the ground in the warm air near the surface. Thus, during a warm, muggy thundershower, it may be refreshing to know that a raging blizzard is occurring a scant three miles away—directly overhead!

8.3.2 Role of ice in the precipitation process

The ice phase complicates the precipitation process enormously, because direct transition from each of the three phases of water to any of the other phases is possible. Thus, frost is not frozen dew, nor snow frozen rain, but they are instead results of a direct conversion of vapor to solid (called *deposition*) at subfreezing temperatures.

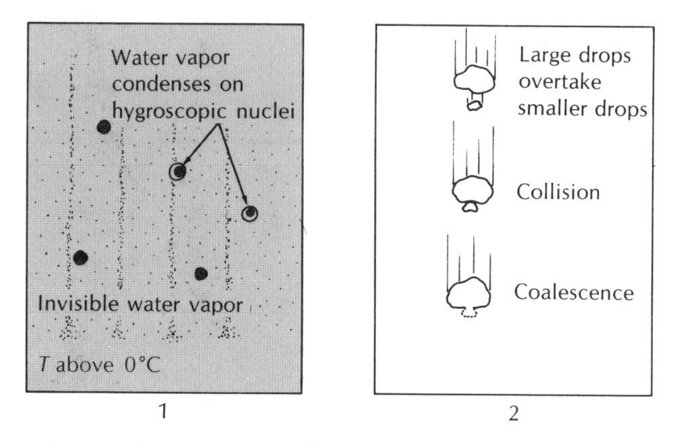

(a) Growth of water droplets by collision and coalescence

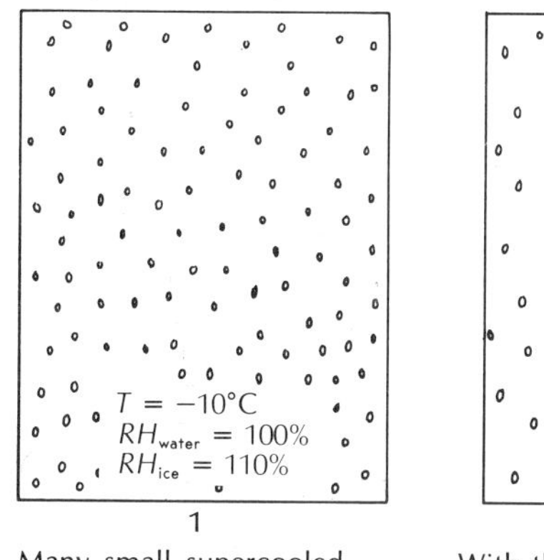

1

Many small supercooled waterdrops falling very slowly.

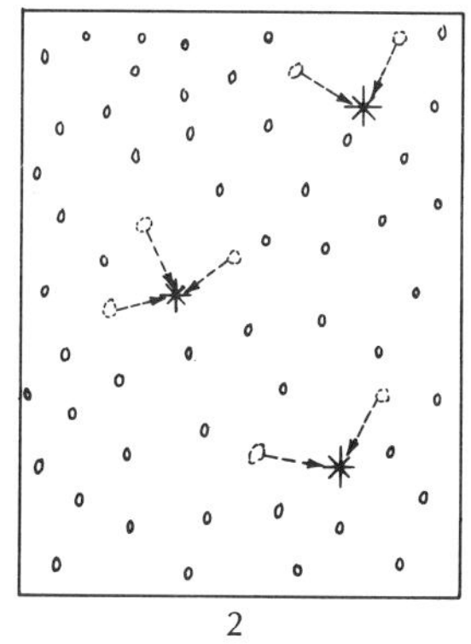

2

With the introduction of freezing nuclei, ice crystals form, which then grow at the expense of water droplets.

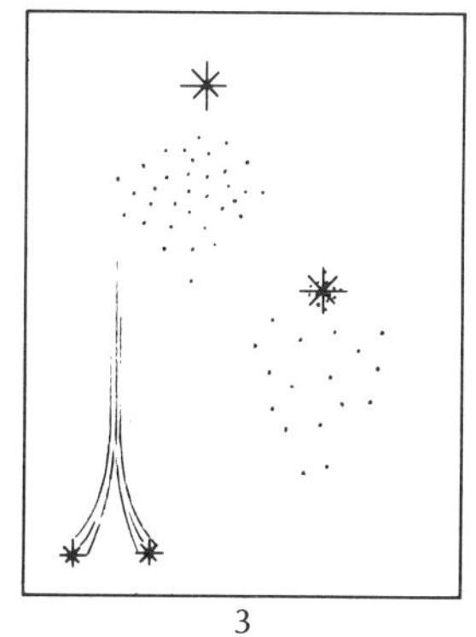

3

Falling ice crystals may directly pick up water droplets, which freeze upon contact, producing riming. Others may fracture, producing additional small ice crystals, which serve as freezing nuclei.

(b) Growth of ice crystals at the expense of water droplets

Figure 8.3 *Precipitation process in (a) warm and (b) cold clouds.*

One of the most important properties of water is that the saturation vapor pressure is less over ice than over water, which means that air of 100 percent relative humidity (calculated with respect to liquid water) has a relative humidity considerably greater than 100 percent when calculated with respect to ice. The difference, which depends on temperature, is shown in Figure 8.4. Thus, if the temperature is −10°C and the humidity is 100 percent with respect to supercooled water, the relative humidity with respect to ice is 110 percent.

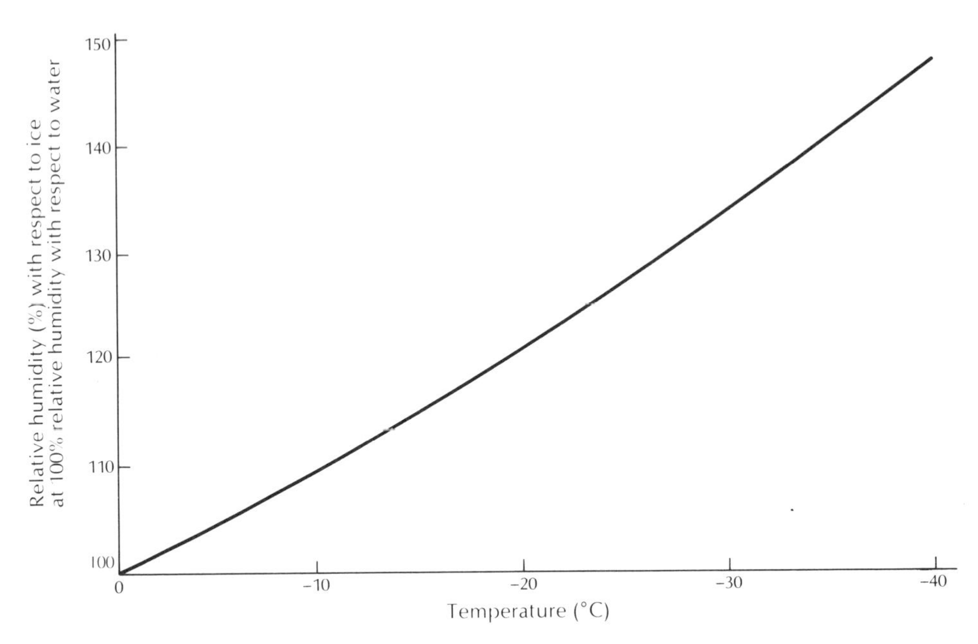

Figure 8.4 *Relative humidity with respect to ice when the air is saturated (100 percent relative humidity) with respect to water.*

The difference in saturation vapor pressure means that water droplets and ice crystals cannot peaceably coexist in the same cloud, for the air will always "appear" more saturated to the ice crystals than to the liquid droplets. Consequently, the ice crystals will continuously grow at the expense of the water droplets, which evaporate and yield the resulting vapor to the hungry ice crystals. The huge supersaturations (up to 150 percent) that are possible with respect to ice mean that the growth of ice crystals by this process can be very rapid, and so sizable ice crystals (large enough to fall to the earth before melting or evaporating) can be created. There is one catch, however; ice crystals seem reluctant to form, even at temperatures as low as −20 or −30°C, unless given a start by *freezing nuclei* (solid particles which have a structure resembling ice crystals). Ice crystals themselves are excellent freezing nuclei. When the temperature reaches −40°C (also −40°F), even pure water freezes, without the benefit of freezing nuclei.

The growth of ice crystals through the process of robbing Peter (water) to pay Paul (ice) is the basis of the most common form of rainmaking—the seeding of clouds which contain supercooled water by freezing nuclei (usually silver iodide crystals, which resemble ice crystals). Because nature frequently does not have enough natural freezing nuclei (in contrast to the usual abundance of condensation nuclei), the precipitation process can frequently be nudged along by introducing artificial nuclei. Once the freezing process is initiated, a feedback process begins, with the first ice crystals growing, fracturing, and serving as freezing nuclei for other ice crystals.

An additional effect of seeding supercooled clouds with freezing nuclei that stimulate additional cloud growth and increased precipitation is that converting supercooled water to ice at 0°C liberates to the air an additional 80 calories of heat beyond the original 600 calories that were released when the vapor condensed into the cloud droplets. This extra heat produces additional buoyancy, which in marginally unstable air can make a tremendous difference in the upward growth of the cloud and the amount of precipitation produced. Explosive growths of 3 kilometers (10,000 feet) by towering cumulus clouds have been observed following seeding, although skeptics would argue that this growth would have occurred anyway. In spite of the skepticism, however, controlled experimentation has indicated that under favorable conditions important increases in precipitation through cloud seeding can be achieved.

The seeding of supercooled water clouds at the ground has also been used successfully to reduce fogs at airports. The seeding initiates a conversion of many suspended water droplets to a few heavy snowflakes, which fall to the ground. The amount of snow produced in this way is negligible, although it may slightly whiten the ground. Needless to say, this technique does not work on warm fogs—fogs which occur when the temperature is above freezing.

The actual form of the precipitation reaching the ground depends on which of the previous processes have occurred during the lifetime of the water particle (liquid, solid, or both) reaching the ground. Rain may occur either as a result of collision and coalescence or from melting of ice in the warm air near the surface. Snow occurs when the ice crystals remain in subfreezing temperatures throughout their tumble to the ground. Rimed flakes (flakes coated with a granular deposit of ice), snow pellets, and graupel (soft hail) are formed when snowflakes collide with supercooled water droplets, which freeze upon contact with the snowflakes, a process called *accretion*. Sleet can be either partially melted snowflakes or partially frozen raindrops. Hail, discussed in Chapter 6 with thunderstorms, results when ice pellets are tossed repeatedly upward through layers of supercooled water, acquiring a ring of ice with each upward thrust.

8.4 Distribution of Water in the United States

The total water supply of the earth, including water in oceans, glaciers, lakes, rivers, soil and rock, and the atmosphere, amounts to 1,500,000,000 cubic kilometers—an exceedingly large figure. If all of this water were readily available to people, there certainly would not be any water shortages in the future. However, water is distributed very unevenly, and most of it has a salt content that makes its direct consumption impossible. As a result, only a small fraction of the total water supply is available for use.

The relative amounts of water in the different portions of the hydrologic cycle are presented in Table 8.1. Over 97 percent of the total water supply is contained in the oceans and salt lakes. This water, unless desalinated, is not usable by people and, as a consequence, they must look elsewhere to satisfy their needs. It is also immediately apparent that approximately three-fourths of the total fresh water supply is relatively inaccessible, being held in glaciers, polar ice, and snowfields. Consequently, the amount of water that is easily available to people is a very small percentage of the total water supply. Atmospheric water only makes up 0.0010 percent of the total water supply at any given time. Even so, this is still a large absolute volume. This atmospheric water content can be expressed as the depth of water over the whole surface of the earth that would result if all the vapor were precipitated out. This depth is 2.5 centimeters (1 inch), which is equivalent to the average precipitation produced in only 10 days. Thus, there is a frequent cycling of moisture through evaporation, condensation, and precipitation. The general climatic factors of a region interact to determine how much atmospheric moisture is removed from the air and reaches the ground as precipitation. Observations show that for some large river basins in the Midwest, only 20 percent of the water vapor passing over them is precipitated.

Table 8.1 *Distribution of the world's estimated supply of water*

Type	Percent
Salt water	97.1370
Fresh water	
Ice and snow	2.2400
Ground waters	0.6129
Freshwater lakes	0.0090
Rivers	0.0001
Atmosphere	0.0010
Total	100.0000

Source: *Water, Earth and Man: A Synthesis of Hydrology, Geomorphology, and Socio-economic Geography*, Richard J. Chorley, ed. (London: Methuen and Co., Ltd., 1969.)

In Table 8.1, we can see some of the reasons why water shortages occur, even in humid areas. Rivers, which are the source from which people commonly extract their water supply, have the least amount of water stored. Lakes, which are also heavily utilized by people, also have a relatively small amount of the total water supply. Because these two components of the world's water supply have the least water, it is not surprising that we sometimes run out of water, especially in the dry, western United States, where the annual precipitation is less than 25 centimeters (10 inches) in most places, and high temperatures and low relative humidities produce high evaporation rates. It is unfortunate that the more abundant fresh water is located in inaccessible places, i.e., beneath the surface as groundwater, and in remote, frozen areas as ice and snow. It is projected that by the year 2000 most of the United States will be experiencing the water shortages the western states currently face, mostly because of an ever-increasing demand for water. We have to start answering questions about how and where we are going to get additional water to satisfy these needs.

The figures in Table 8.1 indicate that the water supply in lakes and rivers will soon be exhausted. One way to extend this supply, however, is to treat water that has already been used once and make it available for reuse. Another way would be to harvest additional water from the atmosphere. There have been many attempts at cloud seeding designed to increase rainfall, but only one type has much promise, at least in the immediate future. That is the seeding of orographic clouds on the sides of mountains to increase the water supply held in the snowpack for subsequent release during the spring melt period. Figure 8.5 shows the orographic uplift situation. When a flow of air

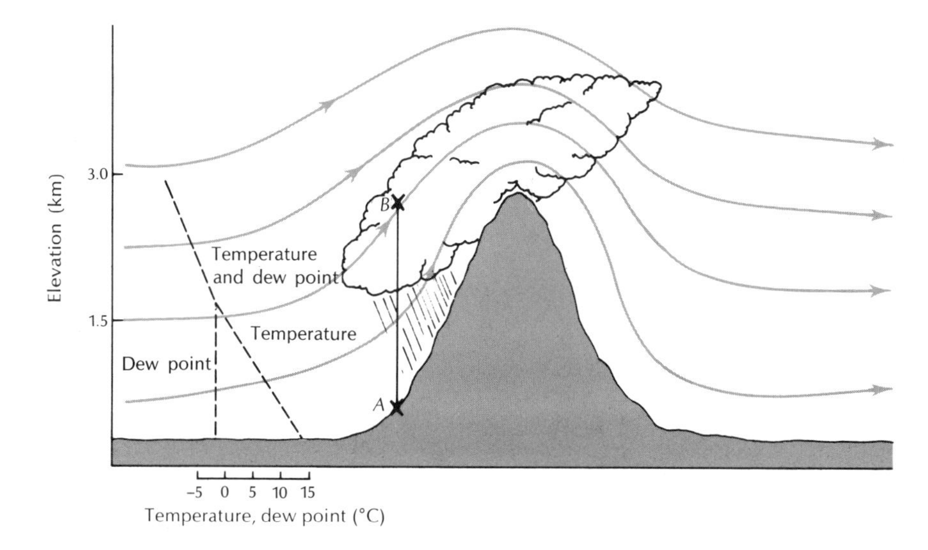

Figure 8.5 *Orographic lifting and precipitation.* Sounding of temperature and dew point (°C) taken along path *A-B.*

encounters a physical barrier such as a mountain, it first attempts to flow around the mountain. But if the mountains are relatively continuous, as are the Rockies, the air has no choice other than to move upslope and over the mountain. When the air rises, it expands, cools, and water vapor condenses, causing clouds to form. Hence, we frequently observe clouds hanging over mountain peaks. Because the physical barrier is stationary, this meteorological situation is repeated many times during the winter snow season, with the clouds approximately in the same place each time. Due to this consistency, cloud-seeding generators can be placed at strategic locations on the windward mountain slope and turned on at appropriate times to release silver iodide particles, which are carried up into the cloud by the wind currents. As we have seen, silver iodide closely resembles ice in its crystal structure and actually "fools" water drops existing in the cloud. When ice crystals and supercooled water coexist, the ice crystals grow at the expense of the water droplets, which evaporate. Further accretion of supercooled droplets on the ice crystals increases their size and causes them to fall out of the cloud in numbers greater than for an unseeded cloud. Such activities are being carried out quasi-operationally by the U. S. Bureau of Reclamation in the San Juan Mountains of Colorado. The main problem with this type of weather modification is that, at present, it is only really efficient in an orographic lifting situation, and these potential orographic seeding sites are limited in number.

As a result, we must turn to other sources of water in flat areas, such as the groundwater supply. The methods for the drilling of groundwater wells are refined and can be easily used in most areas. To allow more efficient use of the groundwater, however, we have to learn more about the location of these subterranean resources through exploratory drilling programs. Environmental considerations are also important. In some areas, the groundwater resource has accumulated over an extremely long time period. Therefore, pumping consists of mining the groundwater, because the supply can only be replenished over centuries. In these areas, a precise knowledge of the amount of water available must be combined with a judicious water development plan to assure enough water for future generations. A more immediate environmental problem exists in arid regions where heavy pumping of groundwater has already occurred, such as in the cities of Phoenix, Arizona and Las Vegas, Nevada. Not only are these cities rapidly running out of water, but they are also sinking. When water is removed from the underlying soil and rock, spaces filled only with air are left. As these cities have increased in size, so has their weight, which has compressed the voids and caused the cities to sink, in some places, on the order of a foot or two per year. In fact, some residents of new office buildings with previously choice first-floor locations are now in the basement!

Realizing all of these problems, it is no wonder that many people have looked longingly at seawater. Desalination is the object of extensive research, and there are already a number of saltwater purification plants operating along the coasts of the United States and in arid countries like Kuwait. The various processes work successfully, but are costly, averaging around $1 per 1000 gallons (1970 data). This price is much greater than the $0.04 per 1000 gallons (1970) we are accustomed to paying. Eventually, we may reach the point where the scarcity of fresh water will make desalination a viable alternative, but for now we must search for other sources.

Another possible alternative is the utilization of the ice and snow storage component, which is the largest, but also most inaccessible, component of fresh water. It has been shown that glaciers and late summer snowfields can be "milked" by spreading carbon-black materials over the surface to decrease reflectivity and consequently increase melt during dry periods. A problem with this scheme is that the operation of evenly spreading these reflectivity-reducing materials is quite formidable. Additionally, a black glacier is not nearly as aesthetically pleasing as a white one.

There remains, however, one major source of freshwater ice that can be exploited, and the plan for accomplishing it is one of the most novel and interesting proposals ever advanced. This scheme advocates the capture and towing of an iceberg (which is made up of fresh water) to a coastal distribution area. Here the iceberg would be grounded and allowed to melt. The meltwater would be collected and pumped into existing water supply systems for use. Amazingly enough, this particular proposal has precedents. In the nineteenth century, ships made trips to Alaska to collect freshwater ice to put into their holds and to transport back to San Francisco. More recently, airplanes stopping in Greenland harvested glacier-derived ice cubes for use in the United States in mixed drinks. Thus, towing icebergs is a logical extension of earlier ideas! Let us look at the iceberg towing theory in more detail.

First of all, although icebergs are cost-free, not all of them are suitable for towing. Most of the ideal large, smooth, disc-shaped icebergs are found in the antarctic. In the arctic, the icebergs are very irregular and very dangerous to tow because of their tendency to capsize. In general, the antarctic would be the best place to "capture" the proper stable iceberg. It has been shown that the Earth Resources Technology Satellite (ERTS-1) can be used to locate these huge, flat icebergs, which may measure 10 kilometers X 20 kilometers and stand 50 meters above the water. Figure 8.6 shows an ERTS-1 view of some likely candidates.

Figure 8.6 *ERTS-1 satellite view of icebergs.*

The problem is the towing of these monster pieces of ice. Actually, rather than tow the icebergs, a future "super tug" or a tandem of three already existing giant tugs would steer the iceberg along favorable ocean currents to its destination. It is suggested that it would cost about $1 million to pay for the tugs and crew for the one year required to locate, capture, and tow the iceberg. to port, and another $500,000 for handling and distribution of the meltwater. Even though up to 70 percent of the original iceberg would melt in transit, enough would remain to supply water to a major city for at least six months. More than likely, countries in the Southern Hemisphere would be the best candidates to utilize these icebergs. Chile's extremely dry Atacama Desert area would be ideal.

Although many of these plans for developing new sources of fresh water seem reasonable, most efforts will be directed toward extracting more water out of our conventional sources—rivers and lakes. As a result, we will have to understand better how the hydrologic cycle works. Another reason for knowing more about the surface water flow is that the energy shortage dictates that we obtain new and additional sources of power. One way to do this would be to construct more hydroelectric power dams. From past experience, we know that dam building is risky and that environmental effects are sometimes drastic. Reservoirs many times collect silt at a rate accelerated by upstream development and fill up much too rapidly. Additionally, most of the good available reservoir sites have already been used. Finally, dam builders must know with certainty that they will get enough water to fill the reservoir. Such knowledge requires a detailed understanding of the hydrologic cycle.

8.5 The Hydrologic Cycle

Figure 8.7 is a schematic diagram of the hydrologic cycle; it is evident that it has no end and that water is continually circulating through it. Water evaporates from the ocean and enters the atmospheric water vapor reservoir. Some of this water vapor condenses to form clouds, and some of these in turn develop into rain clouds through microphysical processes. As precipitation falls toward the earth, a portion of it may evaporate before it reaches the ground surface; when visible as gray streaks trailing beneath clouds, this phenomenon is called *virga*. Another factor that affects precipitation before it strikes the ground is vegetation. Dense vegetation, such as forests of Douglas firs, may intercept as much as 40 percent of falling precipitation and store it until the storm ends, at which time it evaporates back into the atmosphere, a process called *evaporative interception loss*.

Once precipitation reaches the ground, it may follow one of three courses. It may collect in pools and puddles as surface storage and eventually be evaporated into the atmosphere. Second, it may flow over the surface into rills and small channels to become surface runoff which finds it way into streams and lakes. The third alternative is infiltration through the surface soil layer to join existing soil moisture. After infiltration, the water may be held in the soil-moisture reservoir and later evaporated or transpired into the atmosphere.

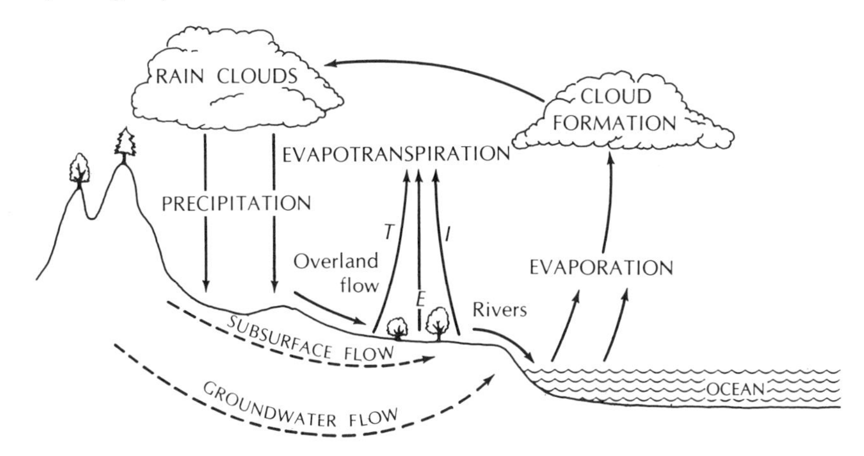

The hydrologic cycle (E = evaporation, T = transpiration, I = evaporative interception loss).

Figure 8.7 *The hydrologic cycle.*

Evaporation occurs directly through the surface pore spaces, whereas transpiration is a more complicated process in which plant roots absorb soil moisture and transfer the water through the stems to the leaves, where plant water is evaporated through the stomata (leaf pores). These two components combine with the previously mentioned evaporative interception loss to make up the term *evapotranspiration*.

8.5.1 The precipitation component of the hydrologic cycle

Meteorologists are mostly interested in the precipitation component of the hydrologic cycle, and for good reason. Precipitation directly or indirectly is the cause of many varied effects, both beneficial and detrimental. On the positive side, precipitation spurs plant growth, increases water supplies, not only for human demands, but also for generation of hydroelectric power, allows navigation on inland waterways, provides nature's part for snow and water sports, cleans the air by removing particulate matter and other pollutants, and washes dust off vegetation and artificial urban surfaces. Conversely, precipitation can be responsible for causing large amounts of economic loss. Excessive precipitation can lead to flooding, especially near urban areas where development has occurred on the floodplains. In mountainous areas, snowstorms lead to avalanches, which can cause considerable damage and loss of life. On soil surfaces not protected by vegetation, the force of rain causes the soil to be eroded, with attendant downstream sedimentation problems. In addition, significant precipitation events very often cause a temporary disruption of transportation and communication.

When considering the influence of precipitation on streamflow, the type of precipitation is of great importance. For example, if precipitation falls in the summer in the form of rain, its effect is felt almost immediately. Conversely, if snow falls during the winter, the water is very often stored in the snowpack and may not reach the stream channel until the spring melt season. We find that this latter process is extremely important in mountainous regions of the United States. It is thus convenient to divide the effects of precipitation into delayed storage (snow) and relatively immediate influences on streamflow (rain). These more immediate effects can only be examined by considering specific characteristics of rainfall.

Rainfall intensity (i.e., the amount of precipitation occurring in a given period of time) is extremely important in determining the rapidity with which runoff reaches the stream channel. If low-intensity rainfall occurs, usually all of the rainwater will infiltrate the soil and take a relatively long period of time to reach the stream. If, however, much rain falls in a short time period (e.g., in thunderstorms), the rainfall intensity may be high enough to exceed the infiltration capacity of the soil. In such a case, not all of the falling water will be absorbed by the soil, and the residual will become surface runoff and reach the stream quickly (usually within an hour of the onset of precipitation). The potential for flooding is very great here. Further complications arise when soil type and attendant infiltration capacities are considered. Sandy soils can usually absorb all the rain even the largest storms can produce and yield no surface runoff. Clay soils, on the other hand, have very low infiltration capacities, and, as a result, even low-intensity rains are likely to produce some surface runoff.

Another important rainfall characteristic to consider is rainfall duration. Duration is important because the infiltration capacity (i.e., the ability of a soil to absorb water) decreases with time during a rain. As rain falls and more water is absorbed by the soil, the total water-holding potential of the soil is decreased, with the result that not as much water is allowed to infiltrate. As a storm continues to produce rain, the infiltration capacity is continually being reduced and may eventually become small enough so that even low-intensity rainfall will exceed it. When this occurs, surface runoff will be produced. In extreme conditions where rains continue over an extended period (e.g., after a hurricane moves inland), enough water percolates through the soil so that the water table may be raised to the surface in low-lying areas. As a result, the infiltration capacity is reduced to zero in these areas, and a serious flood hazard is created.

8.5.2 Floods

In many cases, major floods result, not from widespread rainy conditions, but rather from intense thunderstorms that may only cover a few square miles. An enormous amount of water can be deposited (up to an inch in one minute) in a very confined area over such a short period of time that the soil can only absorb a small portion of it; the rest leaves the watershed as surface runoff. If the same amount of rainfall were distributed over the whole watershed rather than in just one portion, probably no runoff at all would result. This type of local event is often referred to as *flash flooding,* which is the rapid peaking of streamflow and the subsequent rapid recession of flow from the peak. Flash floods are sometimes considered most dangerous because it is relatively difficult to predict when and where they will occur (mainly because it is difficult to predict the occurrence of thunderstorms). Floods resulting from widespread rains are usually easier to predict, but involve a considerably larger volume of flood water (thus, their inherent danger).

Some situations may arise where heavy rains occur frequently over a period of several months so that large-scale flooding conditions result. Such was the case in the winter of 1972–1973 in the Mississippi Valley, where in excess of 200 percent of normal rainfall occurred. The result was the severe 1973 spring floods which inundated land from near St. Louis all the way down the Mississippi River to the Gulf of Mexico. Plate 12 shows an ERTS-1 satellite view of the flooding in the St. Louis area. North of St. Louis, in the scene on the left taken under normal flow conditions, the Mississippi River joins the Missouri River at point *A*, and farther upstream, the confluence of the Illinois and Mississippi rivers is noted by the letter *B*. The photograph on the right of the same region on March 31, 1973, shows areas under water (C) as a result of flooding. In this false-color infrared view, the darkest blue tones indicate areas of deepest water. At the time this picture was taken, the river stage at St. Louis was 38 feet and rising, and already about 300,000 acres were covered by water. Subsequent monitoring of this flood and other floods by space satellites has aided in our efforts in flood relief, flood-damage assessment, and the zoning of flood-prone areas.

8.5.3 Droughts

Perhaps no other agricultural weather disaster is so frustrating to the farmer as a prolonged drought. Hail, tornadoes, and even floods, do their damage quickly and then depart, but drought lingers day after day, gradually taking its toll, first on vegetation, then on livestock, and finally on the earth itself, as hot winds carry away the soil. The several-year drought on the Great Plains in the 1930s which produced the dust-bowl conditions drove thousands of newly settled farmers, lured by the more bountiful rains of the 1920s, from their homes. Nowhere is this tragedy more poetically and poignantly described than in John Steinbeck's Nobel prize-winning novel, *The Grapes of Wrath:*

> *To the red country and part of the gray country of Oklahoma, the last rains came gently, and they did not cut the scarred earth. The last rains lifted the corn quickly and scattered weed colonies and grass along the sides of the roads so that the gray country and the dark red country began to disappear under a green cover. In the last part of May the sky grew pale and the clouds that had hung in high puffs for so long in the spring were dissipated. The sun flared down on the growing corn day after day until a line of brown spread along the edge of each green bayonet. The clouds appeared, and went away and in a while they did not try any more . . . The surface of the earth crusted, a thin hard crust, and as the sky became pale, so the earth became pale, pink in the red country and white in the gray country . . .*

Then it was June, and the sun shone more fiercely. The brown lines on the corn leaves widened and moved in on the central ribs. The air was thin and the sky more pale; and every day the earth paled . . . the dirt crust broke and the dust formed. Every moving thing lifted the dust into the air: a walking man lifted a thin layer as high as his waist, and a wagon lifted the dust as high as the fence tops, and an automobile boiled a cloud behind it. The dust was long in settling back again.

When June was half gone, the big clouds moved up out of Texas and the Gulf, high heavy clouds, rain-heads. The men in the fields looked up at the clouds and sniffed at them and held wet fingers up to sense the wind. And the horses were nervous while the clouds were up. The rain-heads dropped a little spattering and hurried on to some other country. Behind them the sky was pale again and the sun flared. In the dust there were drop craters where the rain had fallen, and there were clean splashes on the corn, and that was all.

A gentle wind followed the rain clouds, driving them on northward, a wind that softly clashed the drying corn. A day went by and the wind increased, steady, unbroken by gusts. The dust from the roads fluffed up and spread out and fell on the weeds beside the fields, and fell into the fields a little way. Now the wind grew strong and hard and it worked at the rain crust in the corn fields. Little by little the sky darkened by the mixing dust, and the wind felt over the earth, loosened the dust, and carried it away. The wind grew stronger. The rain crust broke and the dust lifted up out of the fields and drove gray plumes into the air like sluggish smoke. The corn threshed the wind and made a dry, rushing sound. The finest dust did not settle back to earth now, but disappeared into the darkening sky.

The wind grew stronger, whisked under stones, carried up straws and old leaves, and even little clods, marking its course as it sailed across the fields. The air and the sky darkened and through them the sun shone redly, and there was a raw sting in the air. During a night the wind raced faster over the land, dug cunningly among the rootlets of the corn, and the corn fought the wind with its weakened leaves until the roots were freed by the prying wind and then each stalk settled wearily sideways toward the earth and pointed the direction of the wind.

The dawn came, but no day. In the gray sky a red sun appeared, a dim red circle that gave a little light, like dusk; and as that day advanced, the dusk slipped back toward darkness, and the wind cried and whimpered over the fallen corn.

Men and women huddled in their houses, and they tied handkerchiefs over their noses when they went out, and wore goggles to protect their eyes.

When the night came again it was black night, for the stars could not pierce the dust to get down, and the window lights could not even spread beyond their own yards . . . In the morning the dust hung like fog, and the sun was as red as ripe new blood. All day the dust sifted down from the sky, and the next day it sifted down. An even blanket covered the earth. It settled on the corn, piled up on the tops of the fence posts, piled up on the wires; it settled on roofs, blanketed the weeds and trees.

The people came out of their houses and smelled the hot stinging air and covered their noses from it. And the children came out of the houses, but they did not run or shout as they would have done after a rain. Men stood by their fences and looked at the ruined corn, drying fast now, only a little green showing through the film of dust. The men were silent and they did not move often. And the women came out of the houses to stand beside their men—to feel whether this time the men would break. The women studied the men's faces secretly, for the corn could go, as long as something else remained. The children stood near by, drawing figures in the dust with bare toes, and the children sent exploring senses out to see whether men and women would break. The children peeked at the faces of the men and women, and then drew careful lines in the dust with their toes. Horses came to the watering troughs and nuzzled the water to clear the surface dust. After a while the faces of the watching men lost their bemused perplexity and became hard and angry and resistant. [From *The Grapes of Wrath,* by John Steinbeck. Copyright 1939, copyright © 1967 by John Steinbeck. Reprinted by permission of The Viking Press, Inc., New York.]

Heat and lack of precipitation are the meteorological factors responsible for most droughts. Both of these conditions occur with large-scale subsidence associated with warm anticyclones that extend from the surface through most of the troposphere. If these anticyclones are temporary, lasting only a few days, the drought is short-lived and the principal damage is confined to nonirrigated crops and annual vegetation. Even streams that depend solely on surface runoff will flow 8–12 days without rain.

The far more serious droughts occur over periods of months or years and are associated with major changes in the general circulation—the upper-level winds. As the wet ascending and dry descending portions of the waves in the westerlies reorient themselves in new positions, some regions experience drier-than-normal periods, while others, several thousands of miles away, have more-than-average rainfall. The drought of 1962–1965 over the northeastern United States is a good example. Table 8.2 shows the percentage of normal rainfall during the four seasons of the years 1962–1965 at New York City and Williston, North Dakota. While the East Coast was receiving only 60–70 percent of its normal rainfall, the Great Plains were blessed with above-average precipitation.

Table 8.2 *Percentage of normal rainfall by seasons (1962–1965)*

	Winter	Spring	Summer	Fall
New York	92	58	62	67
Williston, N. Dak.	101	172	125	90

Source: Jerome Namias, *Monthly Weather Review,* Sept. 1966.

The northeast drought was associated with cold temperatures and lower-than-normal pressures off the East Coast during the four-year period. This anomalously deep tropospheric trough, which was especially well developed during the spring, produced a persistent northwesterly flow of dry, cool air over the Northeast. Furthermore, the atmosphere is normally sinking behind troughs (recall Section 4.5.5), and so subsidence contributed to the absence of rainfall. Finally, the location of the trough axis off the coast, and the associated northwesterly winds over the eastern United States, steered the rain-producing cyclones eastward away from the coast, rather than northward along the coast as is typical during wetter times.

During the period of lower-than-average pressure off the East Coast, the pressures were also lower over the Rockies. Therefore, the Great Plains were frequently located on the eastern side of troughs of low pressure, which are favored places for upward motion and, hence, above-normal precipitation.

The reasons for such important, but subtle, changes in the general circulation are not well known, but seem to be related to anomalies in ocean temperatures. For example, the Atlantic water temperatures along the coast from Virginia to New England were 2°C lower than normal during the 1962–1965 drought. Note, however, that we said "related to" and not "caused by," for cause-and-effect relationships in meteorology are very difficult to establish. Thus, although it is tempting to say that the cold ocean waters caused the cold atmospheric temperatures and the deep trough, it would be equally plausible to say that the cold atmosphere and the stronger-than-normal northwesterly flow (which would produce upwelling of cold water) caused the cool sea temperatures.

8.5.4 *Snowfall*

Snow as a form of precipitation is a very important input to the hydrologic cycle. Its importance rests not only in the percentage of mean annual precipitation it comprises, but also in the time lag between its occurrence and when it melts, usually in the spring. Because of this delay, snowfall builds up over an entire winter season and then releases a large amount of water into the active hydrologic cycle over a very short time period, which may lead to flooding situations in some areas. For this reason, hydrometeorologists are less interested in when snow falls as in where it falls, how much has accumulated before the start of the melt period, and how rapidly melting occurs.

Snow also has profound economic and social implications. Heavy snowstorms can sever communication and transportation links and generally cause very hazardous conditions. In individual storms, the relationship between depth of snow and curtailment of human activity is very pronounced. Generally, as the annual accumulation increases at a location, the amount of snow necessary to produce paralysis conditions increases as well. Whereas in Muskegon, Michigan (about 80 inches average annual snowfall) paralyzing disruption occurs with a 14-inch-or-greater snowfall, Greensboro, North Carolina (about 11 inches average annual snowfall) experiences the same type of disruptive effects with a 4½-inch snowfall. The depth of snow is not the only parameter that determines how damaging a snowstorm will be. Snow of lower moisture content produces less difficulty than the wet variety, since it is easier to remove. Also, winds greater than 15 mi/h, when associated with snow, cause drifting of the snow across highways and, therefore, create a greater disruption than do light winds.

8.5.5 *Other aspects of the hydrologic cycle*

It is not sufficient to consider precipitation alone when attempting to evaluate the effects on streamflow. Other climatic variations such as solar radiation, temperature, relative humidity, and wind are important because they affect evaporation from a watershed. This evaporation causes soil moisture deficits which must usually be recharged before significant amounts of runoff will occur. In addition, physiographic features of a watershed interact with the climatic variables to influence the resultant streamflow significantly. Some of these physiographic features are elevation, slope, aspect, soil type, and vegetation.

Although the hydrologic cycle varies considerably from area to area (e.g., the annual rainfall amounts vary from less than 2 inches in Death Valley to greater than 80 inches in the mountains only miles away), it is useful for overall planning to take a look at the average annual water budget for the United States as a whole. This water budget is presented in Figure 8.8. On the average, 30 inches of precipitation fall on the United States, and of that total, 21 inches are evapotranspired back into the atmosphere. The remaining 9 inches

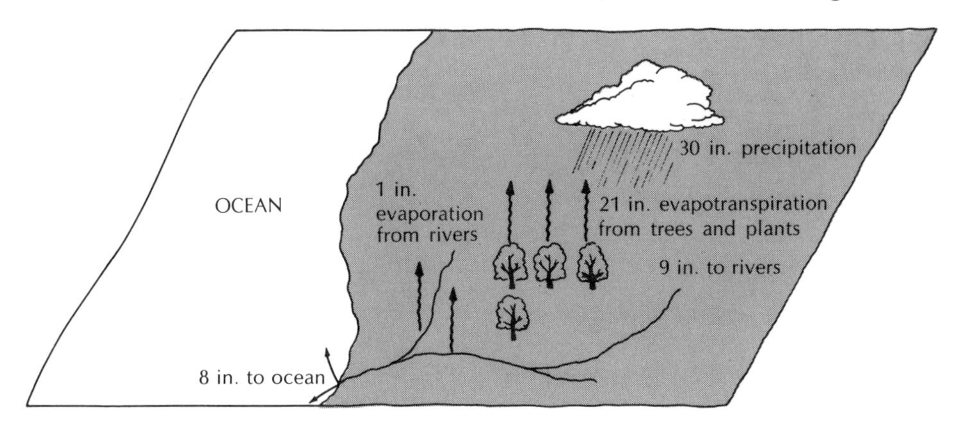

Figure 8.8 *Average annual water budget in the United States.*

make their way through the hydrologic cycle to become streamflow. Presently, about one-third is withdrawn from the stream for use, and two-thirds remains as residual flow. Of the 3 inches withdrawn, 1 inch is evaporated during use and 2 inches are returned to the stream, although they are considerably altered in quality. A total streamflow, then, of 8 inches finally makes it way to the ocean.

8.6 Conservation of Water in the Home

Conserving on the use of water in the home is one way individuals can solve some of our water supply problems. Figure 8.9 summarizes domestic water use in an average U.S.

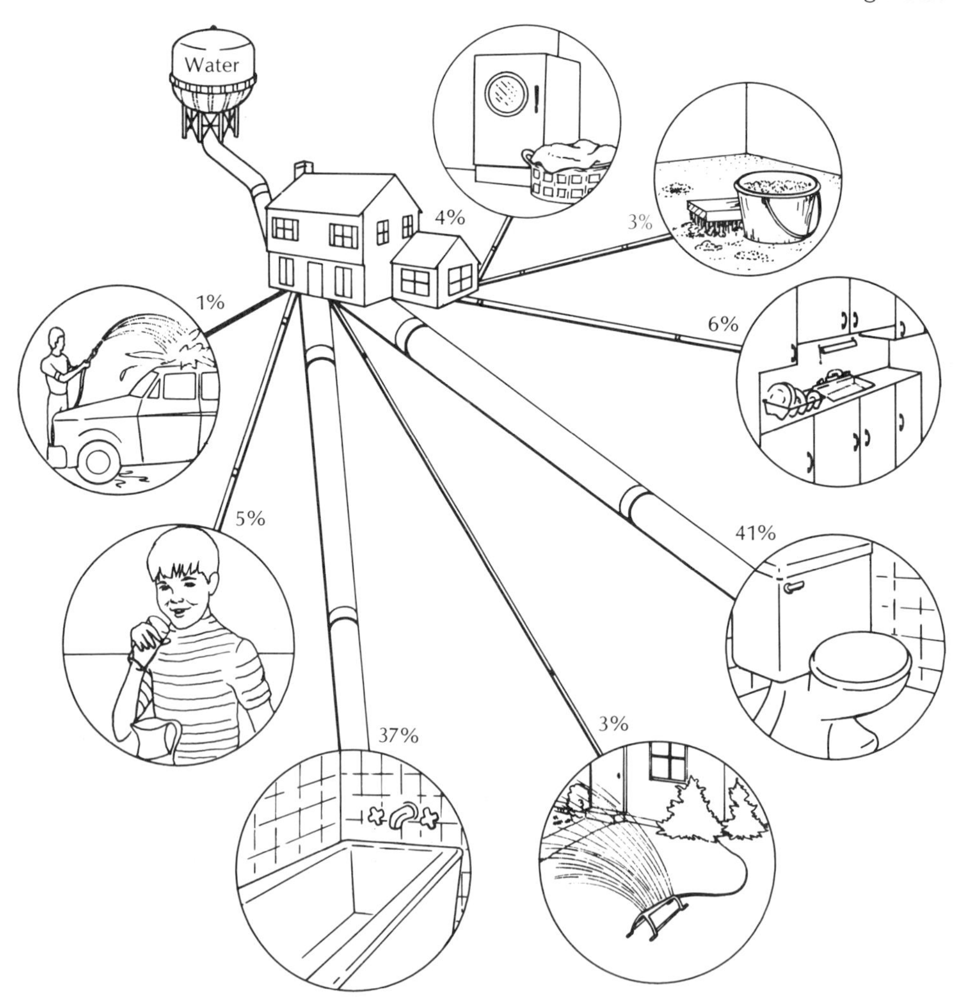

Figure 8.9 *Use of water in an average home in Akron, Ohio (source: U.S. Geological Survey, Water Supply Paper 1812, Washington, D.C., 1964, 164 pp.).*

city, Akron, Ohio. Some ways in which conservation effort could be effective would be to repair leaks and drastically reduce the watering of lawns (which is really unnecessary if the proper type of vegetation is planted). But the greatest savings can be made in the bathroom. A reasonably short shower uses only about one-third as much water as does a bath. Additionally, a tremendous volume of water is used to flush the toilet, and a minimum of flushing would be a great water saver. Insertion of bricks or water bottles into the toilet storage tank also helps.

More importantly, however, is the fact that the toilet is an uneconomical, ancient device. It has not changed in about a century, except that the old copper float is now plastic. This is an area where our engineers should devise a much better solution to waste disposal in order to conserve water. In London about 100 years ago, a contest was held to produce an invention that would conserve the amount of water continually running through homes to carry away waste. The winning entry was the "valveless waste water preventer" that exists in today's toilets. The winning inventor, Thomas Crapper, eventually received knighthood for his effort. In fact, you can read about him in a book entitled *Flushed with Pride: The Story of Thomas Crapper,* by Wallace Reyburn (Englewood Cliffs, New Jersey: Prentice Hall, 1971).

9 Meteorological optics[1]

Meteorology is the study of the atmosphere, and optics is the study of light. It is not surprising, therefore, to find that the modification of light by the atmosphere is as capricious as the weather and as variegated as the rainbow. Halos trace lacy geometric patterns across the dome of the sky, a corona provides an ephemeral splash of eye-blinding color, and a mirage turns distant objects into a carnival hall of mirrors. So rich is the range of observable phenomena that even the most casual observer can often be rewarded.

There are many ways that light shining through our atmosphere can be altered by both the air itself and the cloud and haze particles it contains. These include various combinations of four basic mechanisms: *reflection, refraction, diffraction,* and *scattering* (Section 2.1.4). Examples of each of these mechanisms are easy to recognize when they occur in the atmosphere, so from the point of view of basic physics, they would seem to provide a good basis for classifying the optical phenomena we observe. From the point of view of the meteorologist, though, it is not as important that both a halo and a mirage are caused by refraction as it is that a halo is caused by (refraction through) ice crystals and a mirage is caused by (refraction through) temperature gradients in

[1]This chapter was written by Alistair B. Fraser, Department of Meteorology, The Pennsylvania State University.

the atmosphere. We will therefore divide the optical phenomena into four broad classes, based on whether the light has been modified by ice crystals, water drops, molecules and dust, or temperature gradients.

9.1 Ice-crystal Optics—Halos, Sun Pillars, and Sun Dogs

The four wheels had rims and they had spokes; and their rims were full of eyes round about. [Ezekiel 1:18]

People have long looked with awe at the phenomena that appear in the sky, and they invariably interpret them in terms of their own perceptions of the forces that control their surroundings. Light refracted and reflected by ice crystals in the upper atmosphere is capable of producing a complex of rings, arcs, and bright spots of light, whose appearance is remarkably similar to the description that Ezekiel provides us in his vision.

If this interpretation is correct, it is undoubtedly true that Ezekiel was neither the first nor the last person to deduce divine messages from observing the optics of ice crystals. In A.D. 40, the Roman College of Soothsayers boded good from an observation of three suns that appeared in the sky. Nowadays, we could bode nothing but the presence of hexagonal-plate ice crystals, which, with the help of the sun and refraction, are able to produce two mock suns that sit to either side of the real sun. The Soothsayers, incidently, missed the mark, as the following year was not a good one for Rome.

The term *halo* is generic for all of the rings, arcs, and spots produced by the reflection and refraction of light by ice crystals in the atmosphere. It might be imagined that this cornucopia of patterns is a result of the almost legendary variability of the forms of the ice crystal. Not so, as only two major types of crystals, the hexagonal plate and the hexagonal column, account for almost all the observed halos. One of the reasons that these two crystals can produce so many different halo types is due to their aerodynamic properties. When the crystals are very small (diameter less than 15–20 micrometers), the constant bombardment by the rapidly moving air molecules is able to keep the crystals randomly oriented, rather like peanuts in a bag. This phenomenon, known as *Brownian motion,* ceases to be important when the crystals grow to the larger sizes where aerodynamic forces dominate: large hexagonal plates fall with their flat faces nearly horizontal, like dinner plates set out for guests, while the large hexagonal columns fall with their long axis horizontal, like wooden pencils scattered on the floor. We have, therefore, three different classes of crystal orientations: randomly oriented crystals of either type, or oriented columns, or oriented plates. Each of these crystal groups is capable of producing its own class of halos. In this book, we will discuss only a few of the over two dozen halo types that can be produced.

Reflection is familiar to everybody as the mechanism that enables us to see our face in a mirror or in a calm lake. When light shines on a smooth surface such as water or ice, it bounces off that surface so that it makes the same angle to the surface when leaving as it did when arriving. When large hexagonal-plate ice crystals are settling through the atmosphere, they behave like thousands of tiny horizontal mirrors which can reflect an image of the sun. This image, the *subsun* (illustrated in Figure 9.1), can only be seen from an air-

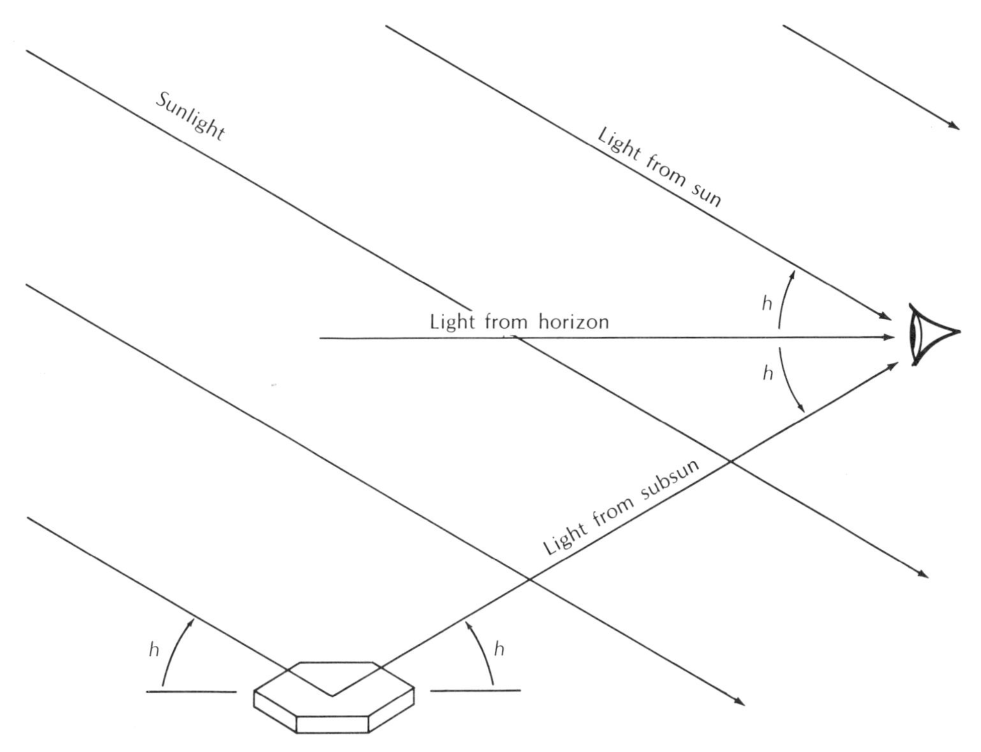

Figure 9.1 *Subsun.* If the sun is at an angular elevation *h* above the horizon, then its image reflected in a horizontally oriented plate will be seen at an angle *h* below the horizon.

plane or a high mountain, for it must appear as far below the horizon as the sun is above the horizon. If instead of plates there are large columns, then reflection off of the columns, which are free to rotate about their horizontal axis, will produce a pillar of light, much like the path of light that extends across a rough ocean surface to meet the setting sun. But in the atmosphere, this *sun pillar* can extend above the sun as well as below it.

When sunlight shines on a water or ice surface, not all of the light will be reflected; some of it passes into the material, and in doing so, is bent so that it travels at a larger angle to the surface. This bending, which is called *refraction* (Figure 9.2), is the reason that underwater objects can appear foreshortened when they are seen from above the surface of the water. (The light rays meet the object at a much smaller angle than if the water were removed.) Because the amount of refractive bending of the light rays depends on its color

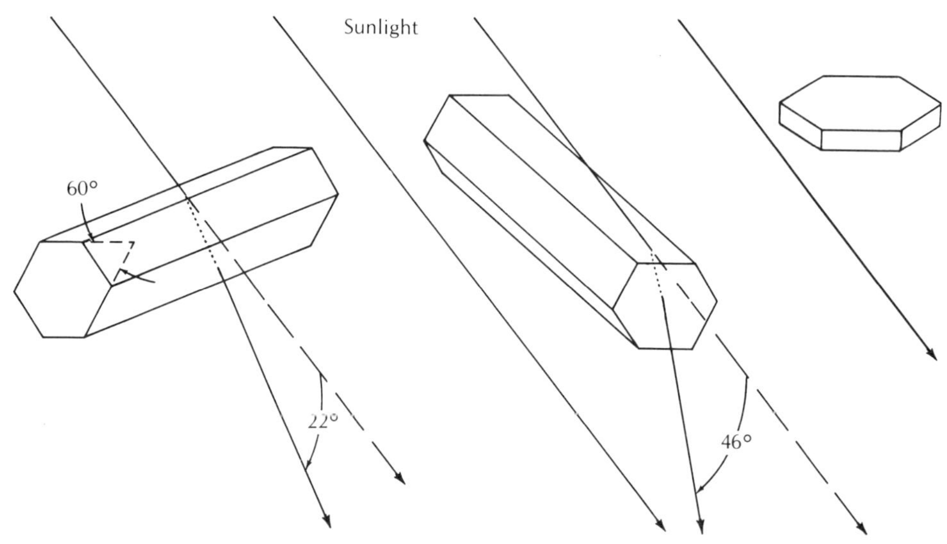

Figure 9.2 *Refraction of light by hexagonal ice crystals.* Refraction is illustrated for the 60-degree and the 90-degree prisms of a hexagonal column only.

(blue is bent more strongly than red), white light can be separated into its component colors. The familiar example of this is the prism. In the atmosphere, tiny hexagonal ice crystals may act as prisms.

A hexagonal ice crystal is really two prisms in one: a 60-degree prism and a 90-degree prism, as illustrated in Figure 9.2. Alternate hexagonal faces make an angle of 60 degrees to each other, and the light that shines through this 60-degree prism gets bent from its original direction by an angle that is equal to or greater than 22 degrees. A sky filled with small, and thus randomly oriented, hexagonal ice crystals is able to produce the *22-degree halo* (Figures 9.3 and 9.4), which is a ring of light that surrounds the sun at an angular distance of 22 degrees (about the width of your hand span on the end of your outstretched arm). Inside the ring, it is relatively dark, as light cannot be bent from the direction of the original sunlight by an amount less than 22 degrees, while outside the halo it is bright, because the crystals can bend the light by more than 22 degrees. As red light is bent less by refraction than the other colors, the inner edge of the 22-degree halo is red.

The sides and the ends of the hexagonal crystals make an angle of 90 degrees to each other. This 90-degree ice prism can bend light from its original direction by an angle that is equal to or greater than 46 degrees. Thus, it is possible to form a *46-degree halo* around the sun (it has a radius of about two hand spans). The 46-degree halo (Figures 9.3 and 9.4) can be observed, at least in part, only about once or twice a year, whereas the 22-degree halo can be seen upwards to a hundred times a year. An analysis that takes into account the optical properties of small prisms, the aerodynamic properties of falling ice

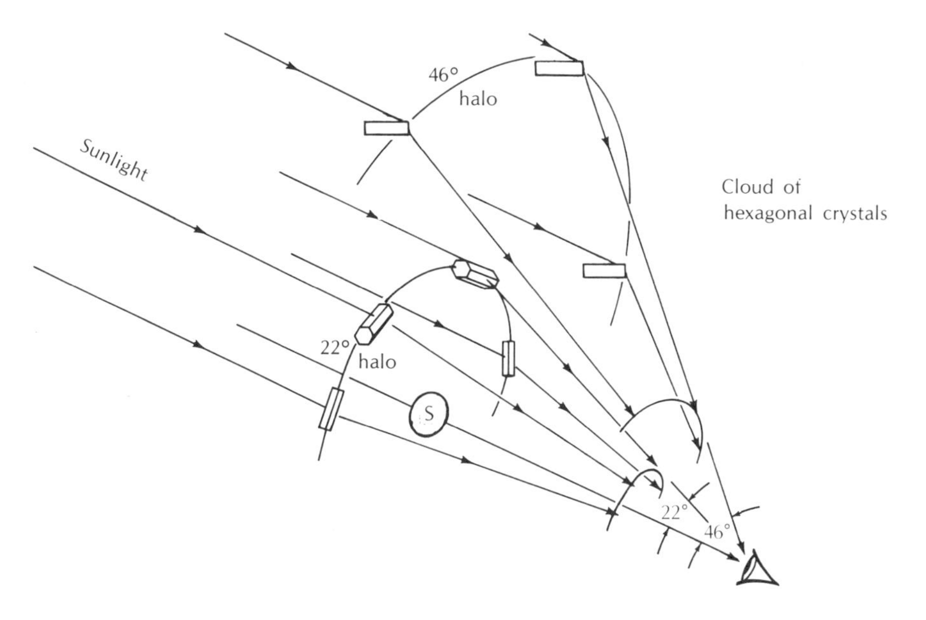

Figure 9.3 *Formation of 22-degree and 46-degree halos.*

Figure 9.4 *Fisheye (180-degree) view of 22-degree and 46-degree halos.* The sun has been blocked out to prevent lens flair (©*Alistair B. Fraser*).

crystals, and the crystallography of growing those crystals in the atmosphere reveals that the 46-degree halo can only be produced by a very narrow size range of hexagonal columns (diameters between about 15 and 25 micrometers) that have been grown very slowly. Neither plates, nor larger or smaller columns, nor rapidly grown columns (the 90-degree prism is not formed well enough) can produce the 46-degree halo. With such restrictions, it is not very surprising that the 46-degree halo is a much rarer sight than the 22-degree halo.

Although halos of other radii are uncommon, they are not impossible, so there is nothing improbable about Ezekiel's observation of "four wheels." Halos with radii of 8, 17, 32, and 90 degrees have been observed and explained.

When hexagonal crystals are very small, and thus randomly oriented, they will produce a 22-degree halo. What happens when the crystals grow large enough to become oriented by the aerodynamic forces depends on whether we are dealing with a column or a plate. Large hexagonal plates can no longer produce the complete halo, but will instead produce two brightly colored spots of light that sit at a distance of about 22 degrees on either side of the sun. Variously called *sun dogs, mock suns,* or *parhelia* (see Figure 9.5), these spots

Figure 9.5 *Parhelia.* Two parhelia and the sun look like three suns in the sky (*photo by Henry Adams. ©Alistair B. Fraser*).

are not as common as the halo, but can be seen about two dozen times a year over most of the United States and Canada.

Large, oriented, hexagonal columns produce two arcs of light which touch the 22-degree halo at the top and bottom. Called *tangent arcs,* their form changes greatly with the height of the sun (see Figure 9.6). When the sun is high in the sky, the upper and lower tangent arcs join to produce the *circumscribed halo* (Figure 9.7).

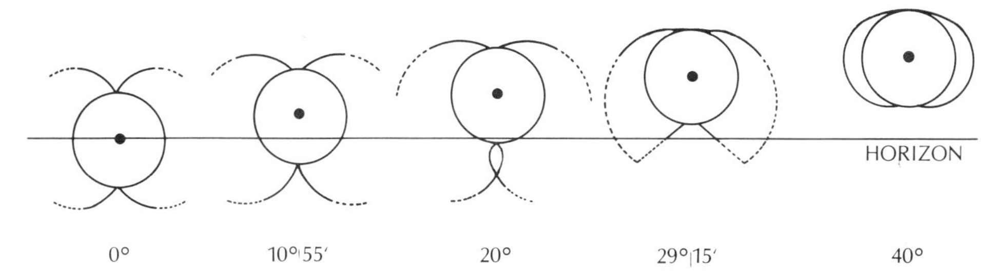

Figure 9.6 *Tangent arcs.* The upper and lower tangent arcs to the 22-degree halo change their shape as the sun climbs higher in the sky.

Figure 9.7 *Circumscribed halo and 22-degree halo when the sun is at elevation of 48 degrees* (*© Alistair B. Fraser*).

Many other halos are possible, some of which can only form when the sun is low in the sky, and others when the sun is high. One of the most beautiful of all the halos, the *circumhorizontal arc* (Plate 13), cannot form when the sun is less than 58 degrees above the horizon. It is most spectacular when the sun is 68 degrees above the horizon, but in the middle latitudes, the sun only gets that high in June and early July, and then just for a few hours around noon. It is not surprising, therefore, that the circumhorizontal arc is a rare spectacle.

9.2 Water-drop Optics—Rainbows, Coronas, and Glories

My heart leaps up when I behold
A rainbow in the sky.

[William Wordsworth, "A Rainbow," 1–2]

Usually seen late in the day after a heavy rainstorm, the rainbow has become a symbol of renewed hope or, in the words of Genesis 9:16, "a covenant between God and every living creature." This optimistic view has not been shared by all cultures. For the ancient Greeks, the rainbow was Iris, the messenger of the gods, who bore news of war and death. Many African and American tribes saw the rainbow as a giant and deadly serpent. This view was extended by the Shoshones to account for the hailstorms that are so common on the high plains of America. While rubbing its back on the icy dome of the sky, the rainbow serpent would chip off small pieces of ice, which would fall to the ground.

The rainbow, undoubtedly the best-known example of meteorological optics short of the blue of the sky, owes its existence to the refraction and reflection of light by raindrops. The light is refracted once as it enters a drop, and is then reflected off the inside back of the drop before being refracted again as it exits (Figure 9.8). Depending on the angle that the light makes to the drop surface as it enters, the whole process of two refractions and one reflection will bend the light through an angle of anywhere between 138 and 180 degrees, filling the sky between these angular distances from the sun with light. The point 180 degrees from the sun is the head of your shadow, and a point 138 degrees from the sun is 42 degrees from the head of your shadow, so that you will see a disk of light centered on the head of your shadow that has a radius of 42 degrees. The rainbow is the brightly colored edge of this disk of light. Outside the rainbow, the sky becomes perceptibly darker, until at 50 degrees from your shadow another fainter rainbow is formed. This *secondary rainbow* is formed when the light is reflected twice inside the drop before it exits. To distinguish these two rainbows, the rainbow that is formed by only one internal reflection is called the *primary rainbow*. A *tertiary rainbow,* resulting from three internal light reflections, would have to form on the side of the sky near the sun, but it would be very faint. In fact, there is no reliable evidence that one has ever been seen.

Because both the rainbow and the halo are caused by refraction, we would expect that the order of the colors would be the same in both, and yet we find that red lies on the *inside* of the halo, but on the *outside* of the primary rainbow. The clue to this apparent contradiction is to be found in remembering

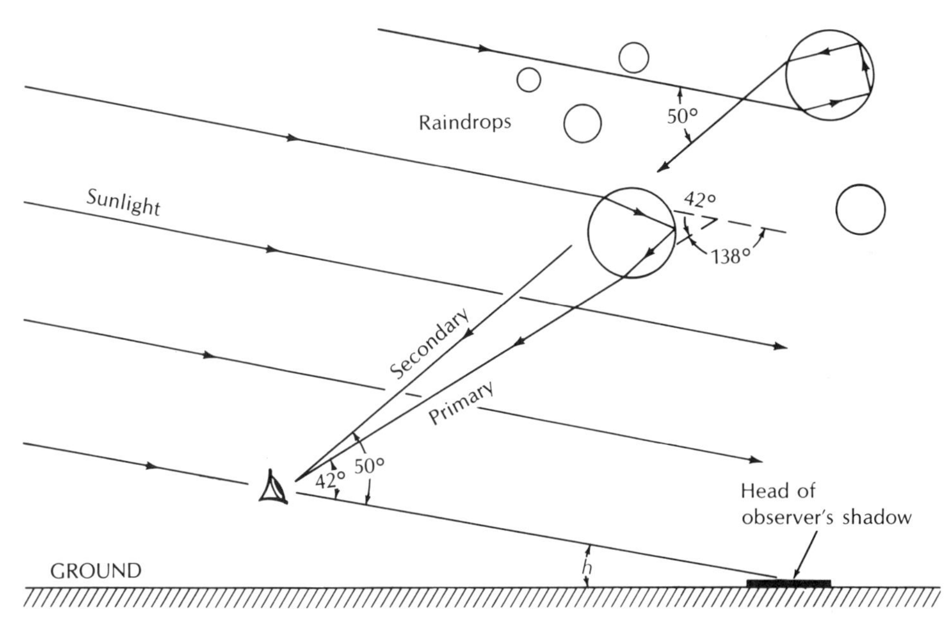

Figure 9.8 *Position of primary and secondary rainbows.*

that, of all the colors, red is bent least by refraction, and will therefore lie closest to the sun—on the inside of the halo, which surrounds the sun, and on the outside of the primary rainbow, which is on the other side of the sky. The secondary rainbow is an interesting illustration of this same principle. Here, the red is again on the inside of the bow, but the light that forms this bow has been bent by refraction through an angle of 230 degrees (180 + 50 degrees), thus seemingly turning it inside out. The red is still the closest color to the sun if we go around the sky in the other direction, passing the head of your shadow on the way.

There is a cliché that speaks of "all the colors of the rainbow," as though the rainbow exhibited all the colors of the spectrum. However, even casual observation will readily convince you that this is not true. Rainbows differ markedly from one to another, and the colors are not even uniform along a particular bow. This peculiar property will be easier to understand after considering a quite different optical phenomenon, the *corona,* which is also caused by waterdrops.

Often at night, irregularly shifting colors play across the surface of ragged clouds as they drift past a full moon. These colors are seen fairly close to the moon, usually at a distance of between 2 and 10 degrees, or about four to twenty moon diameters. (The diameter of both the sun and the moon is about ½ degree, and thus provides a convenient means of estimating angular distances.) On those occasions when the cloud through which the moon is seen appears smooth and uniform rather than ragged, the colors form a series of rings of light around the moon. Called a corona (Plate 14a), the colored rings can also be seen around the sun as it shines through some thin clouds, although it is rarely seen because the light is so dazzlingly brilliant. Newton first mentioned seeing the solar corona by viewing its reflection in still water, and although this method still remains a very good way to cut down on the intensity of the light, it is often easier to view it through a couple of pairs of sunglasses, being careful to block the sun itself from view.

The corona is caused by the diffraction of light around the small drops of water that form the cloud. To explain diffraction, we picture light as a series of waves, much like the waves on the surface of a lake. Imagine some water waves as they pass by a pole sticking out of the water. Each side of the pole causes a disturbance as the waves wash against it, and sends out circular patterns of waves. In some places, the crests of the original waves and the crests of the circular waves combine to produce very large waves, while at other locations, the crests of one group of waves coincide with the troughs of the other, and they cancel out to give calm water. The same thing happens when light passes by the small waterdrops in a cloud (see Figure 9.9). Where the two

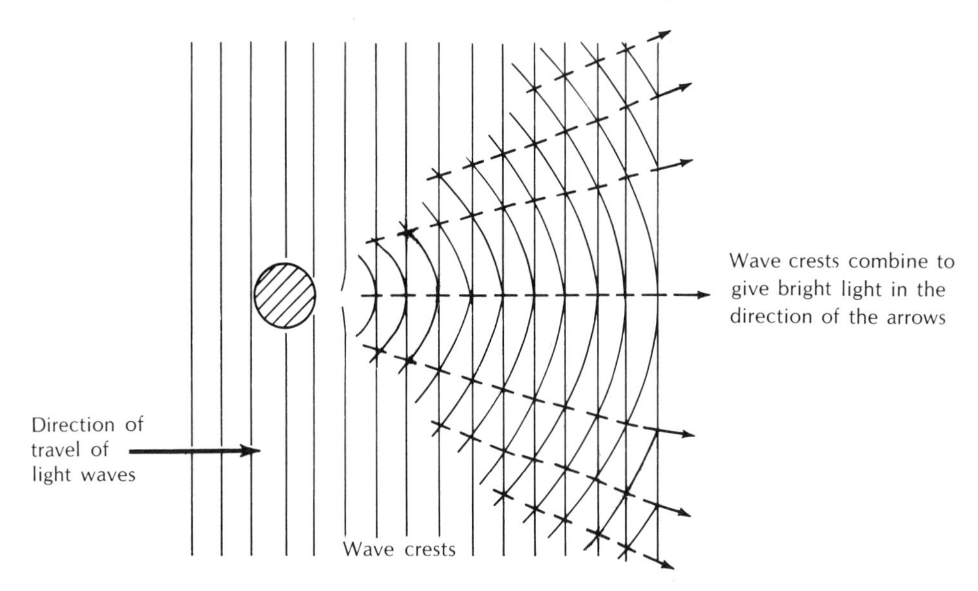

Figure 9.9 *Diffraction.* Circular waves that originate on either side of the cloud drop combine with the original straight waves to cause light to travel out at an angle to the original direction.

series of light waves combine, we see bright light, which travels at an angle to the original light rays; where they interfere, we see darkness. Because the amount of angular bending of the light by diffraction depends on the color of the light, a red ring may appear in the dark region between two blue rings. The result is a sequence of changing colors repeating over and over and getting fainter as we look farther away from the moon or the sun.

In the corona, the size of the rings of light depends on the size of the cloud drops that produce the diffraction; the smaller the drops, the bigger the rings. As a result, we see a perfectly circular corona when the cloud drops are a uniform size throughout the cloud. Drops that vary in size throughout the cloud will produce an irregular-shaped corona, which sometimes bears scant resemblance to a circle. The irregular patches of color that can result are often called *iridescence*.

In this day of frequent air travel, there is a phenomenon caused by diffraction of light by waterdrops in a cloud that is more familiar to most observers than is the corona. Called the *glory* (Plate 14b), it is a series of rings of colored light that can be seen around the shadow of the airplane as the shadow passes over a cloud composed of waterdrops. At first glance, the glory just looks like a fainter version of the corona, but on closer examination, it is found to be much more difficult to explain.

Like the corona, the glory is a diffraction pattern, but unlike the corona, the glory diffraction is caused by a ring of light that seems to emanate from the edges of the waterdrop. This light travels back in a direction toward the sun, and thus can be seen when you stand with your back to the sun and look at the head of your shadow or the shadow of the airplane in which you are riding. This ring of light arises in a peculiar way; first, the sunlight enters one side of the drop and, like the ray that causes the rainbow, is refracted and then reflected off the back side of the drop. It is then refracted again as it exits the other side of the drop. At the end of this process, however, it has not yet been bent through the complete 180 degrees that is required to send it back toward the sun and, thus, toward your eye. It can, at most, have been bent through about 166 degrees, so in order to make up the additional 14 degrees, the light flows along the surface of the drop in what is known as a *surface wave* (Figure 9.10). Having then been bent through the required 180 degrees, the light from all the edges of the drop forms a drop-sized ring that produces the pattern of light which we see as the glory.

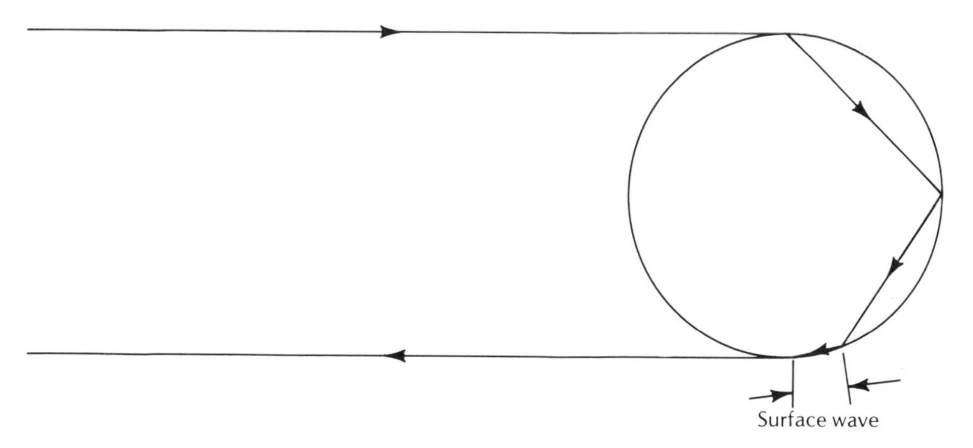

Figure 9.10 *Path a ray of light takes through a waterdrop in producing the glory.*

Both the glory and the corona can only be produced by cloud drops. Because the size of the rings decreases as the drops get bigger, raindrops would produce very small rings indeed. But the light source for these phenomena is the sun or the moon, which have diameters of about ½ degree. Each ring must therefore be at least ½ degree wide. When the diameter of the rings becomes so small that the spacing between them is only ½ degree, we end up with a continuously bright aureole of light, but no distinct rings. When we see a corona or glory, we can be sure that the drops that caused them are less than 50 micrometers in diameter, and thus, are well down in the cloud-drop range (recall Figure 8.2).

We are now in a position to understand why the rainbow does not always appear in the sky just as the simple theory presented earlier would suggest. That theory said that the rainbow arose from light that had been refracted once as it entered the drop, reflected once off the inside back of the drop, and then refracted again as it emerged from the drop. If we now consider that the light that is traveling through the drop is composed of waves, we find that the light that exits the drop has been divided into two different circular wave patterns. Like the two wave patterns that combine to produce the diffraction pattern of the corona, these two wave patterns combine to produce a diffraction pattern at the rainbow. This diffraction pattern produces variations to the simple picture in the rainbow discussed earlier.

At the upper portions of the (primary) rainbow, a number of faint bows can frequently be seen on the inside of the main bow. Called *supernumerary bows,* as if somehow nature had made a mistake by putting them there, these bows are part of the series of diffraction rings, of which the main bow is the brightest. As with all diffraction patterns, when the drops causing the diffraction are very small, the pattern is very broad and the spacing between the supernumerary bows increases. The broadness of the pattern has another effect: as the pattern broadens, the different colors of the main bow begin to overlap and cancel each other, so that when the main bow is caused by small cloud drops, it becomes white. This *cloud bow* (see Plate 15b and compare with the rainbow in Plate 15a), or *white rainbow,* can sometimes be seen from an airplane which is flying over stratus clouds, and can be distinguished from the glory by its very much larger size and lack of color.

The closest approach to "all the colors of the rainbow" is near the foot of the bow, where the largest raindrops can contribute to the brightness of the bow. As a result, the diffraction pattern is very tight, and each color is separated from the others, with no overlap. The large raindrops are flattened by the drag of the air (contrary to the popular myth that raindrops are shaped like tear drops). Near the ground, the orientation of these flattened drops with respect to the observer and the sun is favorable for the production of the rainbow. At higher elevations, however, only spherical drops can produce the proper combination of reflection and refraction that make the rainbow visible. In the upper portion of the rainbow, therefore, the small raindrops (which are constrained to be spherical by surface tension) contribute all the light, and the bow has correspondingly more pallid colors. The small drops also produce a very broad diffraction pattern, which implies a large spacing between the supernumerary bows. Thus the supernumerary bows are most likely to be seen near the top of the rainbow. We see that one of the most beautiful of all natural phenomena, the rainbow, is very complex.

Plate 14a *Portion of solar corona.* A tree branch has been used to block out the sun (© *Alistair B. Fraser*).

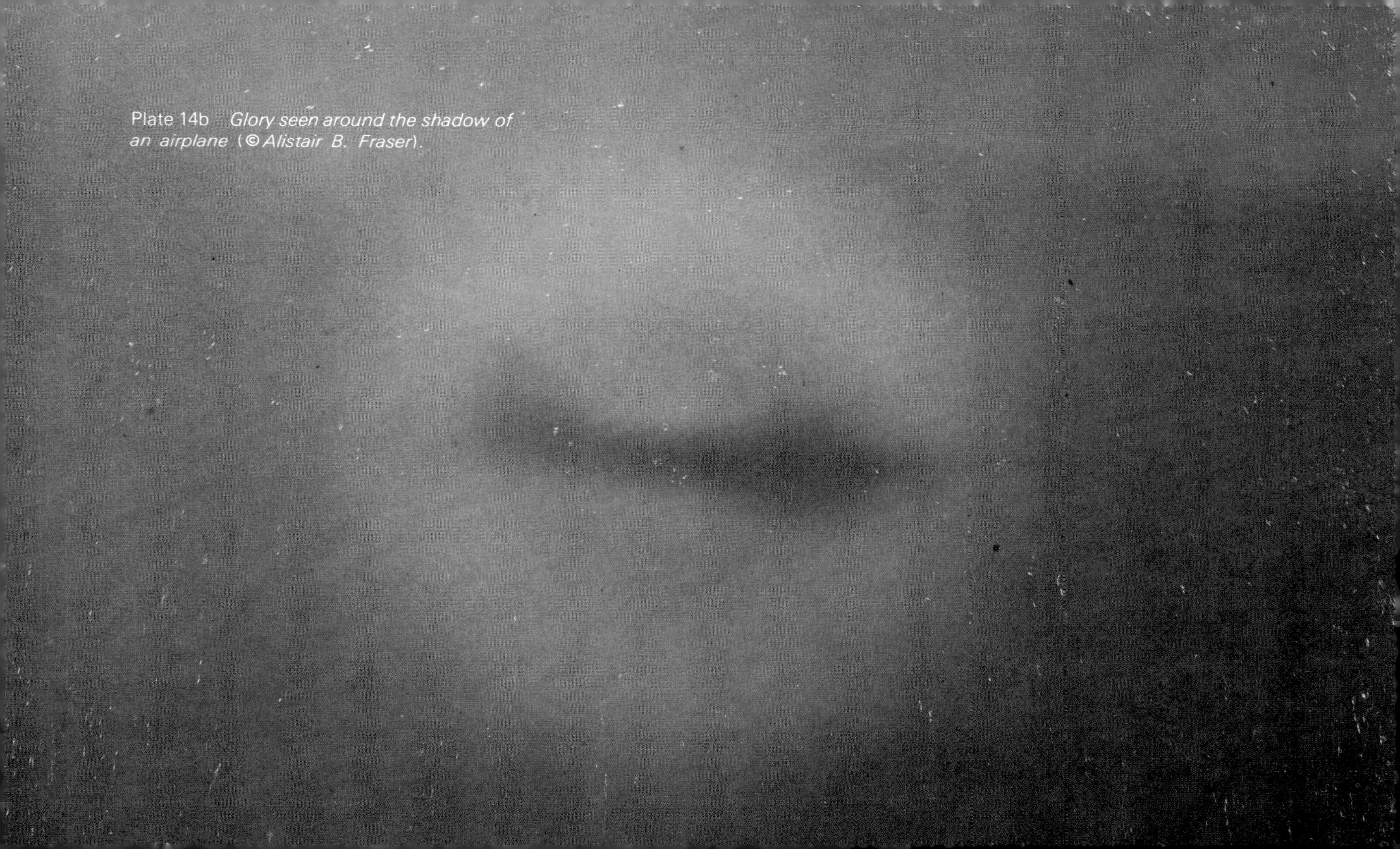

Plate 14b *Glory seen around the shadow of an airplane* (© *Alistair B. Fraser*).

9.3 The Mirage

As for those who disbelieve, their deeds are as a mirage in the desert. The thirsty one supposeth it to be water till he cometh unto it and findeth it naught, and findeth, in the place thereof, Allah, Who payeth him his due; and Allah is swift at reckoning. [The Koran, Surah 24:39]

The above remarkable simile was written in A.D. 628 (the sixth year of the Hijrah), and suggests that the behavior of the desert *mirage* was common knowledge to the Arabs of that day. Indeed, we even find the mirages of North Africa being described as early as the first century B.C. by the historian Diodorus of Sicily. However, serious scientific investigation into the mirage did not begin until the eighteenth century, when navigators and surveyors found that atmospheric refraction greatly inhibits accurate measurements of position. The nineteenth century looked on the mirage chiefly as a curiosity, and it has only been in this century that its potential as a means of remotely determining the temperature structure of the atmosphere has prompted serious investigation again.

The mirage has had the curious fate of being misdescribed as an "illusion" by dictionaries and encyclopedias alike. This is claimed while simultaneously providing the correct explanation, which ascribes the mirage to the refraction of light by the atmosphere. A mirage is a physical reality and, thus if understood, cannot be illusionary. If you as a car driver see a mirage on the road but imagine that you see water, you are deluded, and the water is an illusion (not the mirage). If you recognize it as a mirage, however, you are not deluded, and there is no illusion. When understood, the images seen by the refraction of light through the atmosphere are no more an illusion than are the images seen by the refraction of light through a microscope, a telescope, or, for that matter, a pair of eyeglasses. While it is unquestionably true that the mirage has been the source of many illusions, so have religion and politics, but that does not make either religion or politics an illusion.

The amount of bending of light by refraction in the atmosphere is very small, and so is usually not noticeable. The curved light rays cause an image (the view of a distant boat, for example) to appear slightly displaced from where the object (the boat itself) really is. Whether the image is displaced above or below the position of the object depends on the temperature (strictly speaking, density) structure in the atmosphere. Light rays are always bent so that the colder (denser) air lies on the inside of the curve (Figure 9.11). An image will therefore always be displaced toward the warmer (less dense) air. When the air temperature is greatest near the ground, as it is over a sun-baked road, then the image of a distant object is displaced downward, and the mirage is called an *inferior* (literally, *lower*) *mirage*. When the air temperature increases with height, as is common over lakes on a summer afternoon, the image of a distant object is displaced upward, and is thus called a *superior* (literally, *upper*) *mirage*.

The amount of bending of the light rays, and thus the amount of displacement of the image, depends on the rate of the temperature change with height. If the temperature changes very rapidly with height, then the rays are bent strongly; if the temperature varies only a small amount with height, then the rays

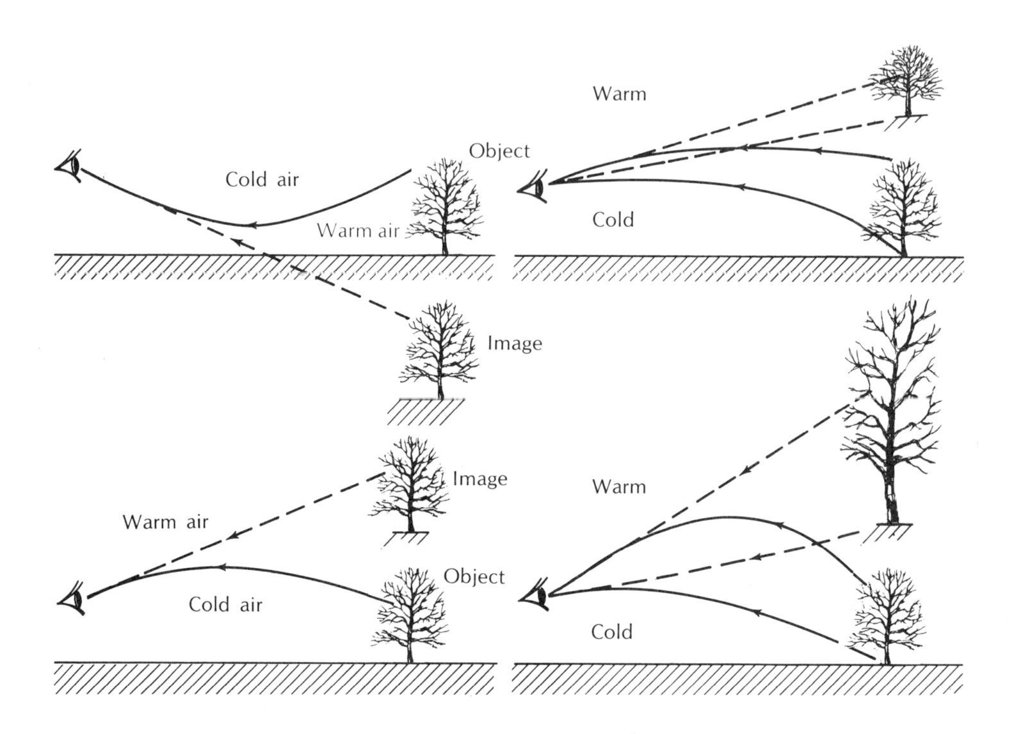

Figure 9.11 *Various forms of mirage.* The vertical scale is greatly exaggerated.

are hardly bent at all. It is possible, therefore, for the top of a distant object to be seen through a stronger vertical temperature gradient than the bottom of the object. The image of the top will be displaced more than the image of the bottom, and the image will appear either magnified or compressed, depending on whether it was displaced up or down. When the image has been magnified, it is said to be *towering*; when it is compressed, it is said to be *stooping* (see Figure 9.11). It is possible for either stooping or towering to occur with either the inferior or the superior mirage. The variety of *imaging* (the formation of images) that can occur in the atmosphere is large, but only a few of the most interesting types will be considered here.

9.3.1 The inferior mirage

Certainly the most familiar example of a mirage to most people is the appearance of what looks like water. Seen in the distance on a paved road, it vanishes as you approach, only to reappear again farther up the road. For this mirage to be seen, the road must be warmer than the air, so that the road will heat the air immediately above it. A ray of light traveling through this air, which is warmer next to the road than higher up, will be concave up (Figure 9.12).

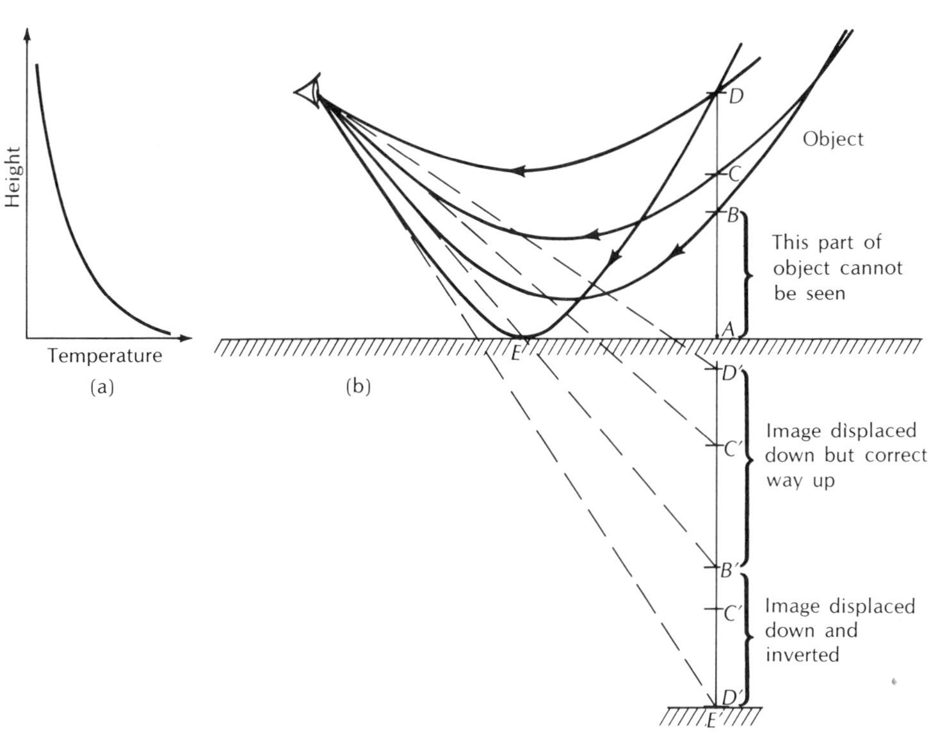

Figure 9.12 (*a*) *Temperature structure that produces a two-image mirage.* (*b*) The rays of light that enter the eye have been bent so that an object *ABCD* will appear to the eye as *D′C′B′C′D′* and the portion between *A* and *B* will have vanished from view.

Because the temperature gradients also are stronger next to the road than higher up, the light rays that approach close to the road are more strongly bent than those that pass higher up. They are bent so strongly that a light ray that enters your eye from the direction of the road can actually have originated much higher up, on a tree, for example, or even in the sky. This image of the tree or sky that is seen on the road is reminiscent of the reflections that can be seen on water. Because the image that is seen is inverted, it strengthens the impression that a reflection has occurred, although no refraction has taken place. This, then, is the *two-image inferior mirage:* a distant object is seen once as it would appear ordinarily, and then again, inverted below the first image. To produce this effect, the temperature and the temperature gradient must increase near the ground.

There is another fascinating aspect to this type of mirage—it can make distant objects vanish. For example, a small portion of the road can no longer be seen, when, in its place, an image of the sky appears. With increasing distance, very large objects such as people, telephone poles, and even small mountains can disappear, even though in the absence of strong temperature gradients these same objects would be readily visible. The objects cannot be seen because the light rays that ordinarily reach your eye from the object get bent so strongly that they pass up over your head instead. The view you get of a person walking off across a desert is striking. The feet vanish first, only to be replaced with an inverted image of a portion of the legs; then the legs vanish and are replaced by an inverted image of the upper body. Penultimately, the whole body is gone, and the head can be seen both normally and inverted. With a little more distance, the person vanishes completely, although he or she may be a kilometer or less from the observer. The whole scene resembles a person walking farther and farther into a lake, and finally drowning. Another example of this type of mirage is shown in Figure 9.13.

Figure 9.13 *Two-image inferior mirage seen over water.* The base of the hull of the fishing boat (left) has vanished. All of the hull of the sailboat (center) in the middle distance has vanished, while the hull and the cabin and the base of the sail of the distant sailboat (right) have vanished from view. In each case, the missing image is replaced by an inverted portion of the correct image (©*Alistair B. Fraser*).

Curiously enough, the same appearance of drowning can be seen if a person stands still at a distance of about a kilometer from the observer while the temperature gradient increases with time. Early in the morning, the person is completely visible, but as the day progresses and the sun warms the ground, it looks as if a wall of water (see Figure 9.14) flows in over the person from a distant sea. A remarkable description that fits precisely how a mirage would behave on a

Figure 9.14 *Inferior mirage.* A two-image inferior mirage over the desert gives the impression that there is a wall of water in the distance (© *Alistair B. Fraser*).

desert can be found in Exodus 14. In verse 21, the account tells how the waters receded at night (the temperature and the temperature gradient will decrease after the sun sets, so that the apparent water will move further away), and verse 22 tells how "the children of Israel went into the midst of the sea upon dry ground, and the waters were a wall unto them on their right hand and on their left" (the apparent water will always remain some distance from the observer on all sides, and as the observer approaches, it will recede). Verse 27 tells us that "the sea returned to his strength when the morning appeared" (the sun heated the ground, the rays were bent more strongly, and the apparent water moved closer). "And the waters returned, and covered the chariots, and the horsemen, and all the host of the Pharaoh . . ." says verse 28 (as the day gets warmer, the rays bend even more strongly, and the apparent waters flow in over the Egyptians; as they vanish from view, the lower inverted image that looks like a reflection provides convincing evidence that they are indeed being inundated by water). To the Egyptians, it would have been the Israelites who had vanished. They would have given up and gone home. A comparably detailed account of a mirage was not made again for over three millenia. Clearly, the inferior mirage is not qualitatively inferior!

9.3.2 *The superior mirage*

On most occasions, the *superior mirage* is not as arresting as the inferior mirage. It has one variety, however, that is not only spectacular, but almost legendary: the *Fata Morgana*. The Fata Morgana is named (in

Italian) for the fairy Morgan, who was the half sister of King Arthur. Morgan was credited with magical powers which enabled her to build castles out of the air. As a result, the residents of Reggio, Italy gave her credit for the strange structures that would appear spontaneously out in the straits to the west. Castles, houses, bridges, and, indeed, whole cities would appear, only to vanish again after a short time. Although comparable sightings have been made in many parts of the world, and many different names exist for the phenomenon, the legendary *Fata Morgana* has become generic.

The Fata Morgana is just an extreme form of towering. A distant object, which might be only one point on the surface of a sea, becomes greatly magnified so that it appears like a high wall (Figure 9.15). The temperature struc-

Figure 9.15 *Fata Morgana.* The cliff on the land to the left and the towers that appear in the strait are both the results of refraction. In fact, the land slopes gently into the water and the "towers" are small boats (© *Alistair B. Fraser*).

ture that produces this type of imaging is what is sometimes called a *lifted inversion* (Figure 9.16). First, the temperature increases slowly with height, then more rapidly, and above that, more slowly again. We thus have a region of strong temperature gradient between two regions of weaker temperature gradients. Because the amount of magnification depends very critically on the relative strengths of these gradients, minor variations will mean that one point will be greatly magnified while another immediately beside it will, by comparison, appear almost normal. The visual impression, then, can be one of a series of towers, or a wall with gaps in it, although, in fact, what is being seen is a smooth

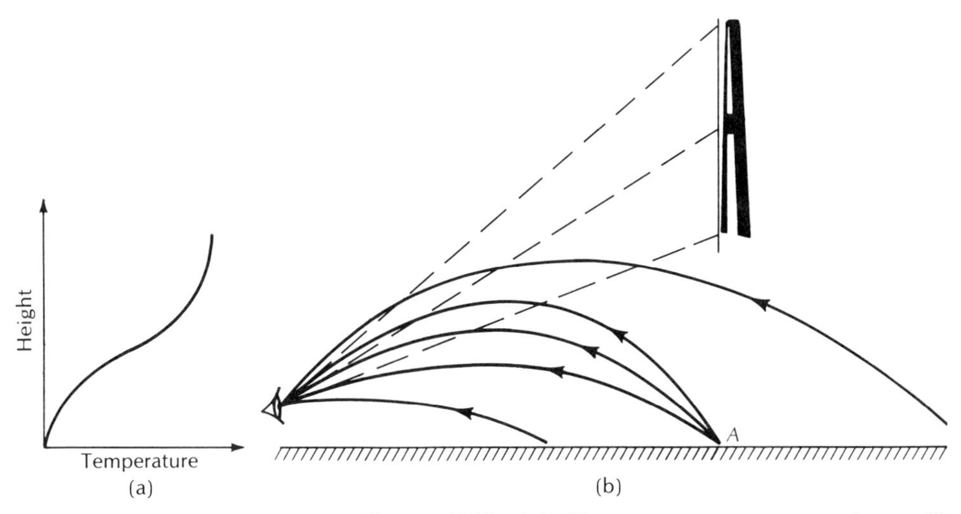

Figure 9.16 (*a*) *Temperature structure that will cause a Fata Morgana.* (*b*) The Fata Morgana occurs when a single point, *A*, is seen through a number of different angles and so appears to be drawn out into a long line.

surface of water. Because the gaps can appear between towers to create the impression of separate buildings, or inside the towers to create the impression of windows, or regularly along the base of a wall to create the impression of a viaduct, a whole city is simulated. As the temperature structure changes, the "city" vanishes as quickly as it came.

9.3.3 The green flash

There is an old Scottish legend that states that those who have seen the *green flash* will never err in matters of love; reason enough, surely, to look for the phenomenon. Occasionally, as the sun is setting or rising, a momentary flash of green light can be seen at the top of the sun (Plate 16). Arriving quite unexpectedly for most people, it seems to contradict normal experience that the low sun is yellow or red.

As the sun's light passes through the atmosphere, it is bent by refraction, so that when the bottom of the sun is seen to be just touching the horizon, the sun itself is actually already below the horizon. Its image, however, has been displaced upward. Red light is refracted less than blue light, so that when the low sun is examined through a telescope, it can be seen to have a red rim on the bottom and a blue rim on the top. Usually, the blue is lost by scattering as it travels through the atmosphere, and so a green upper rim is seen instead (green is next to blue in the spectrum). Under normal conditions, this green rim is so very thin that it is too small for the unaided eye to detect, but under the appropriate mirage conditions, it can sometimes be so greatly magnified that it becomes arrestingly obvious.

10 Impact of weather and climate on man

Weather and climate have directly affected human evolution since the beginning, influencing the rate and direction of civilization, permeating the spiritual side of life, and engraving their many facets and moods into the various forms of our aesthetic expression. Even in today's plastic world of technology, we find our births and deaths tuned to the rhythms of the seasons, and even to the daily variation in temperatures, clouds, and rain. And now, as the exhaustion of many of our traditional forms of energy looms less than a generation in the future, our interest is renewed in the harnessing of alternative sources of energy intimately associated with the weather, such as sunshine, water, and wind.

10.1 Biometeorology

Whoever wishes to pursue properly the science of medicine must proceed thus. First he ought to consider what effects each season of the year can produce: for the seasons are not alike, but differ widely both in themselves and at their changes.

One should be especially on one's guard against the most violent changes of the seasons, and unless compelled, one should neither purge nor apply cautery or knife to the bowels, until at least ten days have passed. [Hippocrates, *Air, Water, Places*]

Since Hippocrates' original ideas, biometeorology has been refined to include the study of the direct and indirect interrelations between the atmosphere and living organisms (plants, animals, and people). *Bioclimatology* is the branch of biometeorology that considers the interactions between weather and life that take place on longer time scales, such as weeks, months, or years. Other subgroups of biometeorology include the study of the influence of weather and climate on the following human aspects:

(1) race and body structure (*ethnological biometeorology*);
(2) mental processes (*psychological biometeorology*);
(3) aesthetic expression—architecture, painting, music, etc.—(*aesthetobiometeorology*);
(4) origin and frequency of disease (*climatological pathology*).

10.1.1 Effect of weather on human health and performance

It is apparent that the weather affects everything we do and that weather forecasts can be used advantageously by many individuals. Besides the direct and dramatic effects that severe weather phenomena have on our activities, weather's effect on life is many times quite subtle. Human performance tests indicate that people work most efficiently in a comfortable environment, i.e., an environment free of temperature extremes that would act as irritants. Telegraphers and typists make more mistakes when the temperature rises above 32°C. Performance of factory workers decreases when they are affected by extreme temperatures. Tests with students in controlled environments show that discomfort associated with excessively warm temperatures interferes with the learning process. The slowing of human response in split-second decision making situations is also well correlated with poor weather conditions.

Weather sometimes has striking effects on human health. Table 10.1 summarizes some of the definite and probable effects that various aspects of air pollution and weather have on life. In general, weather does not cause a particular disease, but rather tends to aggravate existing diseases.

Notable among these are heart and circulatory ailments. Most statistics show that attacks resulting from these diseases follow a very pronounced seasonal course. Usually, there is a peak of deaths in midwinter, partly as a result of strain caused by seasonal work. In addition, during summer heat waves, people with heart trouble are usually more susceptible to heat-related circulatory failure.

Weather not only affects times of death, it affects times of conception as well. Figure 10.1 shows the birth and conception rates by month for 1968 for the entire United States and for Florida. Over the United States, con-

Table 10.1 *Effects of meteorological elements on biological life*

Agent	Definite effects	Possible effects
Sulfur dioxide and related compounds, such as sulfuric-acid rainwater	(1) Aggravate asthma and chronic bronchitis (2) Impair breathing; lead to lung damage (3) Irritate sensory organs (eyes, nose, throat) (4) Damage vegetation	(5) Damage buildings and works of art, e.g., sulfuric acid dissolves marble statues
Sulfur oxides	(1) Increase mortality (death rate) for short term (2) Increase morbidity (incidence of disease) for short term (3) Aggravate bronchitis and cardiovascular disease (4) Contribute to development of chronic bronchitis, emphysema	(5) Contribute to development of lung cancer
Particulate matter not otherwise specified, such as dust	(1) Restricts visibility (2) Dirties surfaces such as cars, laundry, and buildings (3) Reduces sunlight	(4) Leads to increase in chronic respiratory disease
Oxidants (including ozone)	(1) Aggravate emphysema, asthma, and bronchitis (2) Irritate eyes and respiratory tract; impair athletic performance (3) Increase probability of motor vehicle accidents	(4) Alter oxygen consumption (5) Accelerate aging
Carbon monoxide	(1) Impairs oxygen transport function	(2) Increases general mortality and coronary mortality rates (3) Impairs central nervous system function (4) Causal factor in arteriosclerosis
Nitrogen dioxide	(1) Discolors atmosphere	(2) Factor in pulmonary emphysema (3) Impairs lung defenses
Lead	(1) Accumulates in body (2) Proves lethal to animals eating contaminated feed	
Fluorides	(1) Damage vegetation; harm animals	(2) Lead to fluorosis of teeth

Table 10.1 (continued)

Agent	Definite effects	Possible effects
Ethylene	(1) Damages vegetation; hastens ripening of fruit (used deliberately by some tomato growers to artificially ripen tomatoes)	
Chlorinated hydrocarbon pesticides, e.g., DDT	(1) Are stored in body; source usually milk and animal fats (2) Lead to ecological damage	(3) Impair learning and reproduction
Hydrothermal pollutants	(1) Can influence local climate; can interfere with visibility	(2) Have influence on action of hygroscopic pollutants
Airborne microorganisms	(1) Cause airborne infections	
Cold, damp weather	(1) Causes excess mortality from respiratory disease and fatal exposure or frostbite (2) Leads to excess morbidity from respiratory disease and morbidity from frostbite and exposure	(3) Contributes to excess mortality and morbidity from other causes (4) Causes or aggravates rheumatism
Cold, dry weather	(1) Causes mortality from frostbite and exposure (1) Leads to morbidity from frostbite and respiratory disease	(3) Impairs lung function
Hot, dry weather	(1) Causes heat-stroke mortality (2) Causes excess mortality attributed to other causes (3) Leads to morbidity from heat stroke (4) Impairs function of renal and circulatory tracts; aggravates renal and circulatory diseases	
Hot, damp weather	(1) Increases skin infections (2) Leads to heat-exhaustion mortality (3) Causes excess mortality from other causes (4) Causes heat-related morbidity (5) Impairs human performance (6) Aggravates renal and circulatory disease	(7) Increases prevalence of infectious agents and vectors

Table 10.1 (continued)

Agent	Definite effects	Possible effects
Natural sunlight	(1) Leads to fatalities from acute exposure (2) Causes morbidity due to "burns" (3) Leads to skin cancer (4) Interacts with drugs in susceptible individuals	(5) Increases malignant melanoma

Source: Patterns and Perspectives in Environmental Science, report prepared for the National Science Board, National Science Foundation, 1972.

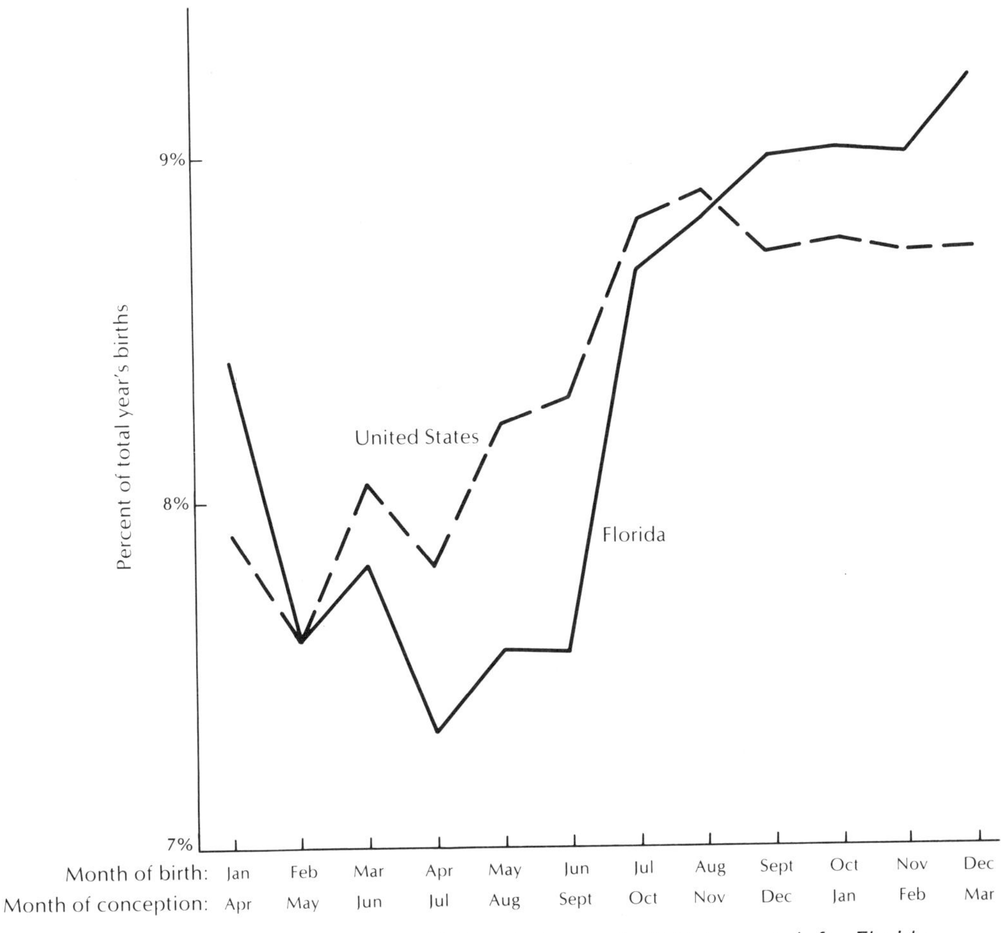

Figure 10.1 *Birth rates in 1968 by month for Florida and entire United States (source: Vital Statistics of the United States, 1968, U.S. Department of Health, Education, and Welfare).*

ception rates were considerably higher during the cold months (October–March) than during the warm months (April–September). Florida, which has extremely warm and humid summers, exhibits even a stronger bias in the rate of conception during the comfortable season. For example, the percentage of conception jumps from 7.56 percent in September, to 8.66 percent in October, the month when cold fronts again affect the state. These graphs suggest that weather significantly affects our sexual lives, as well as other aspects of our health.

Weather conditions can also play an important part in the transmission of disease. Some viruses (especially those that produce colds) are carried in small droplets ejected by sneezing or coughing and move through the air in a fine mist. They are small enough to float in turbulent air currents for long time periods, and, because of certain weather conditions, winter is the best time for transmittal of these viruses. On the average, the outdoor relative humidity tends to be greater in winter than in summer. In the low relative humidities of summer, droplets will evaporate and leave the microorganisms without the base of fluid that they need for survival. In winter, the drops do not evaporate as readily; thus, the chances of encountering cold viruses are increased. In addition, ultraviolet radiation, which is most intense at the earth's surface in summer, is fatal to most microorganisms. During winter, however, ultraviolet radiation is usually blocked by clouds or thick layers of polluted air, and more of the ultraviolet-sensitive pathogens can exist. Finally, in cold winter weather, people tend to congregate closer to one another, thus increasing the chances of transmitting the cold virus.

10.1.2 *Weather and the economy*

The weather in a particular area also determines the type of clothing people wear and the kinds of houses in which they live. Associated with these direct weather-life effects are the effects of weather on people's economic activities. Certain sectors of a nation's economy are very sensitive to variations in the weather.[1] For instance, the energy sales of electric, gas, and oil utilities fluctuate considerably and are strongly dependent on temperature and wind. Proper weather-oriented planning can greatly increase daily operating profit, especially during times of marked weather changes. Also, the impact of weather on retail sales is well known to every merchant, some of whom use storm forecasts to reduce the number of personnel on duty, whereas others attempt to capitalize on shifts in buying moods induced by snowstorms, heat waves, and wet spells. Some plan promotion of weather-sensitive items such as antifreeze, air conditioners, umbrellas, and snow tires. In the investment area, many stock brokers, dealers, and traders on the national exchanges regularly consult private forecasters when planning their strategy in weather-

[1]It is sometimes said that only a fool would try to forecast the weather. But if that is true, then only a fool's fool would attempt to forecast the economy, since the economy depends on the weather, as do the many mysterious and complex aspects of human behavior.

sensitive futures contracts. For example, if a hard freeze is forecast for parts of Florida, it may be profitable to buy orange juice stock a couple of days in advance. Agriculture is also very dependent on weather whims: orchard growers must be forewarned of frost; farmers rely on forecasts to aid in crop planting and harvesting; and in the western United States, temperature and precipitation data enable irrigation companies to deliver only the precise amount of water needed for irrigation of fields. Determine for yourself the weather-dependent nature of the following endeavors: bakeries, airlines, chemical companies, steel mills, highway crews, newspapers, trucking, moviemaking, public transit, construction, education, and the military.

10.1.3 Weather and sports

As an example of how a general activity depends on the weather, sports will be examined. Weather affects the sport itself, the participants, and also the spectators. Baseball is a dry-weather, warm-season game, very sensitive to rainfall and cold temperatures, which cause postponements. A study has shown that one of baseball's greatest players, Babe Ruth, played much better in cool weather, probably because Ruth was rather stout and could not get rid of excess body heat as easily as a thin person in hot weather. Football, conversely, is not usually postponed because of weather conditions, although the players and the fans may sometimes wish it were. Various types of play in football, however, tend to be affected by ground conditions, wind, temperature, and sun angle. In horse racing, the performance of some horses is enhanced by different types of weather and subsequent track conditions. Betting is affected, and some racetracks pay private meteorologists to provide two-hour precipitation forecasts to help in calculating odds before a race.

The major effect of weather on spectators is that adverse weather reduces attendance. Outdoor spectator sports are most severely affected, but fans sometimes cannot make it to an indoor arena because of bad weather. Similarly, amateur sports such as skiing are radically affected by many weather parameters. Rain, wind, and extreme cold are the main deterrents to ski-area attendance, even though an adequate amount of snow exists. But, although heavy amounts of snow may slow traffic and close schools and offices, studies have shown that skiers somehow are able to reach the slopes. Even concessionaires at sporting events are weather sensitive; the most successful operators pay for private weather forecasts to tell them whether to stock cold sodas or hot coffee.

Because of the sensitivity of sports to the weather, various attempts at neutralizing weather effects have been attempted. Underground heating cables have been used to prevent playing fields from freezing in Minnesota. The Houston Astrodome and the Minidome, Idaho have minimized the effects of weather on both the athletes and fans. Artificial turf has solved the ground condition problem in sports, but it has created another one just as serious. Because the artificial surface does not conduct heat very readily, the surface heats up rapidly on sunny days (up to 130°F), creating a severe heat stress on the athletes.

10.1.4 Lightning, plants, and man

One dramatic and direct way in which weather affects the living environment is through lightning-caused fires, especially on forested lands. Over 8000 lightning fires occur annually in the western forests of the United States and Canada. These wildfires create changes that may have profound effects on soil, water, wildlife, vegetation, and human activities in this region.

A generation raised under the friendly admonishing eyes of Smokey the Bear tends to believe that forest fires are invariably damaging and are to be prevented at all costs. However, forest fires are frequently very beneficial to the ecology of the region. They may control undesirable woody plants, allowing grasses to grow on grazing land, in turn reducing soil erosion and runoff. Occasional small fires will lower the chances of a major conflagration by periodically removing vegetative debris that might otherwise accumulate to dangerous levels. In addition, small fires produce soil fertilization from the ashes, control ticks, poisonous snakes, and other undesirable animals, destroy fungi, and create a better habitat for game animals.

There are also some ecological consequences of lightning that are not related to fires. Injury to trees or groups of trees, which may open the way for subsequent insect or disease damage, can be caused by a lightning strike without actual fire. On the positive side, if an old disease-ridden tree is struck, some degree of stand sanitation may be realized. The demise of the older trees may release understory trees and lesser vegetation, affecting both plant and animal life in the immediate area.

There is one direct beneficial aspect of lightning to vegetation. Lightning is a natural producer of fertilizer. The lightning discharge in air produces ozone, ammonia, and oxides of nitrogen, compounds which react with the rainwater to produce ammonia hydroxide and dilute nitric and nitrous acids which serve as soluble fertilizers. Thus, strikes of lightning near (but not too near!) plants are quite desirable.

Lightning's effect on people is very significant also. Lightning is directly responsible for more deaths each year on the average than any other weather phenomenon. Whereas the National Safety Council lists about 150 lightning deaths per year in the United States, the Metropolitan Life Insurance Company lists about 300 killed by lightning.

Lightning has always had a devastating effect on people, as evidenced by the following amazing story, which occurred over a hundred years ago (see Figure 10.2):

> *Mr. Cardan relates that eight harvesters, taking their noonday repast under a maple tree during a thunderstorm, were killed by one stroke of lightning. When approached by their companions, after the storm had cleared away, they seemed to be still at their repast. One was raising a glass to drink, another was in the act of taking a piece of bread, and a third was reaching out his hand to a plate. There they sat as if petrified, in the exact position in which death sur-*

prised them. [Nicholas Camille Flammarion, *The Atmosphere*. Translated from the French; edited by James Glaisher (New York: Harper Brothers, 1874). pp. 439, 440]

Figure 10.2 *"Harvesters killed by lightning."*

It is unfortunate that the vast majority of those people killed by lightning die needlessly. As in the above example, almost all victims are struck while out of doors, frequently while seeking shelter under a tall tree. (Trees are one of the worst things to be near during a thunderstorm, because trees furnish an easy path for lightning to follow.) When struck by lightning, a person is put into a state of suspended animation, as observed in the tale of the unfortunate harvesters. If proper artificial resuscitation is applied, the person may not die. The physiological effect of being struck by lightning is not as complicated as being shocked by a high-voltage wire. After resuscitation of a lightning victim, the heart resumes beating with a normal rhythm, in contrast to the irregular beat of the heart following electric shock. Even before the heart resumes beating and no blood is reaching the brain, degenerative processes seem to be delayed longer than usual (30 minutes in some cases), and the patient will often recover without permanent damage.

10.2 Human Response to Hostile Weather

Because the weather is rarely exactly ideal for human comfort, we have developed ways of adapting to conditions we find uncomfortable. Some of these adaptations are physiological, evolving during people's existence on earth. Many are technological, however, developing over the last several thousands of years as part of the civilization process.

10.2.1 Heat exchange between man and the environment

There are three main physical processes that determine the heat exchange between the body and the environment. The first is *radiation*. People are primarily concerned with the long-wave part of the radiation spectrum. Indoors, this radiation may be absorbed from walls or heating elements if their surface temperatures are higher than the body's. Outdoors, the radiation primarily comes from the sun, sky, and soil. The body itself loses heat by radiation toward objects in its environment that are colder than body temperature. This combination of *absorption* and *emission* yields the *radiation balance,* which is the sum of the radiative heat gained and heat lost. In bright sunshine, the net radiation balance is usually positive, i.e., the body gains heat. In the evening, with a clear sky and cold soil, the balance is negative and the body cools.

The second process in the heat balance of the body is *convection*, which includes heat transferred by air motions, i.e., winds or drafts. If the air temperature is cooler than the body surface, wind will carry heat away from the body. When the air is warmer than the body, such as in the desert or in a boiler room, air flow will carry heat to the body.

Table 10.2 *Average human heat budget*

Heat loss by radiation	40%
Heat loss by convention and conduction	30%
Heat loss by evaporation of perspiration	20%
Heat loss by breathing (evaporation ot water and warming the air)	10%
Total	100%

The third heat exchange process is *evaporation*. To evaporate one gram of water, about 600 calories of heat must be expended. In hot weather, the body perspires, this water evaporates from the skin, heat loss occurs, and the skin is naturally cooled. More water will evaporate if the air is dry rather than humid. The degree of cooling of the skin depends upon the area of skin exposed, the velocity of air, and the relative humidity of the air. Even when the body is not sweating, there is some heat loss from evaporation through the process of breathing, in which relatively dry, cool air is carried into the lungs where evaporation and warming occur. A typical human heat budget is presented in Table 10.2.

10.2.2 Physiological mechanisms man possesses to cope with temperature extremes

When environmental temperatures fall to a relatively low value (say below 14°C), the body temperature starts to drop. Increasing the body metabolism is the natural way to prevent this decrease. One way of increasing the metabolic rate is by eating food, especially proteins; another way is through work, since muscular activity can increase the body temperature 1–2°C, and violent exercise can increase metabolism by 16 times. Just moving around when cold helps keep us warm.

Even if we did not want to be warm, however, there are involuntary reactions of the body which increase the body temperature. Lowering of environmental temperature causes an increase in tensing of the muscles, which is usually followed by synchronous contractions of muscle fibers (shivering). In healthy, relaxed people, there are always microvibrations of the body taking place. The frequency of the vibrations is about 6–12 Hz (cycles/s): the amplitude is 1–5 micrometers. With tensed muscles in a cold environment, the amplitude increases to 50 micrometers, with the same frequency. Again, this involuntary muscular activity tends to increase body metabolism and maintain a high body temperature.

Vasoconstriction is a second physiological process which operates under cold conditions. In this case, a decrease in environmental temperature leads to constriction of peripheral blood vessels in the body; because less blood flows to the outer layers of the skin, the heat loss by radiation is reduced and there is less heat available to be transported away by convection. This mechanism helps to keep the body core temperature high and protects the most vital organs. The extremities such as fingers and toes are most affected, causing a reduction in dexterity on cold days.

It is a familiar experience that windy days feel much colder than calm days of the same temperature. Air is such a poor conductor that if no wind is blowing and we are not moving, a thin layer of warm air forms next to the skin and we may feel quite comfortable. Thus, we may sit on the porch of a ski lodge out of the wind and in the sunshine and feel quite comfortable, even though the air temperature may be 25°F.

The effects of wind and air temperature on the cooling rate of the human body have been combined in the wind-chill index, shown in Figure 10.3.

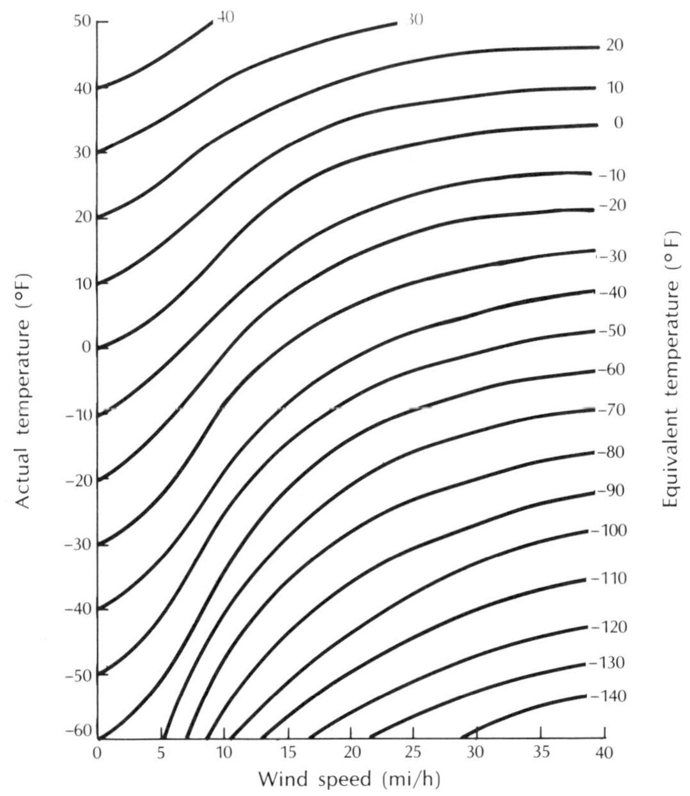

Figure 10.3 *Wind-chill chart. Equivalent "calm-air temperatures" as a function of actual air temperature and wind speed.*

The wind-chill chart translates the cooling power of the atmosphere with wind to a temperature under calm conditions. Thus, a naked body would lose as much heat in a minute with a temperature of 30°F and a wind of 10 mi/h as it would with no wind and a temperature of 16°F.

Newspapers and radio stations are fond of dramatically screeching the equivalent wind-chill temperature to us on cold, windy days. Thus, we are told: "The wind is 30 mi/h, the temperature is 15°F, producing an equivalent temperature of–25°F." But why are we never informed of the other, more optimistic, point of view? For example, on a calm morning with the temperature of 10°F, it might cheer us up to hear "It's 10°F outside, and the wind is calm, producing an equivalent temperature of 40°F with a wind of 40 mi/h."

When the air temperature drops below freezing, exposed parts of the body are in danger of suffering frostbite, which is the actual freezing of the surface tissues. Under frostbite conditions, the vasoconstriction process is so enhanced that very little body heat reaches the extremities, such as the fingers, toes, and ears. Individuals, such as soldiers, that must remain outside on cold, clear nights find that keeping their feet unfrozen is a major problem because of the strong radiational cooling at the ground. The result of this surface heat loss is a strong and shallow inversion, as shown in Figure 10.4a, in which the temperature

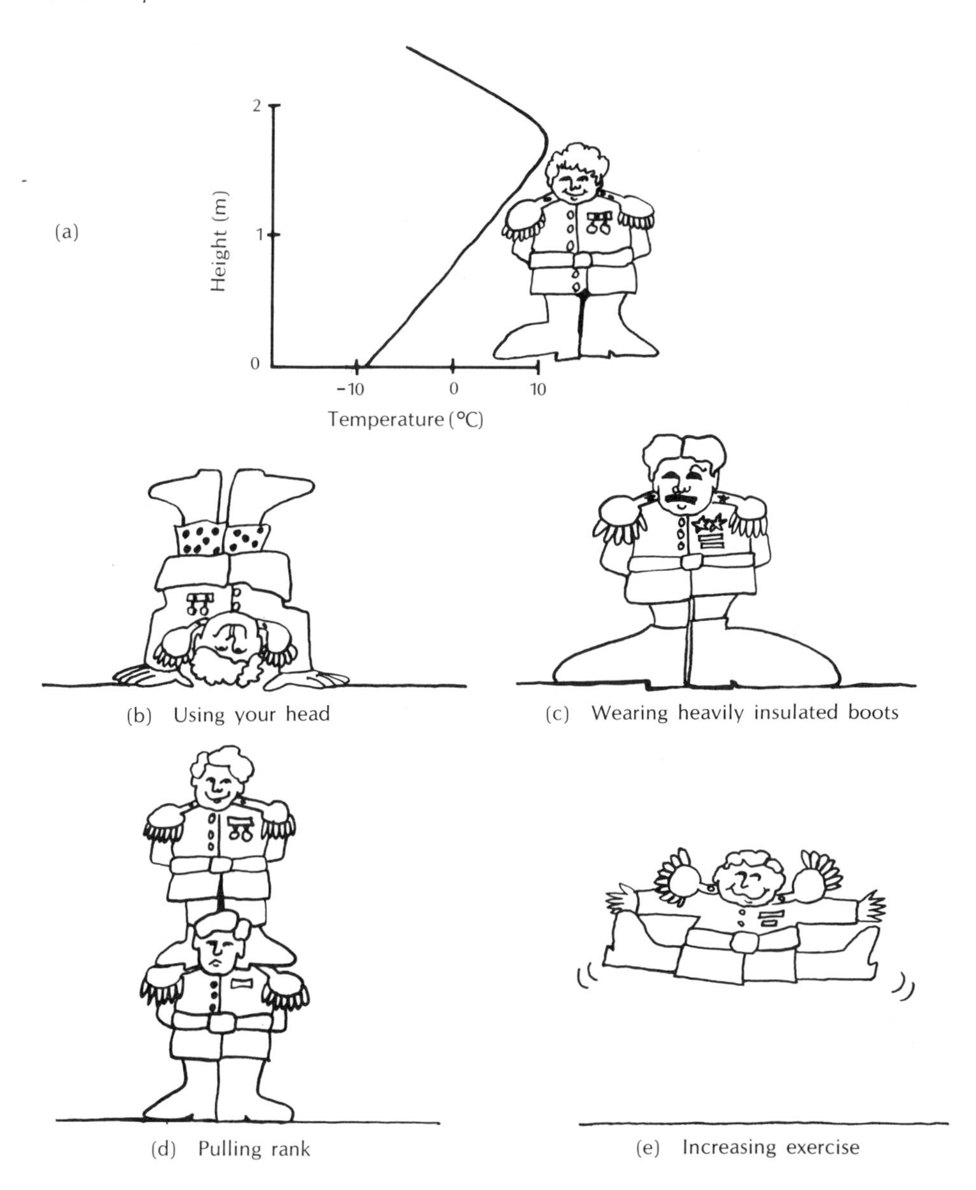

Figure 10.4 *Methods of avoiding radiation-cooling induced frostbite.*

may decrease 10°C from a sentry's head to his boots in extreme cases. Some of the possible solutions to the frostbite problem under such conditions are illustrated in Figure 10.4b–e).

In spite of natural defenses, people do not have a really effective system to combat cold temperature. Even though shivering tends to counteract the effects of vasoconstriction, it also brings more blood to the surface layers by increasing the metabolic rate, which in turn increases the heat loss by radiation and convection from the skin.

People are somewhat better equipped to survive in hot climates rather than cold climates. They have, in particular, the highly effective sweating mechanism that provides evaporative cooling of water in warm surroundings. In each person's skin, there are an estimated two million sweat glands, which can maintain a water loss of about 2 liters/h, provided the water is replaced by drinking an equivalent amount of liquids. In any event, a tremendous heat loss can be maintained at this rate (700,000 cal/m^2/h). Modern civilization, of course, has created a few problems. Socially, it is unacceptable to expose all our sweat glands; clothing tends to absorb perspired water and to restrict free evaporation. In addition, we have developed customs that declare it impolite to sweat. We mop our brows to dry up the sweat and, as a result, defeat the natural evaporation process.

10.2.3 *Artificial minimization of climatic extremes*

Clothing in a cold climate is the human substitute for the fur and feathers of the animal world. The objective, of course, is to interpose an insulating layer between the body and its surroundings, and thus reduce heat loss to the environment. Animal skins worn inside out provide extremely good protection against the cold and wind—the animal skin itself is not readily penetrated by wind, and the fur offers the insulation. It makes considerable difference whether the fur is worn on the inside or outside. With the fur outside, the insulation value drops considerably as the wind speed increases (the trapped air is moved out). However, with the fur inside, protection against the wind remains high. Mink and raccoon coats are really worn with the "wrong" side out.

Modern clothing developed for use in extremely cold environments such as the arctic is designed to make use of the fact that air is a good insulator. It is layered and loose-fitting to provide many air spaces which reduce the conduction of heat away from the body. It is also moisture-proof to prevent vapor loss from the body, which also leads to body cooling.

The head is a small part of the total body surface, but because the skin over most of the face does not vasoconstrict, important amounts of heat may be lost from the head. With proper clothing over the rest of the body, one-half of the total body heat loss occurs from an uncovered head at an air temperature of 5°C (41°F). Therefore, the Eskimo's parka covers most of his face so that it is exposed to the wind only when he faces directly into it.

In hot, humid climates, the clothing should be lightweight, loose-fitting, and porous in order to permit free circulation of air next to the skin. The access of outside air to the skin is important to prevent prickly heat and fungus infection. Also, clothing should be light in color to minimize the absorption of solar radiation.

The climate of the desert is perhaps the most inhospitable of all. Not only does the environment change from hot to cold between day and night, but by day, evaporational cooling is to be avoided because of the lack of water to replenish perspiration. The same general clothes designed for the humid tropics can be used in the desert—loose-fitting, lightweight, porous, and light in color. The fact that clothing necessarily retards natural evaporation to some degree is actually a help in conserving water in the desert. In extreme conditions, desert dwellers wear a lot of clothing to cut down skin surface exposure.

As with clothing, housing in the arctic must insulate the body from extremely cold air. The igloo is an architectural marvel. The hemispheric shape of the structure presents the smallest possible surface area to the environment; at the same time, it encloses the largest possible usable volume for a given amount of material. The hemisphere is also the shape most effectively heated from a small point source of radiant heat, which usually takes the form of an oil lamp in the middle of the floor. The wind-packed snow walls have a small thermal conductivity and, consequently, heat is not easily transmitted through the walls. In addition, the interior heat soon forms a vapor-proof glaze on the walls so that no heat is lost by evaporation.

In the humid tropics, parasol housing is employed. These houses have thatched roofs which shed sun and water, but virtually no walls. Consequently, there is a low thermal resistance to heat loss to the environment. In addition, the good ventilation allows high rates of evaporational cooling. The roof provides shade to the inhabitants and cuts down on direct heat absorption from solar radiation.

In the desert, houses are made with thick mud walls that not only provide a barrier against wind and dry air, but also produce a cavelike variation of temperature within. The temperature inside these structures is moderated and remains comparatively constant. Because of the lag in thermal conductivity, it reaches a maximum during the sleeping hours (when it is cold outside) and a minimum during midday.

10.3 Meteorological Aspects of the Energy Shortage

In recent years, everyone has become aware of the world's shortage of energy, because the effects have extended down to a personal level. One meteorological effect of the energy shortage is that environmental regulations pertaining to air quality will be relaxed to allow certain types of "dirty" fuels, such as coal, to be burned. Air pollution problems, as discussed in Chapter 7, will thus become more important. The increase of air pollution is one effect of the energy shortage on meteorology. On the other hand, meteorology may help relieve the energy shortage, since there are meteorological sources of energy that may be developed to meet local energy needs or help decrease global demands. These sources include solar, water, and wind power.

10.3.1 Solar power

In these days of dwindling supplies of oil, coal, and gas, who has not gazed at the fiery sun on a cold winter day and dreamed of using that energy to heat his house or power his car? The sun is indeed a potential source of safe, clean, and abundant energy. The earth intercepts more solar energy than we could ever conceivably use; in only two weeks, it receives more energy than the entire known global supply of fossil fuels.

The solar constant of two calories per square centimeter per minute is equivalent to 1400 watts (W) per square meter, or 1400 megawatts [a megawatt (MW) is a million watts] per square kilometer.[2] Of course, the amount of radiation reaching the ground is much less, being reduced by clouds and dust in the atmosphere and by low solar elevation angles. Nevertheless, the United States on an annual average receives about 13 percent of this figure, which means that about 180 watts per square meter are available. This rate of heating produces, on the average, about 4 kilowatt-hours (kWh) of energy per square meter each day, which is twice the amount needed to heat and cool the average house. Thus, it is obvious that solar power could make an important contribution to the world's need for power.

Although plenty of solar energy falls on our backyards every day, the technological problems of utilizing this energy on a large-scale basis are substantial. First, no energy arrives at night, which means that supplemental sources or expensive solar energy storage devices are necessary. Variations in the weather, especially in the amount of cloudiness, make the solar energy supply quite variable from day to day and season to season. Furthermore, some means of concentrating the diffuse solar radiation must be utilized. Also, the cost of converting solar energy to electricity by large power plants is not yet competitive with those for conventional power sources, although the costs are projected to become about equal by 1990.

Several means of collecting and concentrating solar energy have been proposed. Arrays of reflectors might be set up over large fields to reflect the radiation to a collector on top of a solar tower in the middle of the array. The collector would absorb the concentrated energy, which would later be used in generating electricity or in heating buildings.

Another way of harvesting solar energy avoids the problems of the vagaries of weather by utilizing solar energy collectors on satellites (Figure 10.5). Here, with the collector always facing the sun, the full solar constant can be intercepted and converted to electricity on the satellite. The electricity can then be converted to microwave energy and transmitted to antennas on earth, where the microwave energy can be safely and efficiently converted back to electricity. It is estimated that one such satellite system could produce 3000–15,000 megawatts.

The preceding power-generating techniques can produce enough electricity to provide important supplements to other power sources, especially during periods of heavy electrical use. Fortunately for the earthbound solar energy systems, the peak in solar radiation received at the ground coincides more or less with the peak in electrical demand, as shown in Figure 10.6.

For many years, solar energy has been used on a very limited basis to heat water for homes or to provide heat for growing plants. Individual solar power systems for air and water heating and air conditioning are feasible and conserve significant amounts of energy. It should be possible to provide

[2] The total U.S. electrical capacity in 1970 was 3.6×10^5 megawatts.

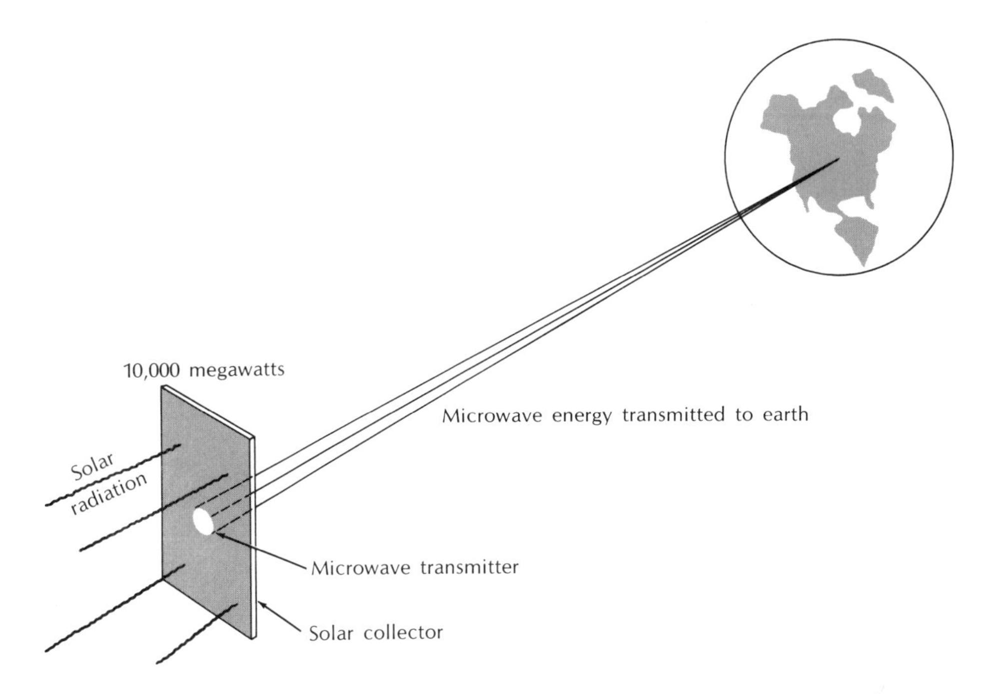

Figure 10.5 *Collection of solar energy by satellite, and relay to earth.*

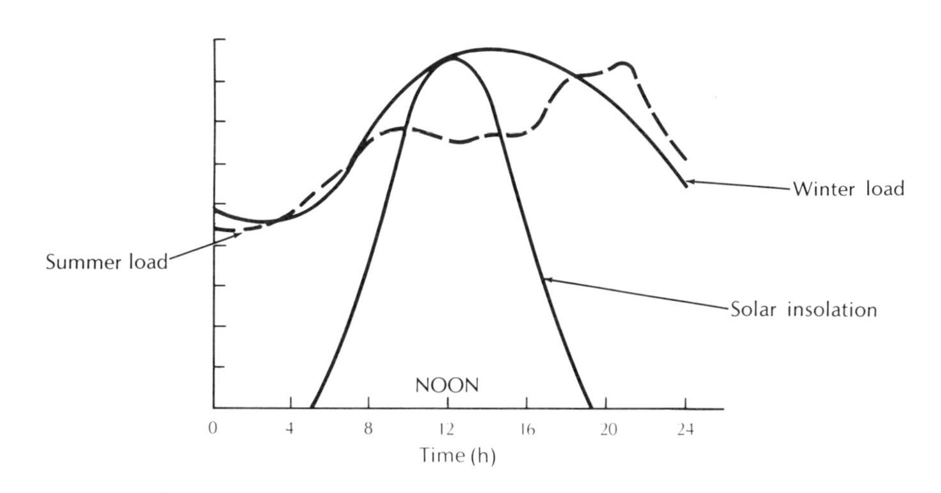

Figure 10.6 *Relative hourly variations in solar insolation and winter and summer demands for electricity (source: Energy, Environment, Productivity, Proceedings of the First Symposium in RANN: Research Applied to National Needs, Washington, D.C., November 18–20, 1973, National Science Foundation, Superintendent of Documents, U.S. Government Printing Office, Washington, D.C. 20402).*

economical home and office solar heating and cooling systems that would generate 60–90 percent of the total heating and cooling requirements for about 75 percent of the United States.

One simple solar heating system for homes, which is available commercially in Australia, Israel, Japan, the Soviet Union, and the United States, is the solar water heater, which utilizes the high heat capacity of water to store the solar energy. A simplified schematic diagram of such a system is shown in Figure 10.7. The sun heats the water in collectors which are mounted on the roof. Then the hot water is pumped into a storage tank where it may be tapped by the hot-water heating system in the house. After being used, the cool water is pumped back into the solar collector on the roof.

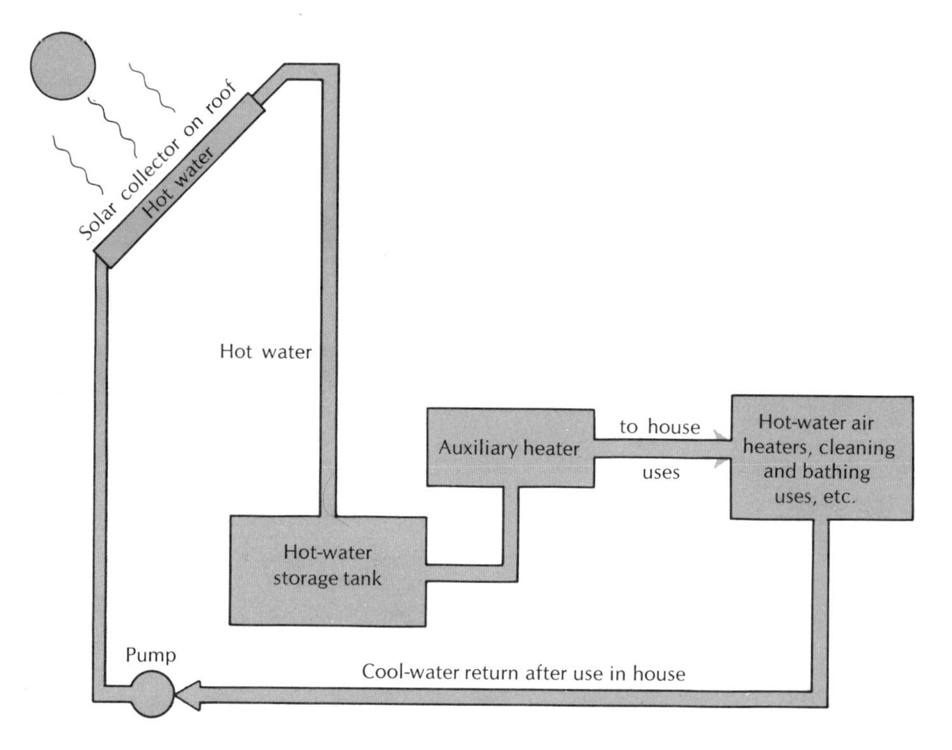

Figure 10.7 *Solar water-heating system.*

The use of solar systems for household heating is not a dream of the future, but a reality of the present. Figure 10.8 compares the range of costs of electric, gas, oil, and solar home heating over a wide variety of U.S. climates. Solar power is competitive right now, and with the cost of the other sources certain to rise as the supplies are depleted, solar energy should soon be the least expensive way to heat a house.

10.3.2 Hydroelectric power

Hydroelectric power is a clean source of energy which is already extensively developed in the United States. There are many potential reservoir sites throughout the world that, if utilized for power development, would approximately double the hydroelectric power output. However, hydro-

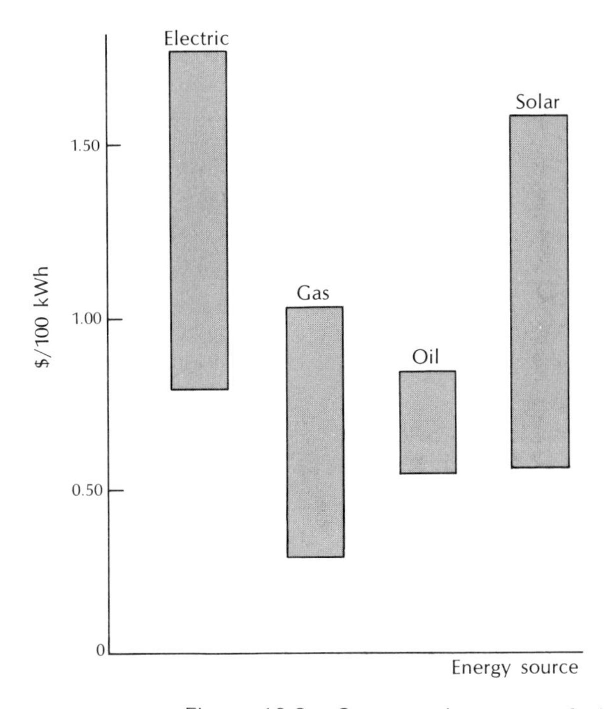

Figure 10.8 *Comparative costs of electric, gas, oil, and solar heating (source: Solar Energy as a National Resource, NSF/NASA Solar Energy Panel, December, 1972, Superintendent of Documents, U.S. Government Printing Office, Washington, D.C. 20402).*

electric power produced only 4.1 percent of the United States' total energy requirements in 1971, and is projected to produce only 3.1 percent in the year 2000 (Table 10.3). Furthermore, in the United States, a large percentage of the favorable reservoir sites have already been used, thus decreasing this nation's hydroelectric power exploitation potential. In the Columbia River Basin, for example, approximately four-fifths of the hydroelectric power storage capacity has been developed.

Table 10.3 *Percentage of total energy consumed in the United States in 1971 and projected percentages for the year 2000*

Type	1971	2000
Coal	18.3%	16.4%
Petroleum	44.2	37.2
Natural gas	32.8	17.7
Nuclear power	0.6	25.6
Hydropower	4.1	3.1

Source: United States Energy through the Year 2000, U.S. Department of the Interior, December, 1972.

Improved management of existing hydroelectric power generation reservoirs through more accurate hydrometeorological forecasting would increase the amount of energy available. Efficient management is dependent upon estimates of the amount of water that will be available for various types of water use. In the western United States, reservoir management is largely based on estimates of the amount of snow available for runoff. In all areas, tradeoffs must be made between optimum reservoir levels for power generation and flood control.

Estimates of runoff to be expected during the snowmelt season are based on widely spaced monthly snow-depth measurements in remote areas. Such estimates involve considerable uncertainty, and runoff prediction errors vary from 7 to 40 percent, depending on the river basin. Such uncertainty makes it impossible to use the water resources to maximum efficiency; i.e., in reservoir management, an overly cautious safety margin must be used, which results in a significant decrease in the hydroelectric power capability. The reduction of the runoff forecast uncertainty will result in greater hydroelectric power capability. This reduction will be realized through a combination of better and more frequent surface snow measurements and the use of high-resolution satellite observations of snow cover to improve both short-term and seasonal runoff predictions.

10.3.3 Wind power

Wind power has been used by people since the early Chinese and Persian civilizations. The wind propelled ships and drove windmills which ground grain, pumped water, and later generated a limited amount of electricity. The windmill improved slowly through the centuries. The greatest improvements in size, amount of power produced, and efficiency occurred from the late 1800s through 1950, when there was great interest in many countries, including Russia, Germany, England, France, Denmark, and the United States, in developing large-scale wind-driven generators. However, although many were built and operated, none were competitive in cost with steam and hydroelectric power plants, mainly because of the abundance of fossil fuels and the uncertainty of wind power due to the variable nature of the wind. As a result, interest in wind power remained at a low level until the early 1970s. The present fuel shortage, however, has stimulated scientists to investigate this potential power source in more detail. Wind power is appealing because it is an energy source, like solar energy, that is inexhaustible and nonpolluting. The windmill has been used to generate small amounts of electricity since 1890, and in the rural United States, thousands of windmill-driven generators have been used to supplement the electrical needs of farms. Most of these small windmills generate about one kilowatt of power. Because *each* person in the United States requires about 1 kilowatt of power, and the projected U.S. requirement in the year 2000 is 1.6×10^9 kilowatts, it would clearly be impractical to rely on windmills of this size.

However, bigger, more powerful windmills are possible, and in fact have already been tested. The most famous example is the "Grandpa's

Knob" windmill, erected in 1941 on a 2000-foot hill in Vermont known as Grandpa's Knob. This windmill consisted of two propeller blades 175 feet in diameter, each weighing 8 tons! The windmill produced energy at an average rate of 1250 kilowatts, before flying apart in 1945, hurling one blade 750 feet into the air.

The Grandpa's Knob windmill did, indeed, prove that significant amounts of energy could be produced by the wind, but a number of problems remain, some meteorological, and some economic. First, large windmills can theoretically convert about 35 percent of available wind energy to electricity, but to reach this maximum, steady wind speeds of over 18 mi/h are required. However, few populated regions of the United States have low-level wind speeds of this magnitude often enough to produce a dependable supply of wind-driven electric energy. Therefore, windmills would have to be located in selected regions where strong, steady winds occur, such as the Great Plains, along high mountain ridges, or on the oceans. Unfortunately, in these places, storage and transport of the electricity would be a problem.

Even in normally windy areas some mechanisms for storing the energy are necessary for the inevitable calm periods. Suggested storage mechanisms include batteries, pumping water back into dams when the wind is blowing for later hydroelectric generation, and electrolyzing water into hydrogen and oxygen, which could later be burned as a clean fuel.

An economic problem with windmills is their cost of producing electricity compared to conventional methods (burning of oil and coal). Crude estimates indicate that the power produced by the most efficient windmills would cost roughly twice as much as power from conventional sources cost in 1970. However, with the rapidly dwindling supplies of fossil fuel and the inevitable rise in prices of the scarcer fuel supplies, this gap does not seem insurmountable.

10.4 The Subtle Influence of Climate on Civilization

It has been argued that the effects of weather are so pervasive that they have affected the development of civilization. In areas where advanced civilization has developed, the climate is characterized by changeable, stimulating weather. Although relatively advanced civilizations are favorably influenced by a moderately challenging climate, for the earliest civilizations to have developed, the climate had to be suitable for primitive living and include mild temperatures and abundant water. The best areas for these conditions are central Africa, tropical Latin America, and Southeast Asia. As civilizations reached a point where they could cope with the more hostile colder climates, migrations occurred and people were subjected to more stimulating conditions.

It has often been claimed that the advances in mathematics in the Near East were encouraged by the ease with which the constellations could be studied in the cloudless sky. Greeks, on the other hand, may have become good sailors because conditions in the Greek islands were favorable for sailing by primitive people. There are numerous harbors and the sailor is seldom out of sight of land. Visibility is nearly always good. Few storm systems pass over the region in summer, and land-sea breezes are often good for sailing.

Finally, the decline of different civilizations has also been examined. It has been suggested that climatic conditions favorable for the proliferation of mosquitoes contributed to the decline of the ancient Greeks by increasing the incidence of malaria. Additionally, some authors partially attribute Rome's collapse to a decline in rainfall.

10.5 Weather and Culture

Let us now focus on how weather affects some of the specific components of human culture. Culture can be defined as the concepts, habits, skills, arts, instruments, and institutions of a given people in a given period.

10.5.1 Weather and religion

Religious rites frequently deal intimately with the weather. People have always sought explanation, through their legends and rites, to explain natural phenomena, which, no doubt, seemed like magic or the supernatural to them. Many times these rites are directed toward some type of weather modification. The witch doctor of some ancient tribe may well have been the first weather wizard to rely on a carefully devised ceremony to bring rain or still the storm. Similarly, the Hopi Indians in the southwestern United States employ rain dancing to beseech ancient gods to end drought conditions. Ironically, at the same time, meteorological airplanes may be overhead seeding clouds with silver iodide. Meteorologists still do not agree which method is more effective.

Many ancient civilizations, such as the early Greek's, attributed the weather to their gods. Jupiter controlled the thunder, Zeus' anger was indicated by lightning, Aeolus caused the wind to blow, and Iris wore a many-colored robe which was the rainbow. Prayers and sacrifices were made to these gods in order to induce them to produce desirable weather and withhold poor weather.

In matters of meteorology, the Judeo-Christian religion is not unlike these ancient sects. Jehova sends the lightning; St. Peter intercedes on our behalf in matters of weather; and in many Catholic rural regions of Europe, holy water is sprinkled on fields during drought.

All over the world, religion adjusts itself to differences in climate. As a specific example, Buddhism in India views hell as a place of intense heat. This is not surprising, for in northern India in the late spring and early summer, it is extremely hot. The maximum temperature may hover around 105 or 110°F day after day. It is understandable then that the Buddhist hell has six levels based on torture by heat. In the mildest level, the people are burned and then allowed to recover before the next burning. In the next level, they are cut to pieces as well as burned, but are allowed a breathing space to recover. And so on, until the lowest hell, where burning continues eternally without relief. When Buddhism migrated northward a few hundred miles to high Tibet, a torrid hell did not seem at all inhospitable to shepherds, whose worst foe is the cold and snow. In order for Buddhism to gain a foothold here, hell had to be remodeled. In Tibet, as in India, hell has six levels, but torture by cold and freezing is the main punishment. In the lowest level of hell, the excruciating ache of fingers that are being warmed after they have been frozen is spread externally to all parts of the body. Again, in contrast to the Judeo-Christian religions, which originated in the hot climate of the Near East and threaten a hell of eternal burning, the Norse and Icelandic myths and legends portray hell as a place of everlasting ice and snow.

10.5.2 Weather and music

Music is another facet of culture where one would expect to find significant influences by weather. Anyone can think of numerous song titles which consist primarily of weather or climate references (e.g., "The Rain in Spain," "Summertime," "Somewhere Over the Rainbow," and "Raindrops Falling on My Head,") In addition, many classical compositions employ musical imitations of weather elements. James Wagner,[3] in his delightful essay "Music to Watch Weather By," summarizes some of the better known classical compositions that feature meteorology. For example, *The Four Seasons,* written in the eighteenth century by Antonio Vivaldi, describes the weather and its effect on human activities in each season of the year. *The Four Seasons* expresses the relief felt on a warm spring day at the end of winter, the violence of a summer thunderstorm, the joy of an invigorating autumn day, and the excitement of the bracing outdoor sports of winter.

Toward the end of the eighteenth century, Franz Joseph Haydn composed *The Seasons,* which depicts a progression of similar weather events to those that Vivaldi described. In his orchestral introduction to winter, Haydn depicts musically the fog and mist which are so prevalent in northern Europe in November and December.

Some of the primal awakenings that accompany the return of spring are present in Igor Stravinsky's *Rites of Spring,* which includes sections describing the pagan orgies conducted by primitive tribes in the spring. This section provoked violent reactions from the audience in its Paris premiere in 1913.

An example of the life cycle of the thunderstorm is found in Beethoven's Sixth Symphony (*Pastoral Symphony*). First, the clouds darken and lower, announced by the soft rumble of distant thunder. After the violence of the storm subsides, the soft harmony of a rainbow appears.

Other familiar classical music references to weather include *The Grand Canyon Suite* by Ferde Grofé, which describes the sunrise, sunset, and a midday thunderstorm; Johann Strauss' "Thunder and Lightning Polka,"

---

[3] *Weatherwise,* August, 1972.

and Moussorgsky's *Night on Bald Mountain,* which describes a raging mountaintop storm.

One form of music in which there are many references to significant weather events is folk music. Blues music is only one type of folk music in which many references to weather or weather-related events can be cited. Titles such as "Texas Tornado Blues," "St. Louis Cyclone Blues," "Nine Below Zero," "Flood and Thunder Blues," and "Mississippi Flood Blues" are just a few examples of weather references in blues music. As a sidelight, the appropriately named Lightning Hopkins has been the blues artist to record the greatest number of weather-related songs in recent years. Numbered among his recordings are such titles as "That Mean Old Twister," "Rainy Day Blues," "Ice Storm Blues," "California Mudslide," and "Hurricane Carla."

Analysis of the titles of all the blues songs recorded from 1920–1966 has shown that the most frequent references are to some form of precipitation, followed closely by flood songs and then, to a lesser degree, by songs about unusually low temperatures. Of the 105 precipitation-related titles, over 90 percent of them are of the mood variety; i.e., the precipitation has produced a depressed feeling in the artist. The flood reference songs occurred primarily in the late 1920s and the 1930s and decreased drastically thereafter. Some of the most devastating floods in history occurred in 1927 and 1937 on the Mississippi River and commentary on the floods rapidly found its way into the blues music, documenting the loss of life and the destruction. As a result of these floods, the federal government embarked on a project of flood control and land reclamation that took many years to complete, but was eventually successful in reducing flood damage. When great volumes of water roared down the Mississippi in the late 1940s and the early 1950s, flooding damage was minimized, and the floods virtually passed unnoticed by the bluesman or folk composer. Finally, because most blues artists lived in the South and were used to high temperatures but not to the relatively rare cold waves, references to the low temperatures in blues are seven times more frequent than high temperature references.

10.5.3 Weather and literature

Meteorological imagery abounds in literature and language. The authors of the Bible, drawn from a stock of weather-conscious farmers, shepherds, and sailors, spoke often of the weather in the Near East. Observant of the peculiarities of the climate in this region of the world, the different writers produced a remarkably consistent collection of weather proverbs. For example, the east wind in Palestine blows off the hot, arid Arabian Desert, and according to the Biblical prophets:

> *The east wind dried up her fruit.* [Ezekiel 19:12]
>
> *An east wind shall come, the wind of the Lord shall come up from the wilderness, and his spring shall become dry and his fountain shall be dried up.* [Hosea 13:15]
>
> *When the east wind toucheth it, it shall wither.* [Ezekiel 17:10]

Plate 15a (above) *Rainbow taken with the same lens as the cloud bow* (© *Alistair B. Fraser*).

Plate 15b (below) *Cloud bow. The cloud bow appears nearly white* (© *Alistair B. Fraser*).

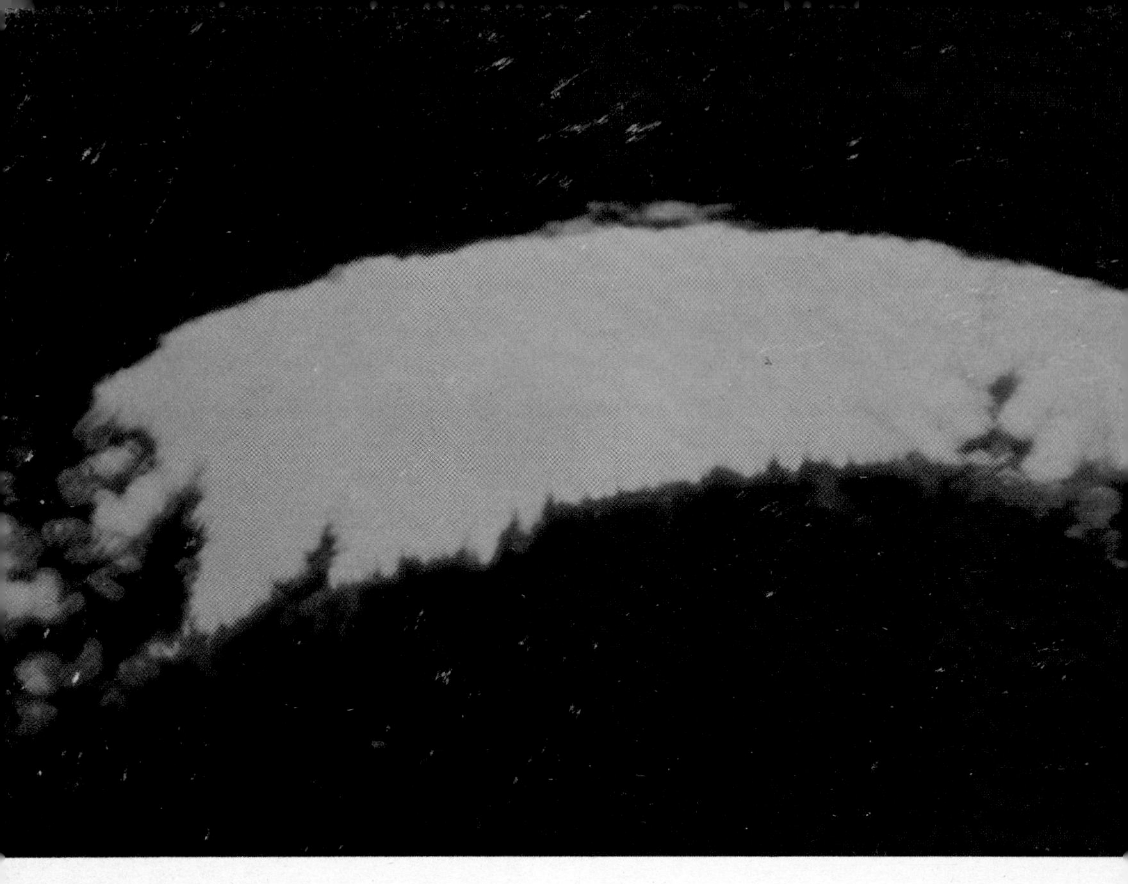

Plate 16 (above) *The green flash.* Occasionally as the sun rises or sets, the upper rim will grow and momentarily produce a brilliant flash of green light (©*Alistair B. Fraser*).

Plate 17 (below) *Fog in the Bald Eagle Valley of Pennsylvania (photo by Richard Anthes).*

And behold, seven thin ears, blasted with the east wind, came up. [Genesis 12:6]

God prepared a vehement east wind [Jonah 4:8]

If east winds produced drought and heat in biblical lands, the west wind carrying moisture from the Mediterranean Sea brought showers, and, according to Luke 12:54:

When ye see a cloud rise out of the west, straightway ye say there cometh a shower; and so it is.

The south wind, on the other hand, produced stormy, unstable weather, and the Bible speaks of whirlwinds coming from this direction:

As whirlwinds in the south [Isaiah 21:1]

And shall go with whirlwinds of the south [Zechariah 9:14]

Out of the south cometh the whirlwind [Job 37:9]

Shakespeare provides us with dozens of examples of the effective use of weather imagery. In England, southerly winds in advance of approaching Atlantic storms bring mild, moist air over the British Isles. Shakespeare, who was very observant of the British climate, writes in *As You Like It,*

Like foggy south, puffing with wind and rain

[3.5.50]

and again, in *King Henry the Fourth, Part II,* he writes

When tempests of commotion like the south,
Born with black vapour, doth begin to melt,
And drop upon our bare, unarmed heads.

[2.4.339–41]

Shakespeare often utilized the imagery of meteorological storms in his dramas of human struggle. In *Hamlet,* Shakespeare sets the ominous mood of the calm before the storm:

But as we often see, against some storm,
A silence in the heavens, the rack stands still,
The bold wind speechless and the orb below
As hush as death: anon the dreadful thunder
Doth rend the region.

[2.2.471–75]

And, speaking of human affairs as well as the weather in *King John,* Shakespeare writes

So foul a sky clears not without a storm.

[4.2.108]

The English language is spiced with idiomatic expressions using weather terms. A few examples are the following: "a restaurant with *atmosphere*"; "a *fair-weather* friend"; "something in the *wind*"; "he *stormed* into the office"; "she didn't have a *foggy* notion"; "he had only a *hazy* recollection"; "they save for a *rainy* day"; and "the student was *snowed*." The reasoning behind some of these examples is somewhat subtle, but easily understood upon reflection.

10.5.4 Weather and art

The visual arts are replete with meteorological images. Dr. Hans Neuberger[4] made an extensive study of the ways in which the climate experienced by the artists is reflected in their paintings. For this study, more than 12,000 paintings in 41 museums in nine countries were analyzed, covering the years 1400 to present. Briefly, here are some of his results:

(1) The frequency of paintings with deep-blue skies is greatest in the clear Mediterranean region, least in the cloudy British Isles. Other writers have commented that English people arriving in the Mediterranean from England in February are aware of a quickening of their mental processes. Italian art, with its vivid coloring, supposedly even stimulates Britishers at home in the same way.
(2) The transparency of the air painted, termed *visibility,* was also estimated; again, Mediterranean paintings had the greatest visibility, British paintings the least.
(3) The frequency of low and convective clouds, which are most often associated with bad weather, is greatest in British paintings, least in Italian paintings.
(4) The frequency of paintings with clear or partly cloudy skies is greatest in the Italian school, least in the British school. As a matter of fact, Dr. Neuberger did not observe a single British painting with a cloudless sky.

The other schools of painting fall between the extremes of the British and Italian schools in the analysis. The statistics developed in this research were even able to delineate the Little Ice Age extending from 1550–1850. In summary, the regional differences in climate are reflected in parallel differences in the paintings of artists having lived in these climates.

---

[4]Hans Neuberger, "Climate in Art," *Weather,* Vol. 25, No. 2, February, 1970, pp. 46–56.

11 A year's weather in the United States

In the previous chapters, we touched on some of the characters in the weather drama and have tried to make the everyday, as well as the extraordinary, weather episodes more understandable to the appreciative observer. The emphasis has been on how the weather affects people, and how people influence the weather and climate, both now and in the future. In keeping with these themes, it is appropriate to summarize the concepts in the book by a month-by-month review of a year's weather pageant over the United States in an effort to illustrate how the physics of the atmosphere unite to produce the dazzling progression of meteorological performances in varied settings.

11.1 January

Jack Frost in Janiveer
Nips the nose of the nascent year[1]

The emphasis in January over the eastern two-thirds of the country is on intense winter storms followed by savage winds and piercing cold. The development of these storms is

[1]Unless noted otherwise, the weather proverbs in this chapter are taken from Richard Inwards, *Weather Lore* (London: Elliot Stock, 1893).

favored by the very strong north-south temperature contrasts and the resulting fast jet streams that are present during January. Winter storms typically begin as weak low-pressure centers that form along an old polar front in Texas or along the Gulf Coast. As a trough in the fast upper-level westerlies sweeps across the Rockies, a favorable upper-level divergence pattern is produced for the deepening of the surface low. Mild and moist air from the Gulf of Mexico is carried northward and overruns the cold, more dense air mass near the ground. As the air is lifted, rain and, farther north, snow, are produced.

The low moves east-northeast as it intensifies, spreading snow as far north as Illinois and Michigan, and drawing warm, unstable air ahead of its path into the south Atlantic states, where thunderstorms break out ahead of the advancing cold front. Behind the developing low, northerly winds carry cold dry air well south of the U.S. border into Mexico and Central America.

The low continues to intensify as it moves up the East Coast, now drawing moist air from the Atlantic as well as from the Gulf of Mexico to feed the heavy rain and snowfall. Paralyzing snows—up to a foot or more—occur to the west of the storm track in the mountains of North Carolina, Virginia, West Virginia, and Pennsylvania. Northeast winds of near hurricane force peak the Atlantic waves into 12-foot crests which crash on lonely New Jersey beaches. Snow spreads into New England, while the winds at Mt. Washington in New Hampshire howl at 150 mi/h (with a temperature of −20°F).

Behind the storm, cold-wave warnings in the upper Midwest signal a fresh blast of arctic air, which only recently covered the icy Yukon. Piercing cold air sweeps southward out of the dark, frozen interior of western Canada. In the North, the snow cover turns dry and hard, squeaking under the feet of the few who must venture outside.

As the polar anticyclone passes southward over Minnesota and Wisconsin and turns eastward to cover Illinois, Indiana, and Kentucky, winds diminish in the East. At night, the quiet of the northern forests is disturbed by trees cracking with cold under skies filled with stars that blaze with a passionate intensity never seen through summer hazes. The earth's surface radiates freely throughout the long night to the cold void of space, unhampered by any water-vapor blanket.

The coldest weather of the year occurs in the calm, nighttime centers of these intense anticyclones. The morning radio calls out the toll:

International Falls, Minn.	−41°F
Bismark, N. Dak.	−36
Minneapolis, Minn.	−21
Madison, Wis.	−15
Chicago, Ill.	−9
Indianapolis, Ind.	−1
Louisville, Ky.	6
Atlanta, Ga.	14
Tallahassee, Fla.	15
Orlando, Fla.	28
Miami, Fla.	37

The emphasis in the East is cold, verifying again one of the truest proverbs:

As the day lengthens, so the cold strengthens.

Only in the Deep South, where the weather is more directly influenced by the solar radiation, do the coldest temperatures occur in December near the solstice.

While the East suffers through the cold waves, the eastern slopes of the Rockies may experience temporary relief from winter by the amazing *chinook,* a strong, warm wind named after the Chinook Indians of the lower Columbia River. The chinook pours down the sides of the Rockies, warming and drying as it is compressed by the higher pressure. It is impossible to describe this wind better than the following account, published in 1896:

> *Picture to yourself a wild waste of snow, wind-beaten and blizzard-furrowed until the vast expanse resembles a billowy white sea. The frigid air, blowing half a gale, is filled with needle-like snow and ice crystals which sting the flesh like the bites of poisonous insects, and sift through the finest crevices. . . . Great herds of range cattle, which roam at will and thrive on the nutritious grasses indigenous to the northern slope, wander aimlessly here and there, or more frequently drift with the wind in vain attempts to find food and shelter; moaning in distress from cold and hunger, their noses hung with bloody icicles, their legs galled and bleeding from breaking the hard snow crust as they travel. . . . Would the chinook never come? The wind veered and backed, now howling as if in derision, and anon becoming calm, as if in contemplation of the desolation on the face of nature, while the poor dumb animals continued their ceaseless tramp, crying with pain and starvation. At last . . . at about the hour of sunset, there was a change which experienced plainsmen interpreted as favorable to the coming of the warm southwest wind. At sunset the temperature was only –13°, the air scarcely in motion, but occasionally seemed to descend from overhead. Over the mountains in the southwest a great bank of black clouds hung, dark and awesome, whose wide expanse was unbroken by line or break; only at the upper edge, the curled and serrated cloud, blown into tatters by wind, was seen to be the advance courier of the long-prayed-for chinook. How eagerly we watched its approach! How we strained our hearing for the first welcome sigh of the gentle breath! But it was not until 11:35 P.M. that the first influence was felt. First, a puff of heat, summer-like in comparison with what had existed for two weeks, and we run to our instrument shelter to observe the temperature. Up goes the mercury, 34° in seven minutes. Now the wind has come with a 25-mile velocity. Now the cattle stop traveling, and with muzzles turned toward the wind, low with satisfaction. Weary with two weeks standing on their feet they lie down in the snow, for*

> *they know that their salvation has come: that now their bodies will not freeze to the ground.*
>
> *The wind increases in strength and warmth; it blows now in one steady roar; the temperature has risen to 38°; the great expanse of snow 30 inches deep on a level is becoming damp and honeycombed by the hot wind.*
>
> *Twelve hours afterward there are bare, brown hills everywhere; the plains are covered with floods of water. In a few days the wind will evaporate the moisture, and the roads will be dry and hard. Were it not for the chinook winds the northern slope country would not be habitable, nor could domestic animals survive the winters.* [A. B. Coe, "How the Chinook Came in 1896," *Monthly Weather Review,* Vol. 24, 1896, p. 413]

Some of the most rapid changes in temperature occur with the onset of the chinook. For example, on January 22, 1943, in Spearfish, South Dakota, the temperature rose from –4°F to 45°F in two minutes! The effects of the chinook are not all beneficial, however. Hurricane-force winds can cause extensive damage in the lee of the Rockies, as the residents of Boulder, Colorado know.

In the Far West, January is a cloudy, stormy month in the north and cool and dry in the south. The precipitation that falls in the Pacific Northwest is caused by a frequent progression of intense Pacific cyclones, whose centers of low pressure usually cross the Canadian coast north of the United States. The moist Pacific origin of these frequent storms and the mountain barriers blocking the onshore winds produce a favorable situation for intense precipitation events—snow in the mountains and cold, dreary rain along the coast. Annual snowfall on the western flanks of the Sierra Nevada and the Cascade mountains exceeds 400 inches in some places.

The southern portion of the Pacific Coast is much drier than the north. Moisture-bearing storms, even in January, usually pass well to the north. The land is also warmer than the cool Pacific, so air from the Pacific warms and the relative humidity decreases, producing day after day the famous southern California sunshine. Occasionally, however, slow-moving Pacific storms move inland far enough south to produce heavy rains and mudslides along the steep slopes of the southern California coast.

One final idiosyncrasy about January weather—in most of the eastern United States, including Washington, D.C., New York, and Boston, around the middle of the month, it is common to experience a temporary respite from the severe cold. After days of subfreezing temperatures, northern residents are awakened by the unfamiliar sound of water dripping off snow-covered roofs. The "January thaw" has arrived. The January thaw is a climatic anomaly in which unseasonably mild weather tends to recur at nearly the same time every year, usually during the period of January 20–23. Although the annual occurrence of the January thaw is not a certainty, temperatures at least 10°F above the seasonal average are more than six times likely to occur in the New York area between January 19 and 24 as between December 4 and 12 or February 7 and 12. We really do not understand why the January thaw occurs, but we appreciate its temporary relief nevertheless.

11.2 February

There is always one fine week in February.

A February spring is not worth a pin. [Proverb of Cornwall, England]

The records show that the groundhog month, February, is nearly as cold as January, but there are usually some hints in most places of warmer days ahead. Toward the end of the month, the sun is above the horizon at latitude 40°N for 11 hours and 7 minutes, compared to a stay of only 9 hours and 20 minutes on December 25. The nearly two extra hours of daylight provide more of a psychological lift than a sensible warming to the winter weary. However, in the South, on days when low pressure systems to the west bring semitropical air from the Gulf of Mexico, daytime temperatures can be quite warm. Along the Gulf Coast, days of 70°F are not uncommon, and early spring flowers such as daffodils appear during this month.

But in the United States, do not be misled by a rare, fine February day, for February can offer as intense cold and deep snowfalls as January, although they usually do not last as long. And, even though February is not the coldest month of the year in the South on the average, some severely cold temperatures have occurred during this month. For example, the cold wave of February 1899 brought subzero temperatures (°F) into Florida.

The ice storm is a particularly damaging, although frequently very beautiful, visitor to the United States during February. Like snowstorms, ice storms are associated with extratropical cyclones and the overrunning precipitation of warm fronts. The difference is in the temperature of the air aloft. In the snowstorm, the temperatures are below freezing through the entire depth of the storm. In ice storms, however, the warm air overrunning the cold surface air remains above freezing through a thick enough layer to melt the snowflakes falling from still higher (and colder) regions. The cold, dense air next to the ground, however, is reluctant to move, especially in mountainous regions, where it remains trapped in the valleys. The rain falls into the subfreezing layer near the ground and freezes upon contact with any surface—trees, telephone wires, roads, or cars. The weight of the accumulated water can be immense: eleven tons of ice have been supported by telegraph poles before snapping; individual wires may accumulate 1000 pounds of ice. Birds are found with their feet frozen to the branches of trees, and deer in the forests have their feet and legs slashed by the edges of the ice-covered snow as they break through the sharp crust of ice. In the United States, only Florida, New Mexico, Arizona, and the southern part of California normally escape ice storms. The region most visited by these quiet, but devastating storms extends from Texas into Kansas, then eastward across the Ohio Valley into the Middle Atlantic states.

February is perhaps the best month for skiing in the United States. Most ski resorts have had two or three months of snowstorms and so have developed a deep base of snow. In the Northeast, the depth may have reached 5 feet; in the West, 20 feet is more likely. Also, the days are noticeably longer and the sun correspondingly brighter in February than in either December or January, making skiing more comfortable.

11.3 March

March was so angry with an old woman for thinking he was a summer month, that he borrowed a day from his brother February, and froze her and her flocks to death. [From island of Kythmos, Greece]

In March, spring gains a firm foothold in the South and occasionally offers tantalizing appetizers of 70°F temperatures to the North. The sun crosses the equator and rises at the North Pole this month, and the days lengthen rapidly with each passing week, providing roughly 15 minutes more sunshine per week. In the North, March temperatures are 10–15°F higher than in February.

The strong March sun in the South and the lingering bitter cold in Canada produce very strong north-south temperature gradients across the United States in March, creating a favorable environment for intense storms. March storms may present a tremendous variety of weather, with blizzard conditions in the cold air over the Dakotas, and near-hot, humid conditions with thunderstorms in the warm sector. The greatest amount of snow in March falls in the Great Plains as the intense lows pass to the south. For example, Rapid City, South Dakota normally receives 22 inches in March, compared to 15 in February and 14 in January.

March is known as the windy month. High surface winds are common in March for two reasons, the first being the frequent passage of the intense cyclones just noted. The second reason is that the atmosphere in March is becoming more unstable. The instability (steep lapse rate) is caused by the strengthening solar radiation warming the ground and lower atmosphere, while the atmosphere aloft retains its wintertime cold temperatures. This instability favors vertical mixing, which transports fast-moving air to the surface from aloft.

Along the Pacific Coast, March is a relatively wet month, although not so wet as January or February. Tempered by the cool Pacific, the temperatures along the West Coast respond slowly to the increasing power of the sun, and March temperatures are only slightly warmer than those in February.

11.4 April

April weather
Rain and sunshine, both together.

The reputation April has for producing showers with sunny interludes is well deserved, for by April, the lower part of the atmosphere is becoming quite warm over most of the United States, while the air at higher levels remains cold, thus producing the instability conducive to shower formation. This instability, coupled with the occasional intense cyclones that persist into April, favors the development of tornadoes, those monsters of nature that are nowhere else so prevalent and intense as they are in the United States.

The atmosphere begins its preparation for tornadoes in a mild enough manner, when a strong anticyclone anchored over the southeastern United States begins swirling hot, humid air across the Texas Gulf Coast and northward into Kansas and Oklahoma, ahead of a developing cyclone in the lee of the Rockies. At first, the southerly winds bring comfortably warm temperatures and pleasant spring weather to the Midwest, but as the jet stream dips southward across the Rockies, funneling cold, dry air over the warm, moist air at the surface, the synoptic stage is set for the sudden release of instability through the mechanisms of the squall line, thunderstorm, and, all too often, the tornado.

First, the southwestern sky darkens; lightning illuminates black clouds, and static fills the radio. The warm, moist wind blows gustily from the southeast, carrying tropical moisture into the developing squall. The line of darkness approaches, and ahead of the rain curtain, the characteristic funnel protrudes from the cloud base, twisting and hissing like an angry serpent. Now, the visible funnel reaches the ground, scouring it clean of trees, buildings, and everything else rising above the ground. Then, a few minutes after its arrival, it is gone, continuing its malevolent track to the northeast. Cold rain washes the air and drenches the devastation as the line of thunderstorms pass. Then, the bright April sunshine returns, and quiet descends over the wasteland.

11.5 May

A swarm of bees in May is worth a load of hay.

By May, the month so rhapsodized by poets, the last frosts of winter have been banished to the extreme northern United States and higher mountains. In most places, May is a superb month, with mild days and cool nights. Warmed by two months of sunlight, the cold of the Canadian prairies has begun to slacken, and with it, the north-south temperature gradient across the United States. With the demise of the huge north-south thermal contrast, the intense winter cyclones move northward into Canada, and in their absence, the likelihood of solid overcast days with steady precipitation diminishes. Rainfall begins to occur in showers and thundershowers, for the air is still relatively unstable. But the showers are fleeting, and the potent sun, now only a month from its zenith, is able to go about its business of warming the land and the atmosphere.

May is a growing and flowering month, as indicated by the above proverb extolling the pollenating virtues of bees. Fruit trees bloom in the North, and gardens in the South.

On the West Coast, the precipitation decreases markedly in May, even in Seattle, where the average cloudiness dips below 70 percent for the first time since the previous September. The semipermanent anticyclone which resides in the Pacific builds northward, and the cyclonic storms are shunted off to the north, away from the U.S. coast.

However, May is not all flowers and light, for the forces of the North do not give up easily. Occasionally, cold pools of air in the upper atmosphere break away from their polar origins and drift slowly over portions of the United States. When these upper-air cold pools anchor themselves over New England, cool, cloudy, windy, and sometimes drizzly weather may persist for three or four consecutive days. If the cold air aloft comes down across the Rockies, severe weather can ensue, including snow in the mountains and tornadoes and hailstorms in the Great Plains along the leading edge of the cold air. But these temporary setbacks do not mar the affection most of us feel for the merry month of May.

11.6 June

It never clouds up in a June night for a rain.

As the proverb states, over most of the United States, rainfall in June is normally of the convective thundershowery variety, and, therefore, likely to occur in the afternoon, when surface temperatures are warmest. Even when weak extratropical cyclones and their attendant warm and cold fronts struggle sluggishly across the continent in the weakened flow aloft (small horizontal temperature differences; therefore, slow upper-level winds), the low-level convergence and associated lifting is more likely to produce transient showers than steady rains. In addition, the proverb has the best chance of being right in June, which has the shortest nights of the year—only eight hours in the extreme northern United States. Thus, there simply is not much time for clouding up at night.

June in the South is the first month of tropical weather; even weak polar fronts rarely penetrate those states south of 35°N. Daytime temperatures are hot, nighttime temperatures warm, and humidities always high. Thunderstorms break out nearly every afternoon in the tropical air mass.

Along the coastal regions of the United States, June is a good month for the development of the sea breeze. The land, under the influence of the strongest solar radiation of the year, heats up more than the surrounding waters, which are still cool from the previous winter. We have seen effects of this differential heating before, on a global scale. Here, during the day, it produces welcome onshore winds which extend 10–50 miles inland. Along the leading edge of this sea breeze (called the sea-breeze front), convergence frequently generates a line of showers and even thundershowers, and many places near the southeast coast can expect a daily shower, more or less at the same time each day, as the sea-breeze front passes. An example of these sea-breeze thundershowers located a few kilometers inland from the Florida coast is shown in Plate 4.

11.7 July

Whatever July and August do not boil, September cannot fry.

In most of the United States, the hottest weather of the year occurs in July, with August running a close second. Even though the sun has started its retreat toward southern latitudes, the net energy budget over the United States is still positive; that is, the atmosphere receives more heat than it loses during most of July.

By coincidence, a typical heat wave might begin as Sirius, the Dog star, rises and sets with the sun, an event which occurs sometime in mid-July and is the reason for the term *dog days*. The heat wave begins slowly as an old polar high ceases its southward drift and settles down for a lengthy visit over the south Atlantic states. At first, the air is warm, comfortably dry, and clear, but as the atmosphere aloft slowly subsides and warms by compression, extreme stability is produced, even with hot surface temperatures. As the warm high aloft increases in intensity, the winds through a deep layer of the troposphere become light and variable and finally die away altogether. Pollution, both natural (pollen and dust) and man-made (gases and particles), is neither carried away horizontally by the winds, nor mixed vertically in the stable air, and so accumulates in the lower mile of the atmosphere. A yellowish haze deepens with each hour over the eastern United States. The sun turns from brilliant white to yellow, then a brassy orange, and finally, a burnt crimson as the heat wave and the haze reach a peak. The warm high-pressure system is now firmly entrenched, maintaining in part its foothold on the U. S. continent by shunting cooler air and cyclonic disturbances clockwise around the edge of its huge circulation into Canada.

Daytime temperatures reach 100°F as far north as Maine, exceeding even Miami's 90° F reading, from which sea breezes provide a little relief. Nighttime temperatures fail to break the 80°F level; brownouts occur in the cities as air conditioners labor 24 hours a day in the torrid heat. Absolute humidities reach their maximum values of the year as the heat wave enters its second week.

Finally, an imbalance in the upper-level circulation permits some cool Canadian air to start southward, driven by the northerly winds associated with a vigorous wave in the upper-level westerlies over Canada. As this wave intensifies, the stagnant high over the United States weakens slightly, then moves grudgingly off the coast, permitting cool, clear Canadian air to replace the muddy air of the heat wave. Ahead of the advancing relief, thunderstorms erupt, providing welcome moisture for parched lands. Behind the Canadian front, the sun sparkles in a cool, deep-blue sky. The heat wave is broken.

11.8 August

August sunshine and bright nights ripen the grapes.

August weather over most of the United States is similar to July's, although the days are noticeably shorter and the temperature drops a degree or so from the July maximum. The exception to this slight cooling is along the Pacific coast, where water temperatures are still rising. Hence, in Los Angeles and San Francisco, August is slightly warmer than July.

In the East, August days continue hot and humid. During the longer nights, local fogs become more common. These *radiation fogs* form under conditions of weak pressure gradients (light winds), clear skies (strong cooling by radiation), and long nights (long time for cooling to occur). They are concentrated in the valleys where cold air drains from the higher surrounding land and collects into stagnant pools of cool, moist air, as shown by the early morning fog in the Bald Eagle Valley of Pennsylvania (Plate 17). Local differences in the early morning temperatures under such conditions can be quite large, and are strongly correlated with topography. Thus, at night, the temperature on top of a 300-foot hill under clear skies might be 60°F, while in the valley, there is a dense fog and a temperature of 50°F. These valley fogs are not very thick, and tend to evaporate soon after sunrise.

It is noteworthy that fogs "burn off" from below, not from above. Very little of the rising sun's heat is absorbed by the top of the fog deck. Part is reflected from the top of the fog; the rest penetrates through the fog and warms the ground and the air in contact with the ground. The lowest fog droplets evaporate in this warmer air. Furthermore, as time-lapse photography shows, local hot spots develop underneath the fog and generate thermals which rise through the fog deck and penetrate the top, mixing drier air downward and helping dissipate the fog. Because such radiation fogs are associated with calm anticyclonic conditions, they are strong indicators of fair weather; hence the proverb:

When the fog falls, fair weather follows.

In contrast to the inland radiation fogs, the *marine fogs* of the West Coast are caused by the cooling of low-level air by cold water. San Francisco is perhaps most famous for its fogs, great banks which, in the summer, pour inward through the Golden Gate in response to the sea breeze. These fogs do not survive more than a few miles inland during the day, as they are eroded from below by surface heating. They have a beneficial aspect near the coast in that they frequently contribute moisture—as much as 0.1 centimeters—to vegetation during this otherwise dry season. For example, the needles of the redwood tree strain the tiny water droplets from the air and combine them into larger droplets, which then drip off the trees, providing necessary water. It has also been shown that redwood needles absorb directly the equivalent of 0.04 inches of rain on foggy nights. These redwoods grow only on the coast. Forty miles inland, where the fogs rarely reach, redwoods cannot survive.

11.9 September

June—too soon;
July—stand by;
August—look out you must
September—remember
October—all over

[Captain Nares]

Then up and spake an old Sailor,
Had sailed to the Spanish Main,

'I pray thee, put into yonder port,
For I fear a hurricane.

Last night, the moon had a golden ring,
And to-night no moon we see!'
The skipper, he blew a whiff from his pipe
And a scornful laugh laughed he.

[Henry Wadsworth Longfellow, "The Wreck of the Hesperus," *The Complete Poetical Works of Henry Wadsworth Longfellow* (Boston: Houghton, Mifflin and Co., 1903), p. 17]

The unwary visitor from the north would hardly suspect anything amiss as he says farewell to the golden red sun on a hot, still September evening in the Florida Keys. Not a cloud disturbs the brassy sky, not even the usual evening clouds that form over the Gulf Stream. And even the usual southeast breeze has died away completely, leaving a strangely tranquil twilight.

The night begins uncomfortably warm; no wind moves the humid air. Walking outside, searching in vain for relief, he sees the gibbous face of the waxing moon shrouded by fine, powdery clouds which cast a faint orange ring in the night sky. Left unsatisfied by even a whisper of cool air, he sinks restlessly back on his bed.

Later at night, he is conscious of an unusual sound—soft at first, but gradually increasing in volume. The gentle lapping of small wavelets on the beach is drowned out by a rhythmic cadence of swells breaking over the sandy shoals far out in the bay.

The morning dawns without a sun; thickening and lowering cirrostratus clouds infect the eastern sky. At last, the wind has returned, but this time from the northwest, an unusual direction for September, but welcome nevertheless. The needle of an old barometer on the wall suddenly becomes unstuck from its rusted position at 1012 millibars and drops to 1001 millibars.

Later in the afternoon, high and middle clouds cover the entire sky, and low, dark clouds appear on the northeast horizon. The wind is blowing noisily in from the northwest now, with brief gusts reaching 30 mi/h. Whitecaps cover the longer, high swells that roll relentlessly toward the coast from the northeast.

The first rain falls an hour later as an outer band of the hurricane sweeps across the islands. A half an inch falls in 15 minutes; then, the rain abruptly ceases as the dark clouds race on. But other nimbus clouds scud in across the frenzied sea, bringing more rain and ever-increasing winds. The top of a thirty-year-old coconut palm snaps under a brief gust reaching 80 mi/h.

The gray sea is now completely covered with spindrift, which merges imperceptibly with the raindrops streaking almost horizontally in the howling wind. Quickly now, the wind and water rise to the climax. The western coast and the landscape are rearranged under the fury of the wind and tons of water; new islands are created, old ones covered forever. Salt spray is driven everywhere, even tens of miles inland, where it damages salt-sensitive plants.

Then, as if by supernatural intervention, the 100-mi/h winds subside within a few minutes, the low clouds open, and a dim sun appears through a thin overcast. Towering white clouds, illuminated by the sun, wall the eye of the hurricane. The wind blows fitfully, first one way, then another. The barometer, now at its ebb, reads 950 millibars.

But the other side of the wall is approaching, and quickly the winds resume their battering, this time from the southeast. Now, the east coast of the island is disfigured, bearing the full intensity of the wind-driven storm tide. Trees which had barely managed to withstand the northwest winds by leaning toward the southeast are snapped back by blows from the opposite direction.

The hurricane leaves as gradually as it came—the squalls become less frequent and less violent, occasional breaks in the clouds uncover the moon, and the winds subside little by little. The sea slowly returns to normal, adjusting gradually to the lighter winds. The hurricane has crossed the Keys and moved into the Gulf of Mexico.

Hurricanes, dramatic as they can be, are rare refugees from the tropics, even in September along the Florida coast, where they are most likely to occur. For most of the United States, September is a superb outdoor recreation month for those who like sunshine. It is remarkable that in nearly every part of the country, September is one of the clearest months of the year. (Note the annual variation of cloudiness in the climatic graphs in Appendix 2.) Also, water temperatures reach a maximum in September, making this month ideal for beach vacations almost everywhere along both coasts.

The reason for the minimum in cloudiness is related to the stability of the air and the infrequent appearance of extratropical cyclones. The autumn atmosphere is generally stable, because, in contrast to spring, the air aloft is warm after the summer months, and the ground is beginning to cool under the lengthening nights. These cooler surface temperatures and warm air aloft are also responsible for the frequent formations of fog during September evenings. Thunderstorms become less frequent in September for the same reason—increasing stability of the air.

Although the arctic region begins to cool substantially in September as it bids farewell to the sun, the temperatures in southern Canada are still mild. Therefore, the north-south temperature contrast over the United States remains small during September, and extratropical storm systems remain weak.

An unwelcome September visitor to California, a bad cousin of the beloved Great Plains chinook, is the dessicating *norther,* or *Santa Ana*. This hot northeast wind sweeps downward out of the Sierra Nevadas, producing temperatures over 100°F and humidities as low as 5 percent, burning and shriveling any unprotected vegetation (and people) in its path, and frequently causing an extreme fire hazard. The cause of the Santa Ana is similar to that of the chinook—air descending steep slopes, warming and drying as it is compressed. Santa Ana winds are favored when high pressure builds over the Pacific Northwest states with low pressure to the south over Mexico, producing a strong north-south pressure gradient and associated geostrophic east winds.

11.10 October

Dry your barley in October, or you'll always be sober. (Because if this is not done, there will be no malt.) [The Reverend C. Swainson]

By October, the forces of the North clearly mean business. Snow covers the dark arctic once again, and hour after hour, the earth radiates the hard-earned warmth of summer to space. The polar front intensifies and thrusts farther and farther south, pushing the warm air back toward the tropics with each passing wave in the strengthening westerlies. Smoke curls from the chimneys again in houses across the land as the first frosts of autumn march southward. The first cold front of the season sweeps past Miami, dropping the dew point into the 40s and ending five months of tropical reign.

October along the Pacific coast, particularly the northwest coast, is the transition month between dry, sunny summer weather and the cold, wet winter season. The anticyclonic nose of the Pacific High, which has extended over the Northwest for the past four months, begins to weaken, and vigorous Pacific storms sweep inland on the now unprotected coast. Seattle's precipitation, for example, doubles from about four centimeters in September to over eight centimeters in October, while the clouds once again cover the sky over 70 percent of the time.

The first widespread snowfalls of the season whiten the aspens of the Rockies in October, each day extending to lower elevations. These early-season snowfalls occur when cold upper-level troughs of low pressure dip southward from Canada along the Rockies, and easterly flow near the surface carries moisture from the Gulf of Mexico up the eastern slopes of the mountains.

October is a changeable month, and it is possible in the East to have frost and 80°F temperatures in the same week. Frequently, an intense polar anticyclone carries early morning frosty weather as far south as Virginia, but then slows down over the Carolinas, and finally stalls over Georgia. The first morning is decidedly cold, and even during the day, sweaters are comfortable in the 50°F temperatures. However, on the second night, the warm air in the south begins its long return northward around the western perimeter of the high, and temperatures fall only into the 40s.

On the second day, the sun and the increasing southerly flow combine to produce 60°F weather as far north as New York. Indian summer has arrived.

The synoptic conditions for Indian summer are very similar to those which produce the August heat waves, only now, the mean temperature is 20°F lower, and the extra warmth associated with the sunshine, sinking of dry air, and light winds is welcome. High pressure aloft and at the surface produces light winds throughout the troposphere, and cool, foggy mornings give way to warm, hazy afternoons. Smoke from burning leaves adds to the natural haze and reddens the evening sky.

As we noted previously, the configuration of a warm high-pressure system at the surface and aloft tends to be slowly changing, with cold air isolated in Canada and cyclone families traveling around the perimeter of the anticyclone. Thus, the pleasant Indian-summer weather may linger for a week or so, allowing the last of the grapes to ripen and the farmers to complete the autumn harvest.

Indian summer—the name itself suggests the full-bodied flavor of a mellow wine. Indian summer—nature's last banquet before grimly settling down to the austere business of winter. Indian summer—the time of the year when the last of the autumn sun's rays strike gold and red in the forests. Indian summer—frost on the pumpkins and fresh apple cider along the roadways. Drink deeply, but slowly; savor every drop, and remember.

11.11 November

No warmth, no cheerfulness, no healthful ease,
No comfortable feel in any member, no shade, no shine, no butterflies, no bees,
No fruits, no flowers, no leaves, no birds. No-Vember

[From *The Works of Thomas Hood,* Epes Sargent, ed. (New York: Putnam, 1865), p. 332]

November slashes most of the country like a cold blade of steel. There are no more reprieves like Indian summer, no more pleasant nights of sitting on the porch, no more running barefoot down dusty paths, and no more late-season bonuses from the garden. The colorful leaves of last week are ripped away in an angry north wind, and the golden harvest moon turns white with cold.

In no other month is there such a change of weather. Nearly everywhere the cloudiness increases dramatically to near wintertime levels. Look at the climatic charts (Appendix 2) of Chicago, for example, where the cloudiness jumps nearly 20 percentage points in November. While the clouds increase, the mean temperatures tumble precipitously—from 55°F to 39°F at Chicago, from 59°F to 44°F in St. Louis, and from 48°F to 32°F in Minneapolis.

Significant snows reappear in St. Louis, Boston, Washington, D.C., Minneapolis, Minnesota, and Rapid City, South Dakota. Only in the extreme south does the weather improve or remain hospitable. In Miami, for example, the precipitation decreases from 21 centimeters in October to 7 centimeters in November, and the mean temperature drops to a comfortable 73°F.

In November, the famous Great Lake snowstorms begin to assault the leeward shores of Lakes Superior, Huron, Michigan, Erie, and Ontario, and turn the lakes themselves into nightmares for those commercial ships seeking to make one more run. In contrast to the large-scale heavy snows associated with extratropical cyclones, the Great Lakes storms produce their maximum snowfall after the low and cold front have passed. An ideal synoptic situation occurs after a deep cyclone (very low pressure) moves northward along the eastern seaboard, with a strong Canadian high situated over the Dakotas. The large west-east pressure gradient accelerates frigid, dry air along a trajectory from the Canadian ice fields across the Great Lakes. Pouring over the relatively warm waters, the air in the lowest mile or two quickly becomes saturated. As the

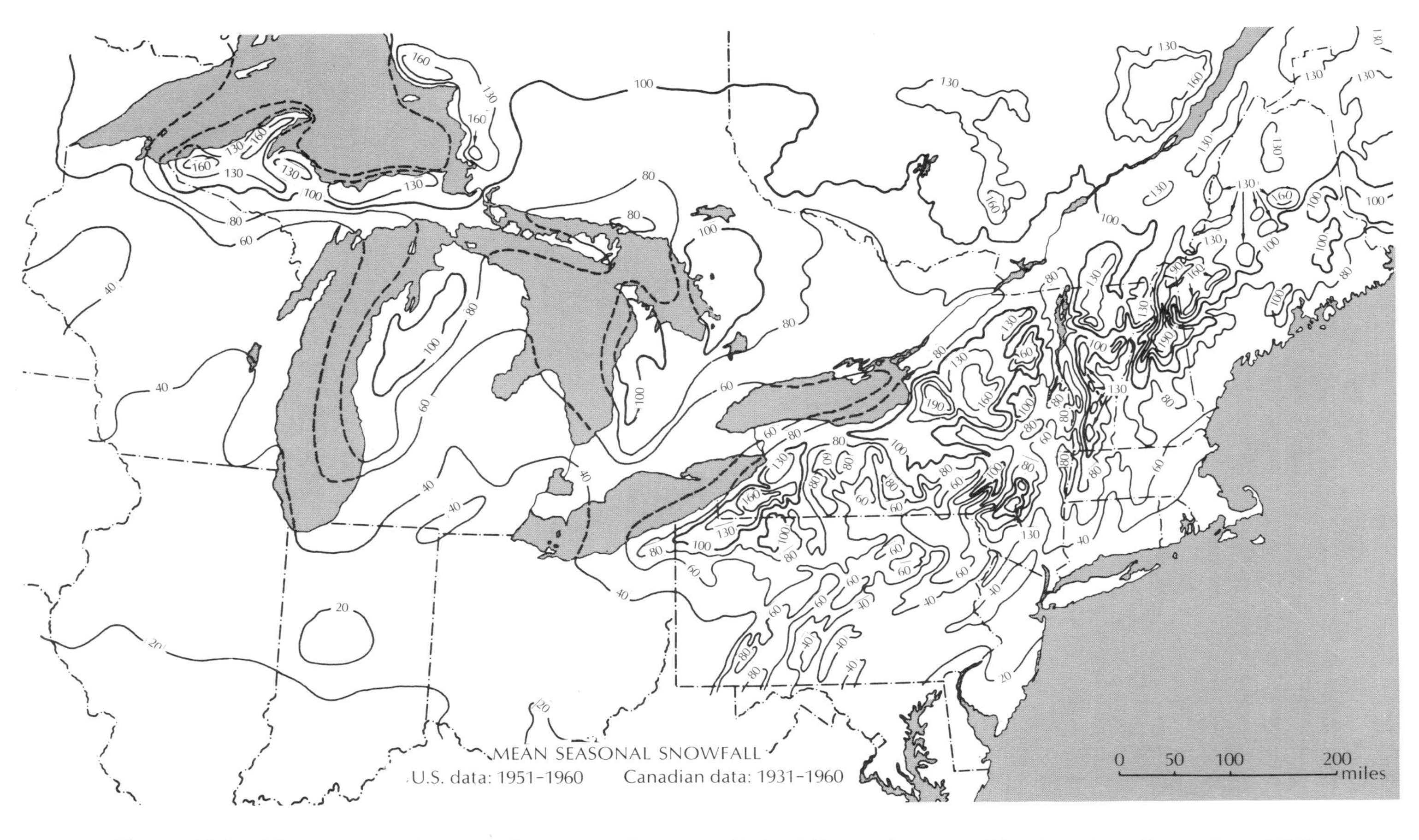

Figure 11.1 *Mean seasonal snowfall over northeastern United States (source: Weatherwise, December, 1966).*

moistened air reaches the opposite shore, amazing (to those who have never seen a Great Lakes snow squall) amounts of snow can be dropped. The efficiency of the lakes as snow producers can be seen in Figure 11.1, which shows the mean seasonal snowfall over the northeastern United States. Maxima in snowfall exist along the southern and eastern shores of the lakes, which are downwind of the prevailing west and northwest flow. For a specific example of the downwind Great Lake effect, consider Milwaukee, Wisconsin and Muskegon, Michigan, which are located only 60 miles apart on opposite shores of Lake Michigan. Milwaukee receives 109 centimeters (43 inches), mostly from extratropical storm systems, but Muskegon, with its extra bounty from the lake, receives double that amount (220 centimeters, or 87 inches).

The heat added from the Great Lakes to the atmosphere also contributes to the heavy lake snowstorms. Because warm air is less dense than cold air, a trough of lower pressure tends to form over the Great Lakes whenever the water is significantly warmer than the surrounding air. This trough of low pressure produces covergence of surface air, which then rises and contributes to the precipitation process.

The Great Lakes snowstorms continue through the winter until the lakes freeze and ice stops their role as a heat and moisture source. Although most of the Great Lakes do not freeze until late December or January, it is interesting to mention here how the freezing occurs. One of water's peculiar properties is that it reaches a maximum density at a temperature of 4°C (39°F). The water near the bottom of deep freshwater lakes in the northern United States (such as Lake Superior) has a temperature of 39°F year around. During the summer, the surface waters reach a temperature of 60–70°F, and therefore float on top of the denser water below. As the subfreezing blasts of the November winds blow across the lakes, the water at the surface is chilled, becomes dense, and sinks, very much like convection currents in the atmosphere. This vertical mixing through cooling from above is efficient until the entire lake temperature equals 4°C (39°F). Then, further cooling produces lighter water, vertical mixing is inhibited, and freezing of the uppermost layer may readily occur.

11.12 December

December cold with snow, good for rye.

A green Christmas makes a fat church yard.

December is the darkest month of the year everywhere in terms of the time the sun spends above the horizon. In Fairbanks, Alaska, the sun appears for only 3½ hours on December 21, and, then, only if the skies are clear. Adding to the dreariness of a 21½-hour night are the frequent ice fogs in Fairbanks. In most parts of Alaska, the winter air is dry enough so that even in the coldest weather only a few ice crystals precipitate, and visibility remains good. In cities, however, significant amounts of water vapor are emitted by cars, aircraft engines, and the combustion of fuels. Here, under the calm anticyclonic conditions of December, where radiative cooling frequently drops the surface temperature to –50°F, dense fogs of ice crystals hover over the city. The inversion during these episodes, which may last for a week or more, is extremely well developed. Temperatures may rise by 30°F from a surface value of −50°F in only 1000 feet, making it desirable to live on the mountain slopes rather than in the valleys.

The cold polar atmosphere can sometimes play strange tricks on the senses. Imagine being situated in the remote outpost of Barrow, Alaska, which is located on the shore of the Arctic Ocean at a latitude of 71°N. At noon on November 21, the sun sets for an advertised two months, until its scheduled reappearance on January 21. But two days later, the southern horizon brightens again, and a distorted, yet unmistakable image of the sun reappears, casting a feeble yellow glow over the icy sea. A wild hope springs to the heart. Are the astronomers wrong? Has the tilt of the earth's axis suddenly changed? Will the sun steadily climb in the sky, erasing the clouds and snow like a bad dream? But even as we eagerly watch, our hopes are shattered as the sun scintillates, bends, wobbles, and plunges a final time below the frozen horizon. The whole scene has been a cruel hoax, caused by refraction. The cold, dense atmosphere has bent the sun's rays over the horizon for one brief moment.

Over the continental United States, December is dark for meteorological as well as astronomical reasons, the brightest times being indoors during the holiday season. December is one of the cloudiest months of the year, particularly in the Pacific Northwest and along a band extending from the mountainous areas of Pennsylvania northeastward into New England. For example, cloudiness reaches 81 percent in Seattle, 90 percent in Portland, Oregon, and 74 percent in Caribou, Maine. Thus, it must have been in December when the frustrated poet from Maine wrote the following (note that *dirty* was formerly a synonym for *cloudy*):

Dirty days hath September,
April, June and November;
From January up to May,
The rain it raineth every day.
All the rest have thirty-one,
Without a blessed gleam of sun;
And if any of them had two-and-thirty,
They'd be just as wet and twice as dirty.

December begins the snowfall season over much of the United States, as cyclones sweep across the middle and southern sections of the country with increasing frequency. December is the month when many people, not yet tired of cold weather, appreciate a snowstorm, especially if the snowfall occurs during the Christmas season. Still, the chances of snow cover on Christmas day are small, as shown in Figure 11.2, except in the northern third of the country and in the mountainous regions. Thus, most locations wait through green Christmas after green Christmas, dreaming in vain for that elusive magic morning of white.

As a final note, it is interesting to see throughout the centuries of weather wisdom how people come to expect and trust "normal", or what they consider "proper," weather for the season, such as hot weather in July, or snow at Christmas. Exceptions to these ideas of what is "right" for the season are

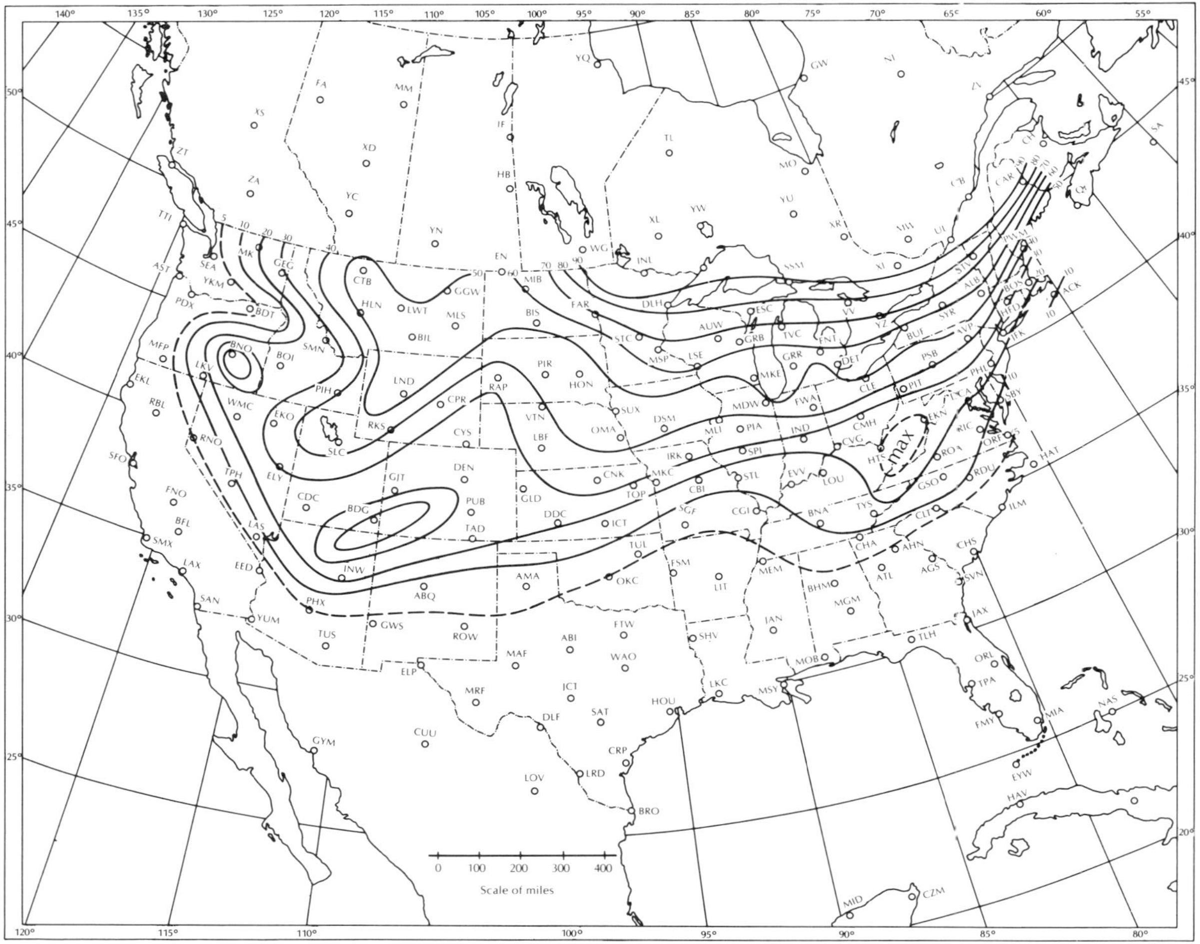

Figure 11.2 *Probability of a "White Christmas" (one inch or more of snow on ground) (source: "Statistical Probabilities for a White Christmas," U.S. Department of Commerce news release, Washington, D.C., December 15, 1971).*

commonly greeted with suspicion, and even as portenders of evil, as the proverb relating green Christmases to fat churchyards (many graves), or the following odes to January indicate:

"Normal" (to be trusted)	*When oak trees bend with snow in January, good crops may be expected.*
"Abnormal" (to be regarded with suspicion)	*In January should sun appear, March and April pay full deal.*
	A January spring is worth naething. [Scotch proverb]

These forebodings are, for the most part, misapprehensions of what is really quite characteristic of the atmosphere, which rarely behaves normally or follows the neat climate tables or the smoothly drawn curves on the annual graphs of temperature and rainfall. These "normals" are really made up of an average of many "abnormal" days. As an extreme example, consider a mythical climate in which half of the days have temperatures of 100°F, and the other half have temperatures of 50°F. The average temperature of 75°F would never occur! Thus, the "normal" weather may be actually less likely than the so-called atmospheric freaks such as April snows or January thaws. As the Norwegian proverb tells it:

> *There are many weathers in five days, and more in a month.*

Appendix 1

Weather observations, units, and orders of magnitude

One of the barriers of communication between scientists and laymen is the use by scientists of units that are unfamiliar to nonscientists. This difficulty is especially pronounced in the United States, where the scientific community generally uses metric units (e.g., meters, kilometers, degrees Celsius), while the general public utilizes the more unwieldy English system of units (e.g., feet, miles, degrees Fahrenheit).[1]

In writing a book for people who are accustomed to the English system, compromises must be made between a rigid adherence to the more rational metric system (which would lead to many unfamiliar units and hamper communication) and the complete use of the familiar English system (which would perpetuate the unwieldy system). Added to the dilemma is the fact that some meteorological variables are commonly expressed in both units; for example, surface temperatures on weather maps are plotted in degrees Fahrenheit (°F), but temperatures aloft are expressed in degrees Celsius (°C). In this

[1] It is ironic that the English have recently joined the majority of nations in adopting the metric system, so that the United States now stands alone among major nations in its use of the archaic and inconvenient English system. Perhaps it should be renamed the "U.S. system" until the United States finally abandons it in favor of the much easier metric system.

book, the metric system is emphasized, but, for convenience, English units are frequently given also. An attempt is made to introduce the metric system without requiring memorization of a set of conversion factors.

Besides the difficulties they have with units, many people frequently have only vague ideas of the representative distance scales associated with meteorological phenomena or of the typical and extreme values of many meteorological variables. Unless you know the characteristic order of magnitude of a given variable, you may be unable to appreciate the meaning or significance of the numerical values given in the text. For example, you may not be impressed by the surface pressure value of 908 millibars (mb) recorded in Hurricane Camille unless you know that the average sea level pressure on the earth is about 1013 mb, and that pressures lower than 910 mb have only been recorded a few times in history. Therefore, in addition to discussing some of the commonly used units in meteorology, this appendix presents typical values and extremes of several meteorological phenomena to give you a feeling for what is normal and abnormal in the weather.

A.1 Temperature

Temperatures on surface weather maps are plotted in degrees Fahrenheit (°F); on upper-level maps, degrees Celsius (°C) are used. The conversion from °F to °C is given by

$$T(°C) = 5/9[T(°F) - 32]$$

One degree Celsius is 1.8 times bigger than a degree Fahrenheit; for example, an increase in temperature of 1°C equals an increase of 1.8°F. Therefore, if the temperatures on weather broadcasts are reported in °C rather than °F, some precision is lost. For example, a temperature of 32°F represents a range from 31.50 to 32.49°F. The corresponding temperature of 0°C, however, represents a range from 31.11 to 32.88°F. When the temperature is close to freezing, this loss of information can be important. Figure A.1 compares the Celsius and Fahrenheit scales, and gives several benchmark temperatures for reference. In theoretical work, degrees absolute (°A) or kelvins (K) are used. The temperatures in degrees absolute or kelvins are the same, and are related to the temperature in degrees Celsius by the expression

$$T(\text{K or °A}) = T(°C) + 273$$

A.2 Length

The units of length commonly used to express horizontal distances are kilometers (km), statute miles (stat. mi), nautical miles (naut. mi), and degrees of latitude (° lat.).

Relationships among these are the following:

1 naut. mi = 1.15 stat. mi
1° lat. = 60 naut. mi
1 km = 5/8 stat. mi
1° lat. = 111 km

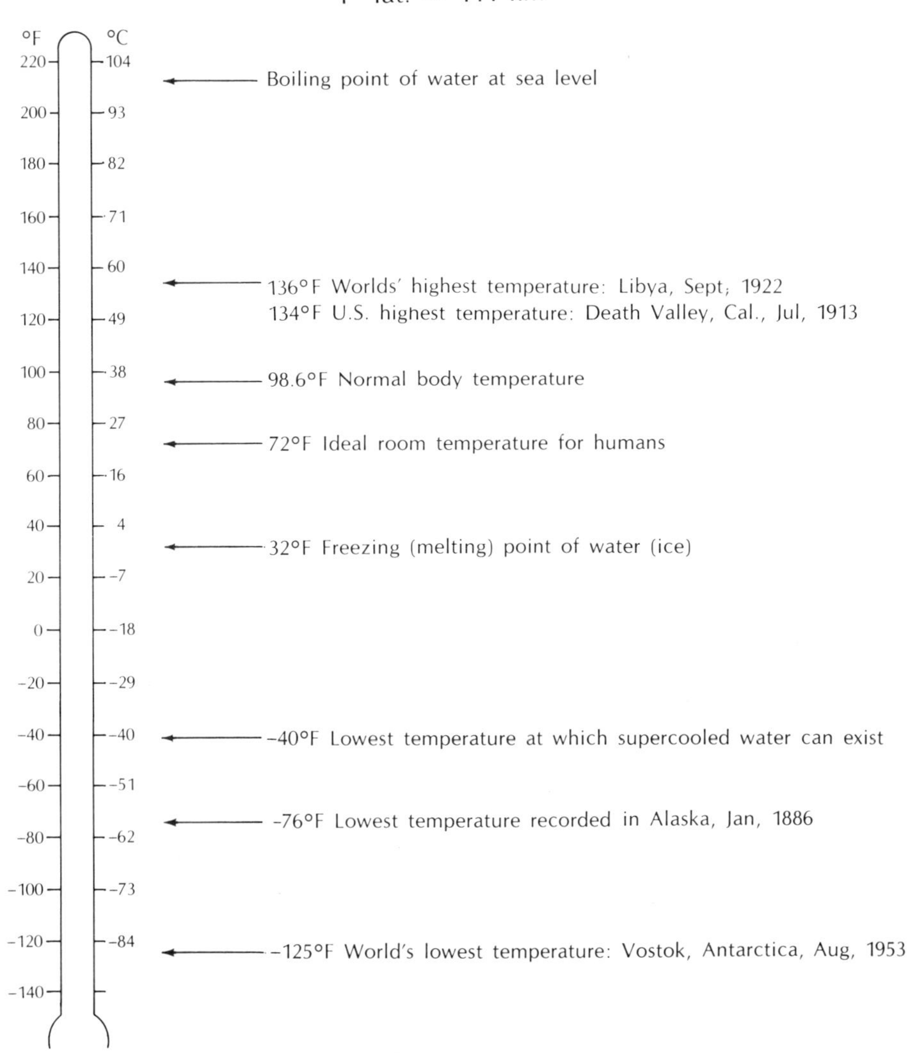

Figure A.1 *Comparison of Fahrenheit and Celsius temperature scales.*

Degrees of latitude are useful when working with weather maps. Thus, the great circle distance between New York and San Francisco is 2571 stat. mi, 2233 naut. mi, 4138 km, or 37.2° lat. (Figure A.2). Vertical distances are usually expressed in

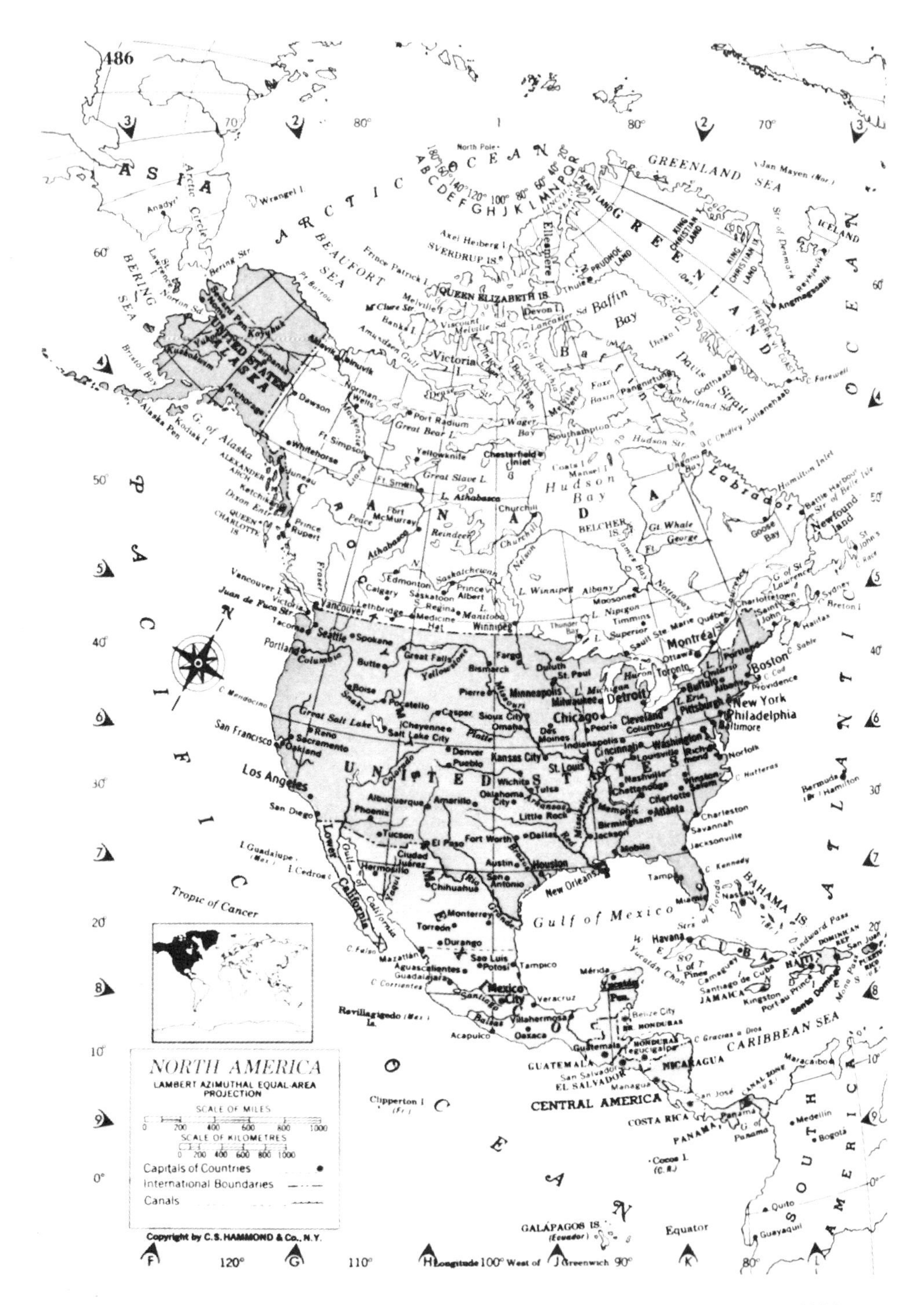

Figure A.2 *Map of North America. © Hammond Incorporated # 10399.*

thousands of feet, kilometers, or miles:

$$1 \text{ km} = 3280.8 \text{ ft}$$
$$1 \text{ km} = 0.621 \text{ stat. mi}$$

The heights of typical atmospheric phenomena are given in Figure A.3.

A.3 Horizontal Wind Velocity

Wind speeds are plotted on weather maps in knots (kt) and reported over the radio in miles per hour (mi/h). A knot is a nautical mile per hour, and is slightly faster than a statute mile per hour (mi/h). One knot equals 1.15 mi/h. For most order of magnitude purposes, they may be considered the same. The effect of wind velocities that may be observed on land is given by the Beaufort scale (Figure A.4), which was devised in 1806 by Sir Francis Beaufort. Beaufort originally described the effects of the wind on a full-rigged man-of-war; the effects of the same wind speeds on land-based phenomena were added later.

A.4 Vertical Wind Velocities

The up-or-down component of air motion is the vertical velocity, which is usually much smaller than the horizontal velocity. For example, a typical value for vertical velocity in the atmosphere outside of thunderstorms is 5 cm/s (0.1 mi/h), truly a snail's pace. In strong thunderstorms, or in the vicinity of steep terrain, the vertical velocities may be larger, say 40 mi/h.

A.5 Pressure

Pressure is the force exerted by the atmosphere per unit area of surface. In meteorology, the most commonly used units of pressure are millibars (mb), but inches of mercury (in. Hg) are also reported. Note that "inches of mercury" are not really units of pressure, because inches are not the correct dimensions for force per unit area. When we say that the pressure is 30 inches of mercury, we mean that the atmospheric pressure is sufficient to support a column of mercury 30 inches high. The conversion from inches of mercury to millibars is

$$1 \text{ in. Hg} = 33.86 \text{ mb}$$

Some typical and extreme values of sea level pressure are given in Figure A.5.

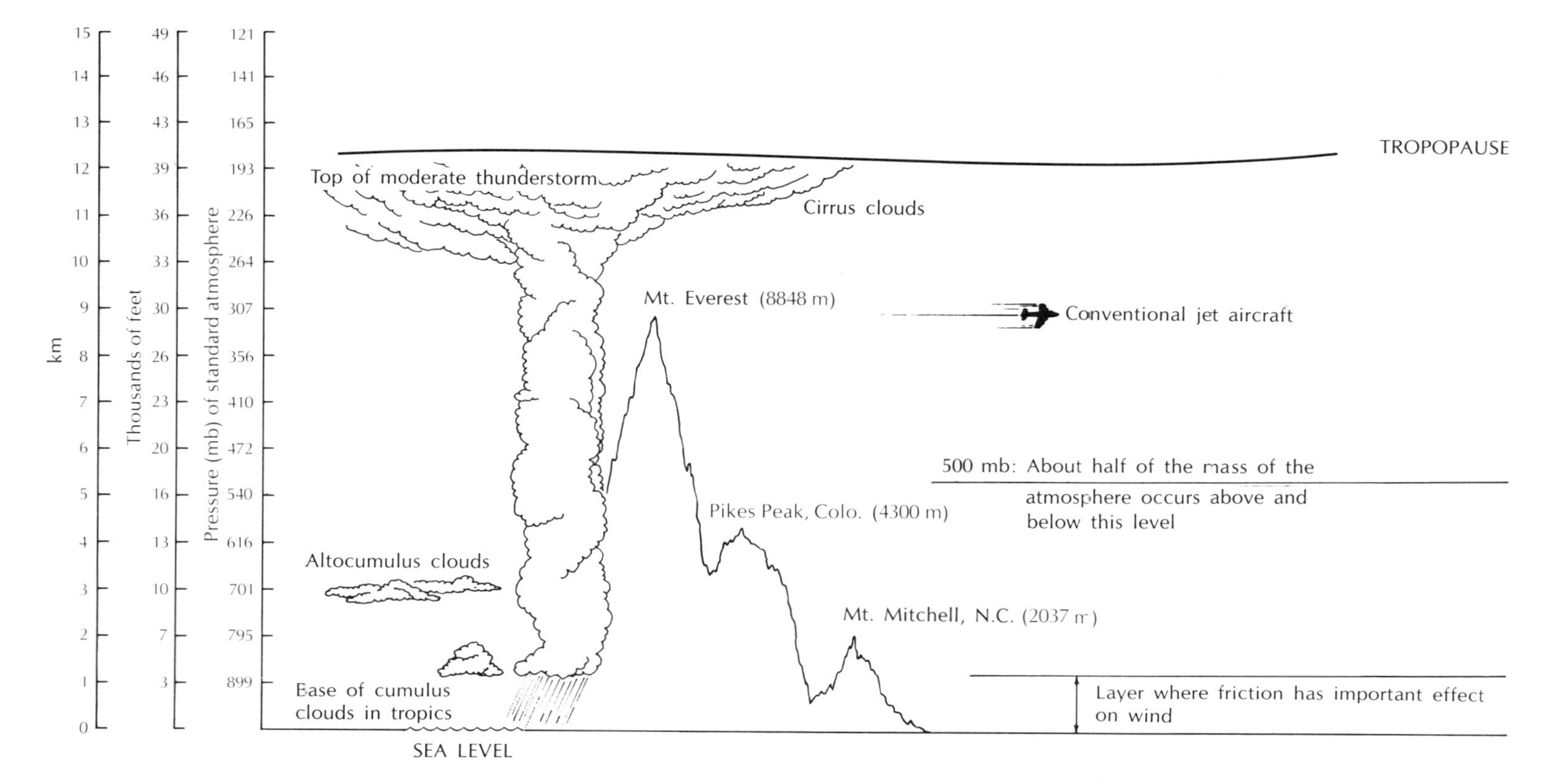

Figure A.3 *Vertical distances in the atmosphere.*

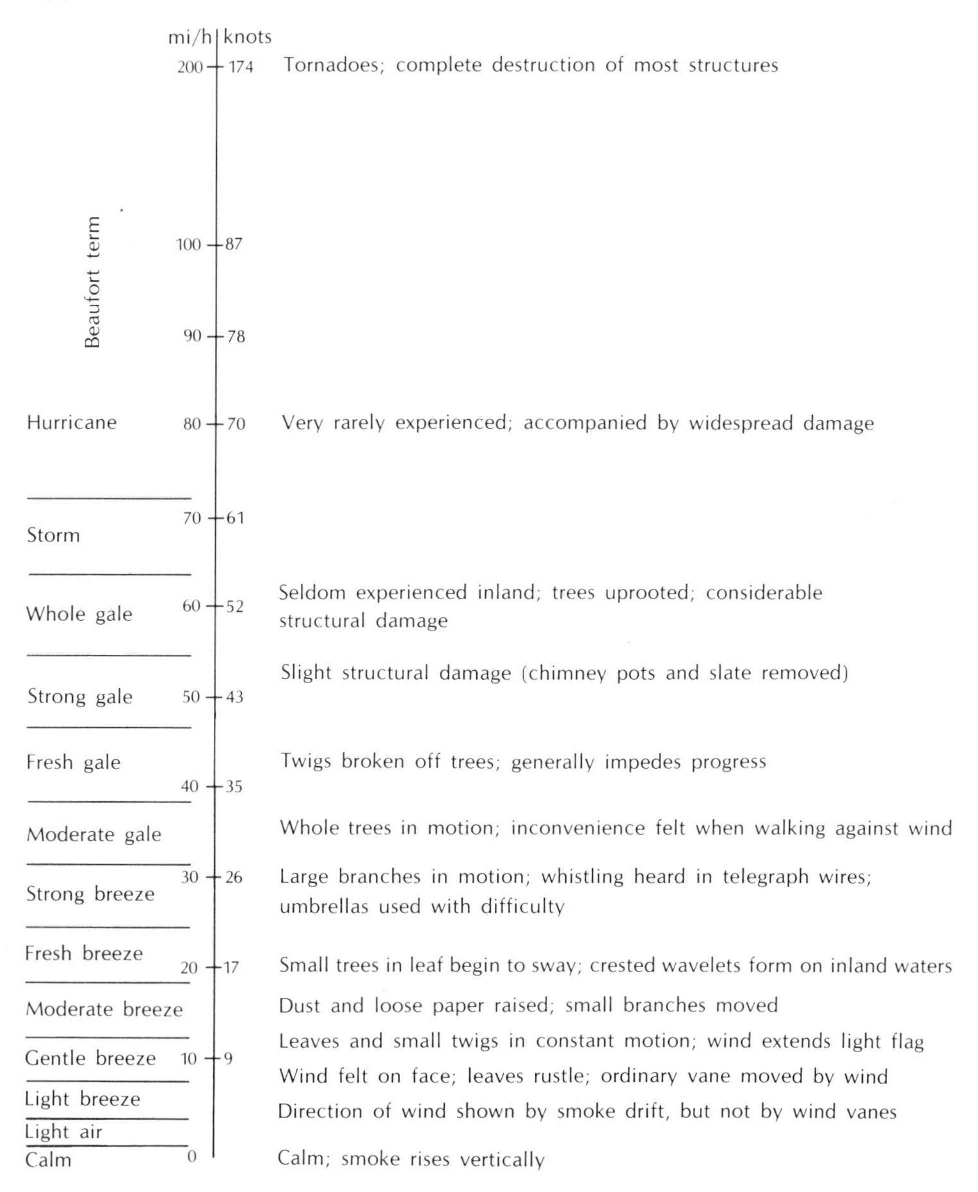

Figure A.4 *Beaufort's description of wind effects.*

A.6 Weather Observations

Almost everyone has used thermometers and barometers. Meteorologists use the same types of instruments at regular intervals to determine the temperature and pressure at many surface locations. Some weather stations report their observations (usually on a teletype network) every hour on the hour, but others report less frequently, perhaps only once a day. In addition to the temperature and pressure, many

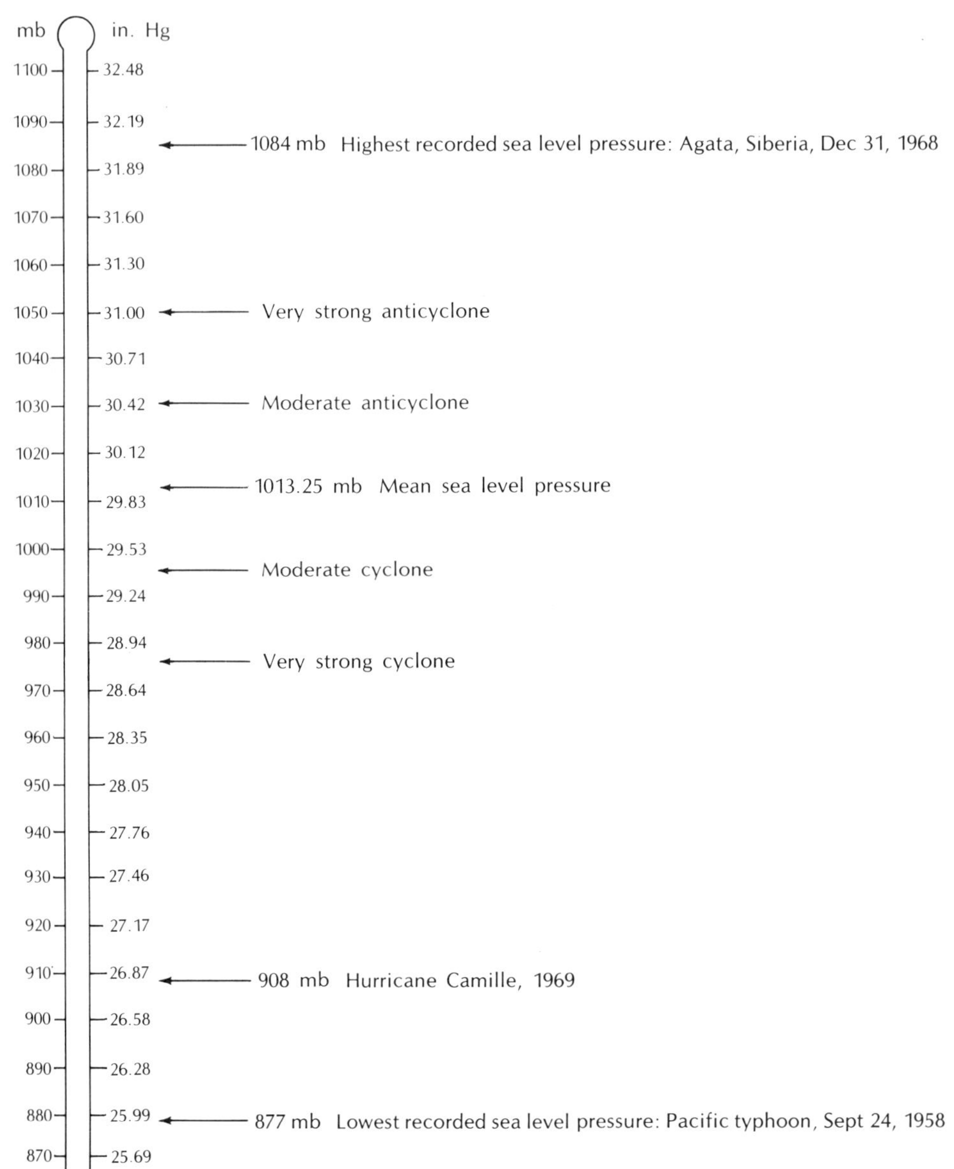

Figure A.5 *Representative barometric pressures in the atmosphere expressed in millibars and inches of mercury.*

stations report humidity, wind speed and direction, cloud amount and type, and amount and type of precipitation that has fallen since the previous observation. The surface weather is well observed over the continents of the world, but not nearly so well over the oceans.

Surface observations are not sufficient to predict the future weather; for predictions, observations are needed to heights of about 30 kilometers. Many of these observations are obtained from a balloon system called the *rawinsonde*. As the balloon and its instrument package rise, information about pressure, temperature, humidity, and wind at the height of the balloon are transmitted back to ground by radio. However, because rawinsondes are expensive and can be used only once, only the richest nations have a dense network of rawinsonde stations. Over the United States, rawinsondes are launched twice a day—at 0000 GMT and 1200 GMT. There are a few weather ships on some of the oceans that are equipped with rawinsondes, but upper-air data over most of the oceans are insufficient for adequate resolution of the upper atmosphere, especially over the Southern Hemisphere, where the oceans cover more than 80 percent of the area.

Meteorologists often blame inaccurate weather forecasts on insufficient data, especially those missing from the upper atmosphere. The weather satellites have shown considerable promise in eliminating this excuse by receiving weather information over the entire atmosphere. The early satellites showed mainly cloud photographs, which were useful in understanding the weather qualitatively, but could not provide much quantitative information about the temperature, humidity, or wind. Recently, however, cloud motions (and, therefore, the winds, since clouds tend to move with the wind) have been calculated from pictures taken by the "stationary" weather satellites, which orbit at about 35,000 kilometers from the earth. At this distance, the orbital period of the satellite equals the rotation period of the earth, so that the satellite remains above the same point on earth all the time. From such satellites, the same cloud patterns can be photographed at frequent intervals (for example, every half hour), and the speed of the clouds calculated by measuring displacements over these intervals.

The second important innovation in weather satellites has been the remote sensing of atmospheric temperatures by measuring the infrared radiation emitted by the atmosphere. Some of the gases of the atmosphere emit infrared radiation, as we saw in Chapter 2. The amount of radiation emitted is a measure of the temperature at the level of radiation. If the radiation is measured at a wavelength where the absorption is large, radiation emitted from low in the atmosphere cannot get out into space. Instead, the satellite senses radiation from a high level in the atmosphere, and the intensity of the radiation received yields a measurement of the temperature at this level. On the other hand, if the satellite receives radiation at a wavelength where the atmospheric absorptivity is small, the radiation will come from far down in the atmosphere. The intensity of radiation at this wavelength will then be a measure of the lower atmospheric temperature. Thus, temperatures at many levels can be inferred by measuring the radiation at different wavelengths.

Appendix 2

Climatic summaries of selected U.S. cities

The following graphical summaries of the climate at selected U.S. cities were prepared from the data presented in *World Survey of Climatology: Climates of North America,* Vol. 11, Reid Bryson and Kenneth Hare, eds. (Amsterdam: Elsevier Scientific Publishing Company, 1974):

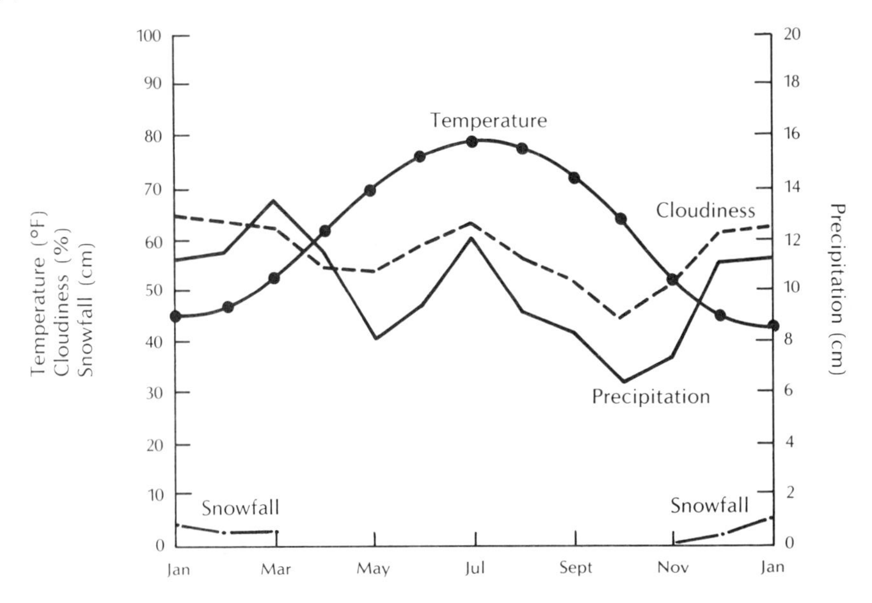

Atlanta, Georgia: 34°N 84°W;
elev. 308 meters

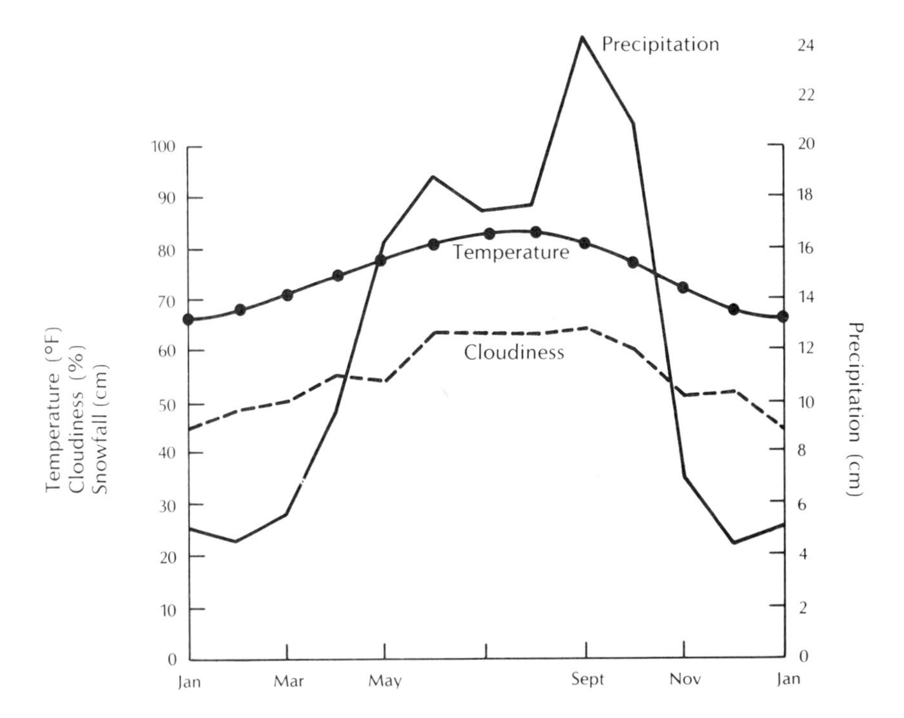

Miami, Florida: 26°N 80°W;
elev. 2 meters

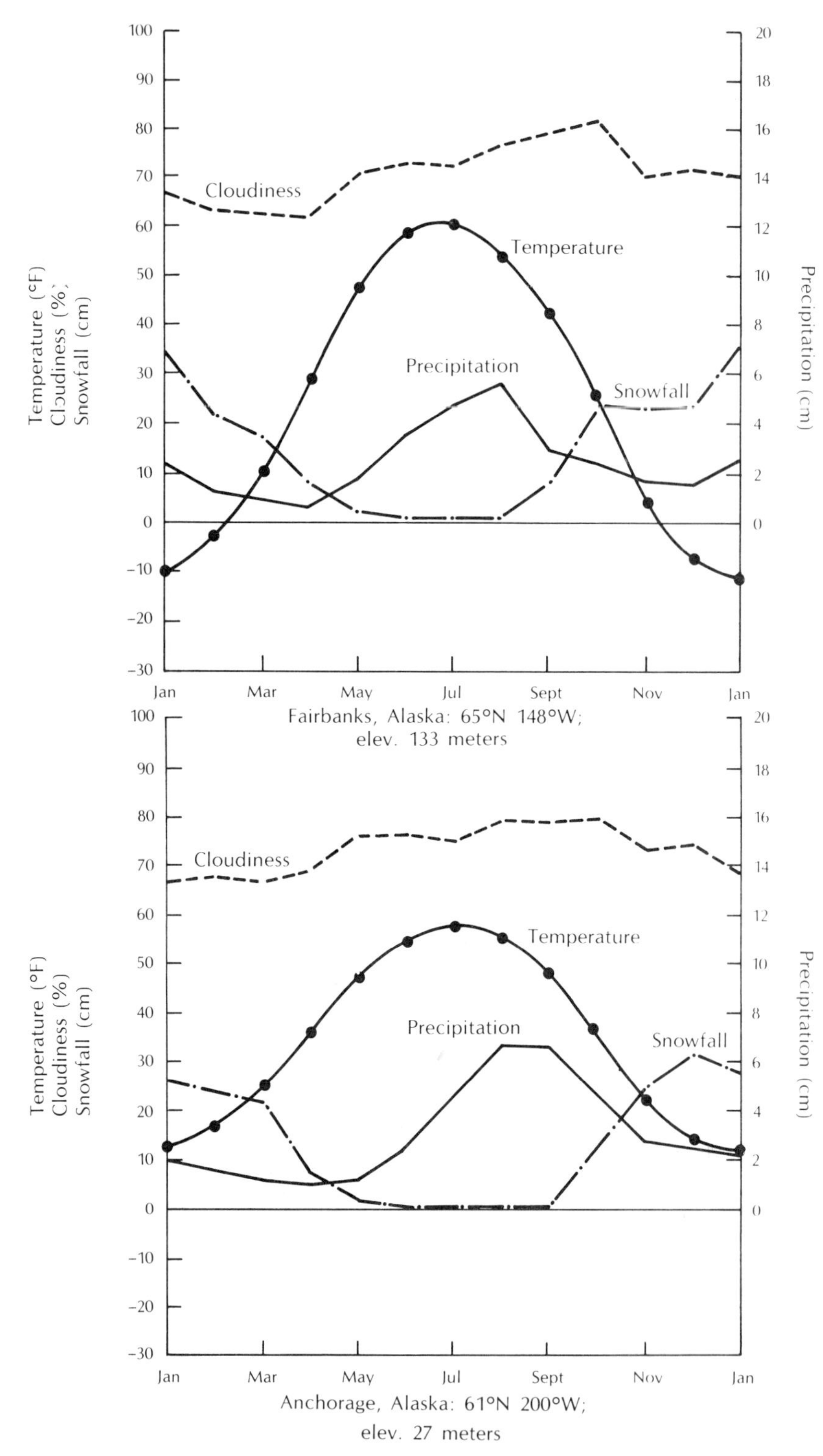

100
90
80
70
60
50
40
30
20
10
0
-10
-20
-30
20
18
16
14
12
10
8
6
4
2
0
Cloudiness
Temperature
Precipitation
Snowfall
Temperature (°F)
Cloudiness (%)
Snowfall (cm)
Precipitation (cm)
Jan
Mar
May
Jul
Sept
Nov
Jan
Fairbanks, Alaska: 65°N 148°W;
elev. 133 meters
Cloudiness
Temperature
Precipitation
Snowfall
Temperature (°F)
Cloudiness (%)
Snowfall (cm)
Precipitation (cm)
Jan
Mar
May
Jul
Sept
Nov
Jan
Anchorage, Alaska: 61°N 200°W;
elev. 27 meters

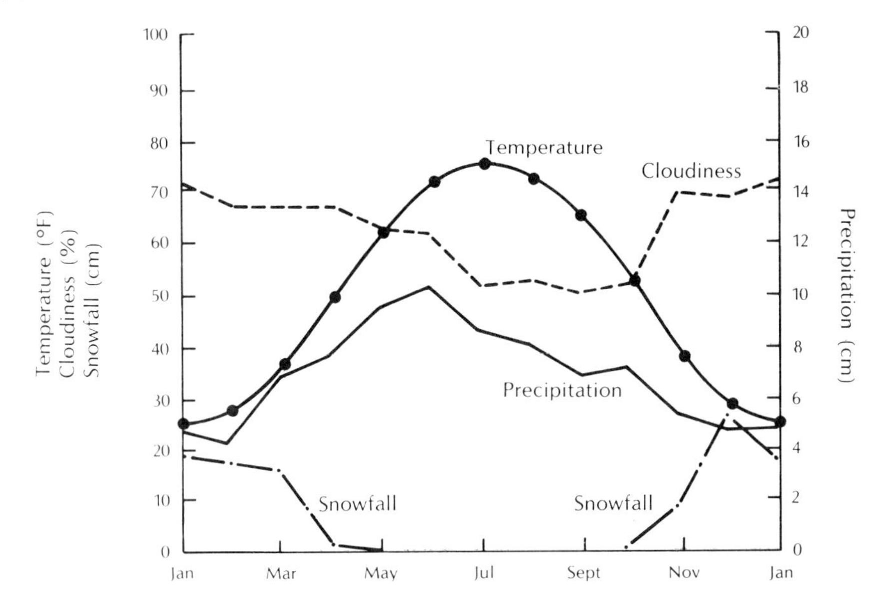

Chicago, Illinois: 42°N 88°W;
elev. 185 meters

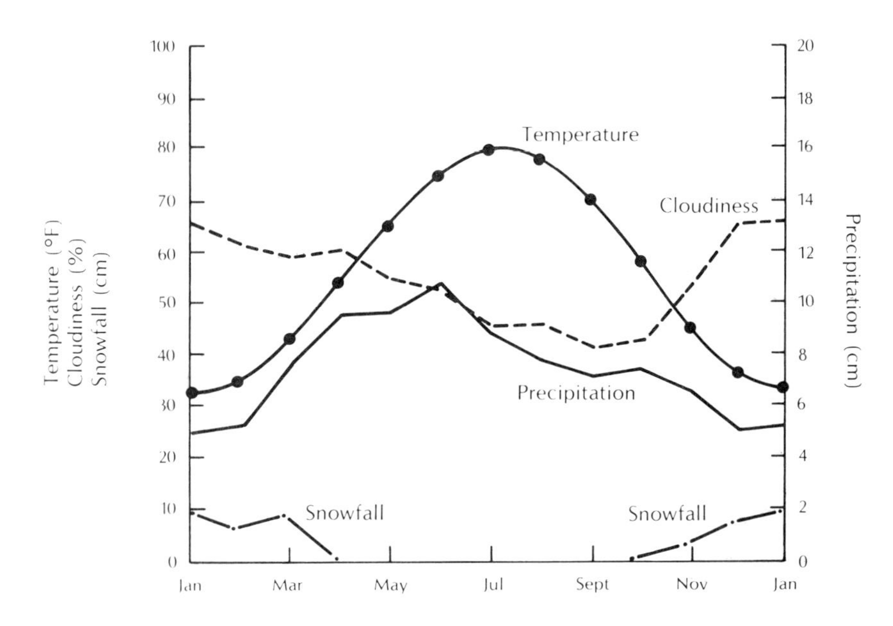

St. Louis, Missouri: 39°N 90°W;
elev. 142 meters

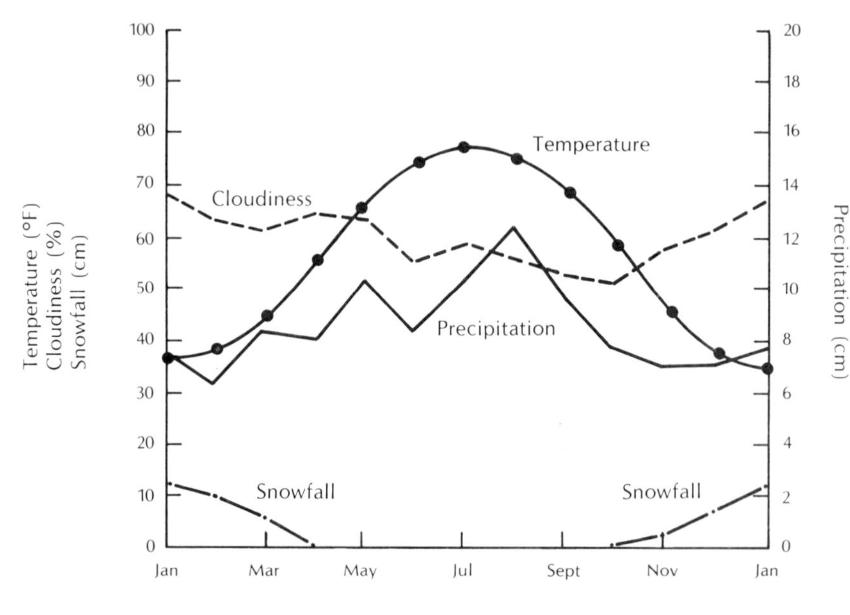

Washington, D.C.: 39°N 77°W;
elev. 4 meters

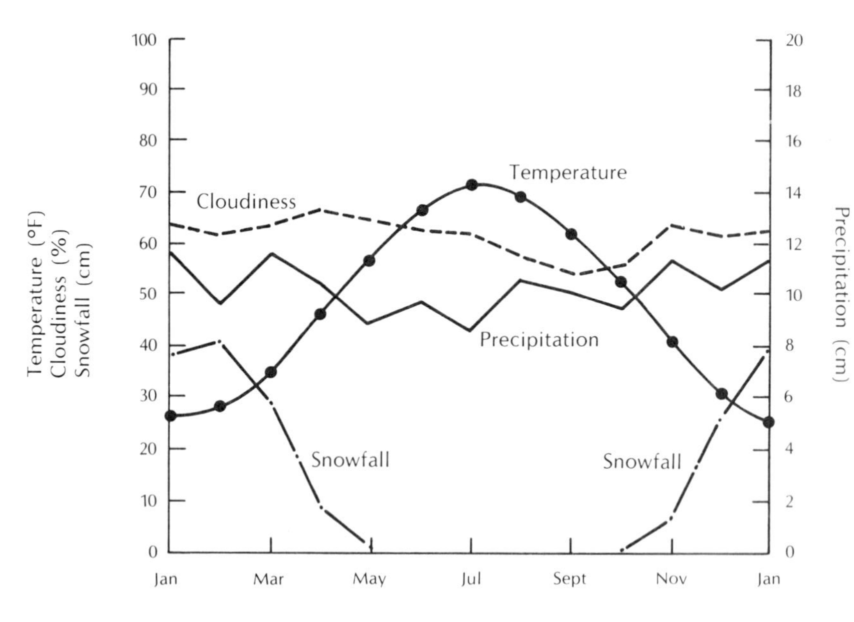

Boston, Massachusetts: 42°N 71°W;
elev. 192 meters

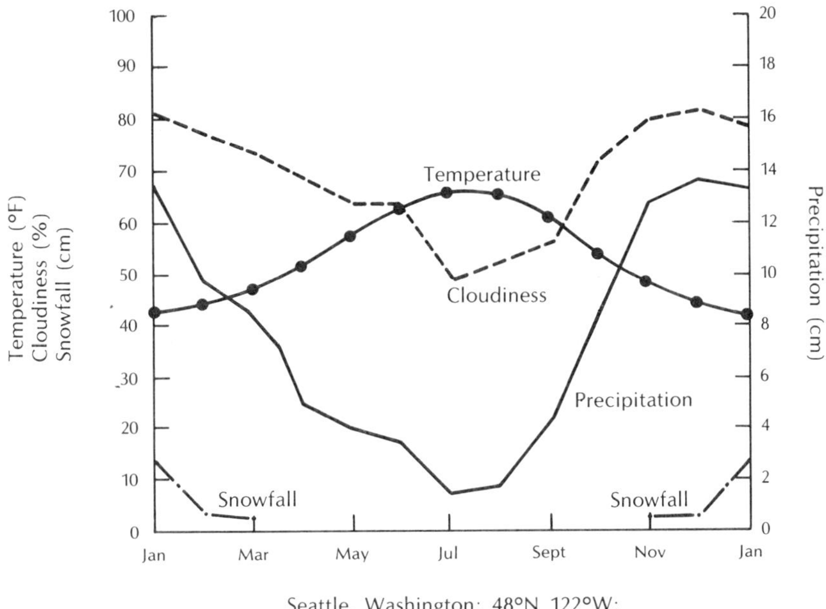

Seattle, Washington: 48°N 122°W;
elev. 4 meters

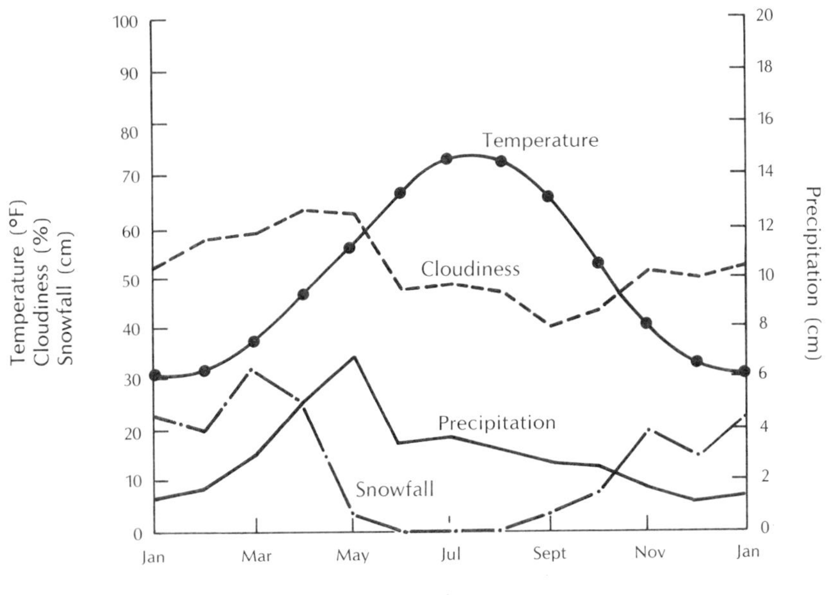

Denver, Colorado: 40°N 105°W;
elev. 1610 meters

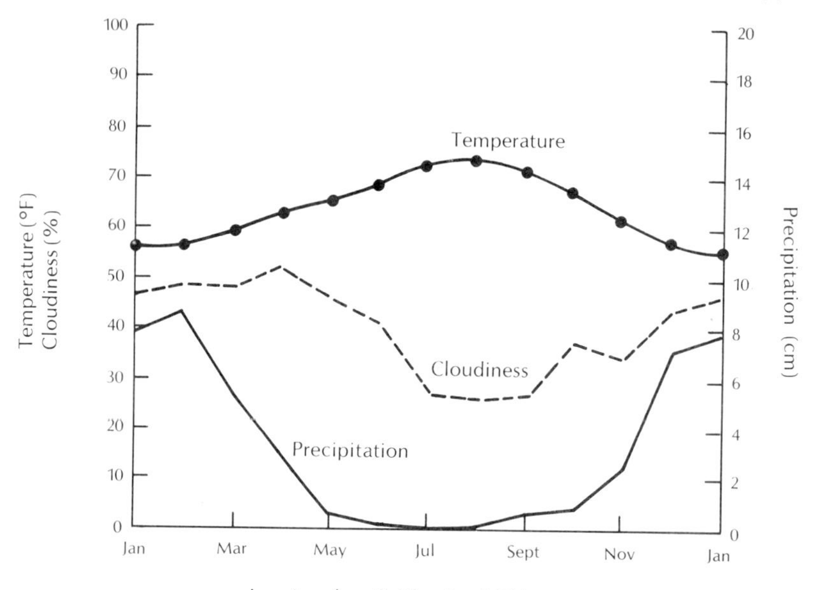

Los Angeles, California: 34°N 118°W;
elev. 104 meters

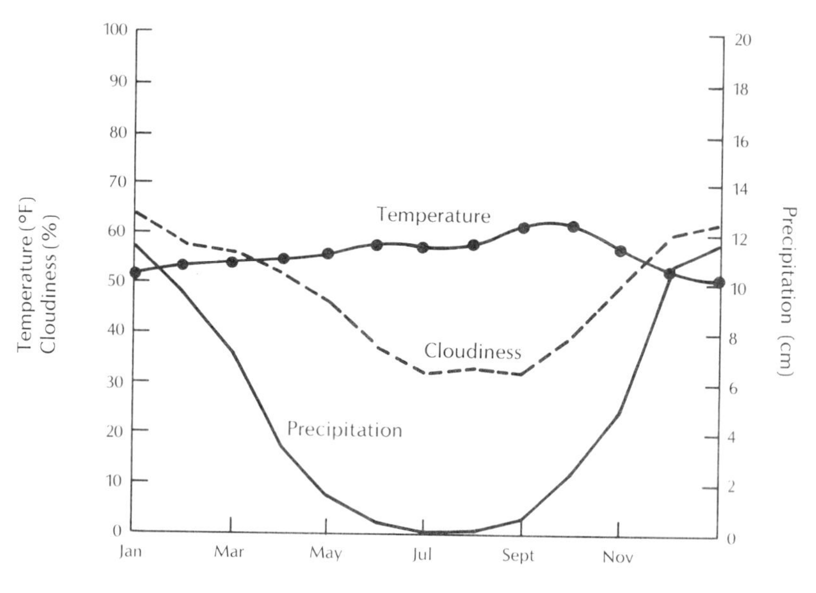

San Francisco, California: 38°N 122°W;
elev. 20 meters

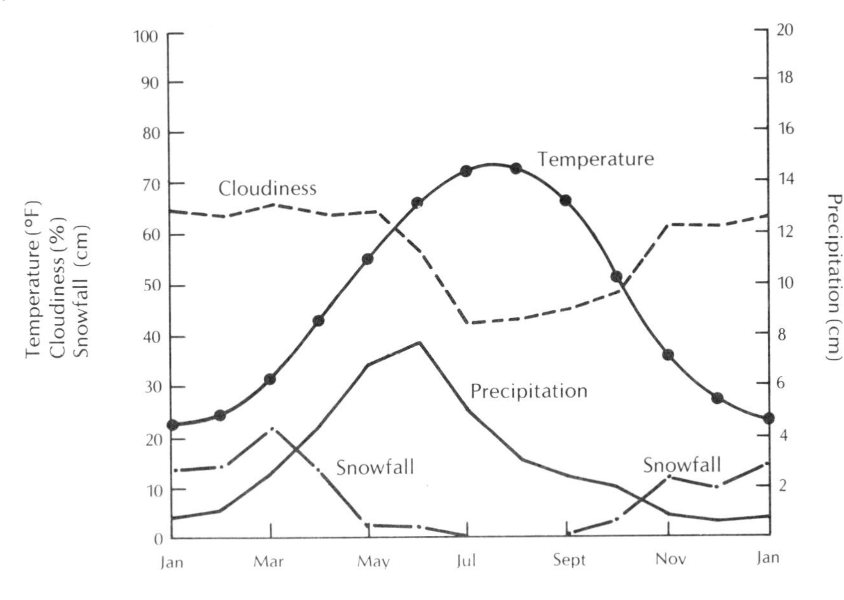

Rapid City, South Dakota: 44°N 103°W;
elev. 993 meters

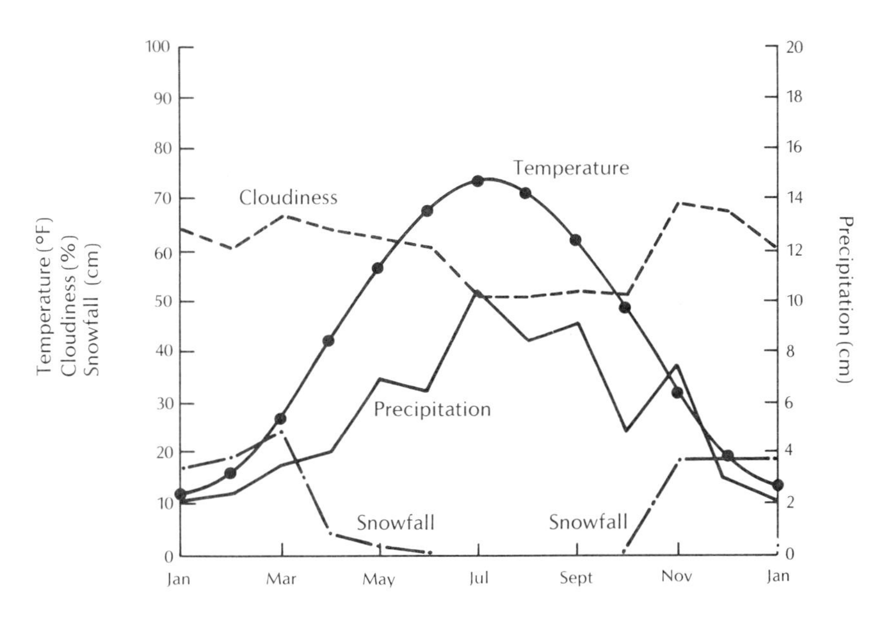

Minneapolis-St. Paul, Minnesota: 45°N 93°W;
elev. 254 meters

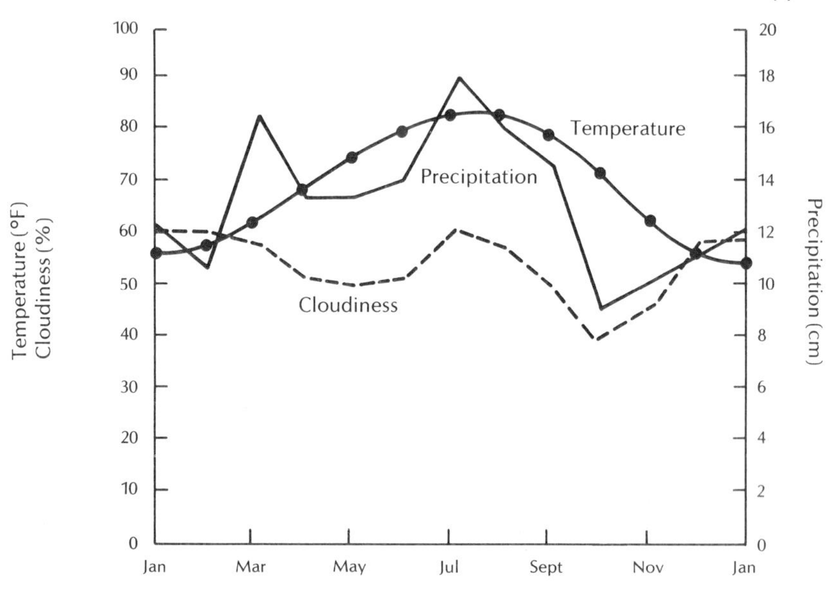

New Orleans, Louisiana: 30°N 90°W;
elev. 3 meters

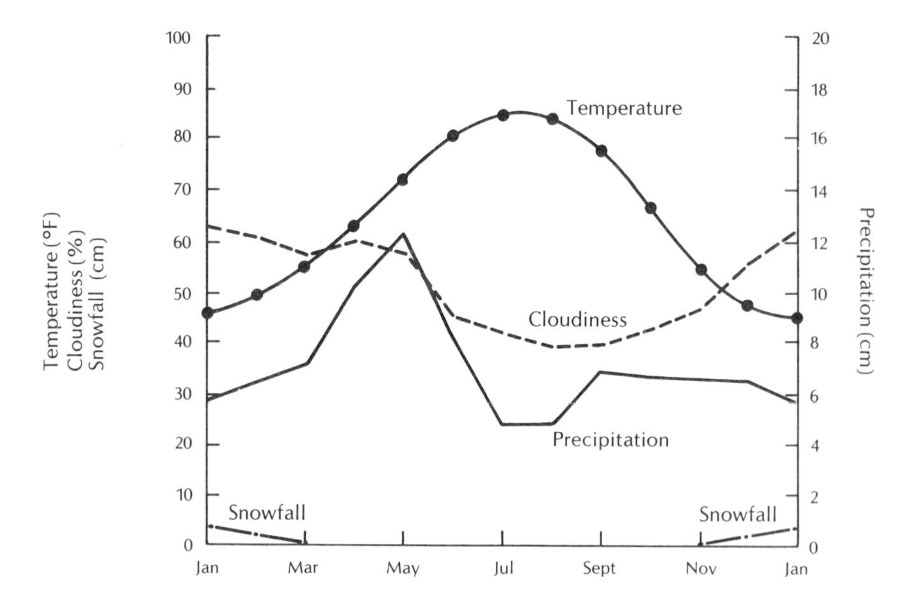

Dallas, Texas: 33°N 97°W;
elev. 146 meters

Appendix 3

Additional source materials

This appendix contains a sample selected from literally hundreds of additional sources for information on meteorology or related fields.

ADDRESSES:

For air pollution information:

Air Pollution Control Association
4400 Fifth Avenue
Pittsburgh, Pennsylvania 15213

The National Center for Atmospheric Research, sponsored by the National Science Foundation, and the National Weather Service are good sources for meteorological information:

National Center for Atmospheric Research
Boulder, Colorado 80303
Attn: Public Information Office

Public Information Publications
National Weather Service
National Oceanic and Atmospheric Administration
Rockville, Maryland 20852

Also, much climatology data can be obtained from local National Weather Service offices or the

National Climatic Center
Asheville, North Carolina 28801

Another useful address for World Meteorological Organization publications on meteorology is:

WMO Publication Center
UNIPUB, Inc.
P.O. Box 433
New York, New York 10016

BOOKS:

Battan, Louis J.: *Radar Meteorology,* 161 pp., Chicago: The University of Chicago Press, 1959. A nontechnical description of the use of radar in studying the atmosphere.

_____: *Nature of Violent Storms,* 160 pp., Garden City, New York: Doubleday and Co., Inc., 1961. Discussion of thunderstorms, hurricanes, and tornadoes.

_____: *Harvesting the Clouds,* 148 pp., Garden City, New York: Anchor Books, Doubleday and Co., Inc., 1969. A lively descriptive book on the various aspects of weather modification, including the LaPorte anomaly, cloud seeding, and hurricane modification.

Bentley, W.A. and W.J. Humphreys: *Snow Crystals,* 226 pp., 180 Varich Street, New York, New York: Dover Publications, 1962. A remarkable and beautiful book first published in 1931, containing 2453 illustrations, most of them striking photographs of the many varieties of ice and snow crystals.

Byers, H.R.: *General Meteorology,* 540 pp., New York: McGraw-Hill Book Company, 1959. A comprehensive, elementary meteorology textbook.

Dunn, G.E. and B.I. Miller: *Atlantic Hurricanes,* 377 pp., Baton Rouge, Louisiana: Louisiana State University Press, 1964. A very readable book on the many aspects of tropical storms.

Edinger, J.C.: *Watching for the Wind,* 148 pp., Garden City, New York: Anchor Books, Doubleday and Co., Inc., 1967. Enjoyable style with emphasis on local meteorology.

Flora, S.D.: *Hailstorms of the United States,* 201 pp., Norman, Oklahoma: University of Oklahoma Press, 1956. A description of these damaging storms in the United States.

Hodges, L.: *Environmental Pollution,* 370 pp., New York: Holt, Rinehart and Winston, 1973. Complete treatment of all aspects of pollution, including air, water, noise, thermal, pesticide, etc. pollution.

Inadvertent Climate Modification, 308 pp., Report of the Study of Man's Impact on Climate (SMIC), Cambridge, Massachusetts: The MIT Press, 1971. Fascinating review of the many aspects of the effect of human activities on climate.

LaChapelle, E.R.: *Field Guide to Snow Crystals,* 101 pp., Seattle, Washington: University of Washington Press, 1969. Discussion of the meteorological factors producing different types of snow crystals; 57 photographs.

Landsberg, H.E.: *Weather and Health,* 148 pp., Garden City, New York: Doubleday and Co., Inc., 1969. An introduction to the fascinating link between weather and human life.

Lowry, W.D.: *Weather and Life: An Introduction to Biometeorology,* Corvalis, Oregon: Oregon State University Book Store, Inc., 1967. A look at the many effects of weather on life.

Mather, J.R.: *Climatology: Fundamentals and Applications,* 411 pp., New York: McGraw-Hill Book Company, 1974. Includes many interesting applications of climatology, such as those in health, architecture, and industry.

Miller, Albert and J.C. Thompson: *Elements of Meteorology,* 2nd ed., 384 pp., Columbus, Ohio: Charles E. Merrill Publishing Company, 1975. An elementary meteorology text.

Neuberger, H. and J. Cahir: *Principles of Climatology,* 178 pp., New York: Holt, Rinehart and Winston, 1969. Brief treatment of factors behind the climate; numerous examples.

Riehl, Herbert: *Introduction to the Atmosphere,* 516 pp., New York: McGraw-Hill Book Company, 1972. An elementary meteorology textbook.

Sloane, Eric: *Folklore of American Weather,* 63 pp., New York: Meredith Press, 1963. A collection of weather proverbs with a brief explanation regarding the scientific basis (if any) of each saying.

Stewart, G.R.: *Storm,* 349 pp., New York: Modern Library, Inc., 1947. A well-written novel of the birth, life, and death of a Pacific storm and its dramatic effect on the personal lives of people in the western United States.

Trewartha, G.T.: *An Introduction to Climate,* 4th ed., 408 pp., New York: McGraw-Hill Book Company, 1968. A general, thorough treatment of regional climatology.

CLOUD CHARTS:

Cloud charts can be obtained from PI Publications, National Oceanic and Atmospheric Administration, Rockville, Maryland, or from:

Science Associates
Nassau Street
Princeton, New Jersey

FILMS:

Most films are available for a modest rental. You may order distribution information from:

Modern Learning Aids
P.O. Box 302
Rochester, New York 14603

or

Universal Education and Visual Arts
221 Park Avenue South
New York, New York 10003

Above the Horizon (general meteorology)
Formation of Raindrops
Solar Radiation I (the best one)
Solar Radiation II
Atmospheric Electricity (interesting equipment)
Convective Clouds
Sea Surface Meteorology (electrical phenomena emphasized)
Planetary Circulation of the Atmosphere (may be a bit specialized for the nonmeteorologist)
It's an Ill Wind (air pollution)

McGraw-Hill Text Films
330 West 42nd Street
New York, New York

The Inconstant Air (excellent general film)
The Flaming Sky

Two excellent films from your local Weather Service Office:

Tornado: Approaching the Unapproachable
Hurricane!

Also, a number of films on air pollution are available from:

Audio-visual Facility
Public Health Service
U.S. Department of Health, Education and Welfare
Atlanta, Georgia 30333

INSTRUMENTS:

You can get an instrument catalog from Science Associates (see address under CLOUD CHARTS). Weather instruments tend to be expensive, so it pays to shop. Two other places are:

Taylor Instruments
Rochester, New York

Weather Measure Corporation
San Jose, California

LIBRARIES:

University, college, and public libraries have a wealth of popular books on meteorology. Many older libraries contain interesting (and sometimes amazing) early books on weather.

PAMPHLETS:

For government pamphlets which are reasonably good and inexpensive, write for PL48 (Price List 48 - Weather, Astronomy, Meteorology), Superintendent of Documents, Government Printing Office, Washington, D.C. 20402.

PERIODICALS:

Nontechnical

Bulletin of the American Meteorological Society: published monthly by the American Meteorological Society, 45 Beacon Street, Boston, Massachusetts 02108.

Weather: published monthly by the Royal Meteorological Society, Cromwell House, High Street, Bracknell, Berks, United Kingdom.

NOAA: published quarterly by the National Oceanic and Atmospheric Administration, Public Information Publications, Rockville, Maryland 20852.

Meteorological Magazine: published monthly by the British Meteorological Office, British Information Services, 845 Third Avenue, New York, New York.

Weatherwise: published bimonthly by the American Meteorological Society, 45 Beacon Street, Boston, Massachussetts 02108.

Technical

Ecology Today

Journal of Applied Ecology

Journal of Applied Meteorology

Journal of the Atmospheric Sciences

Quarterly Journal of the Royal Meteorological Society

Tellus (geophysics)

Water and Pollution Control

Water Resources Research

WEATHER MAPS AND SATELLITE PICTURES:

The *Daily Weather Map* is published in weekly sets by the Public Documents Department, Government Printing Office, Washington, D.C. 20402. The series contains a good readable surface chart, upper air chart, and temperature and precipitation maps for each day. The January 1975 price was $16.50 per year. Composite global satellite pictures, such as Figure 4.1, are published monthly by the National Environmental Satellite Service (NESS) under the title Environmental Satellite Imagery. These are available from the National Technical Information Service (NTIS), Sills Building, 5285 Port Royal Road, Springfield, Virginia 22151.

In addition, glossy prints of images obtained from both polar orbiting and stationary equatorial satellites are available from the National Climatic Center, Asheville, North Carolina 28801. A typical price for a glossy print is $2.75. However, the National Weather Service Office in a major city could be a veritable silver mine of prints and fascimiles for the asking. Finally, the National Aeronautics and Space Administration has Educational Offices at Greenbelt, Maryland; Hampton, Virginia; and Cleveland, Ohio, which can be helpful.

index

index

References to figures are printed in **boldface** type. References to plates are noted in *italic* type.